Metal Complexes *in* Biomimetic Chemical Reactions

N_2 Fixation in Solution, Activation and Oxidation of Alkanes, Chemical Models of Photosynthesis

A. E. Shilov, Ph.D., D.Sc.
Institute of Biochemical Physics
Russian Academy of Science, Moscow
and
Institute of Chemical Physics, Chernogolovka
Russia

CRC Press
Boca Raton New York

Library of Congress Cataloging-in-Publication Data

Shilov, A. E. (Aleksandr Evgen'evich), 1930-
Metal complexes in biomimetic chemical reactions : N_2 fixation in solution, activation, and oxidation of alkanes, chemical models of photosynthesis / A.E. Shilov.
p. cm.
Includes bibliographical references and index.
ISBN 0-8493-4953-2 (alk. paper)
1. Nitrogen--Fixation. 2. Biomimetics. 3. Metal complexes.
I. Title.
QR89.7.S47 1996
589.9'0133--dc20 96-3959
CIP

This book contains information obtained from authentic and highly regarded sources. Reprinted material is quoted with permission, and sources are indicated. A wide variety of references are listed. Reasonable efforts have been made to publish reliable data and information, but the author and the publisher cannot assume responsibility for the validity of all materials or for the consequences of their use.

Direct all inquiries to CRC Press LLC 2000 Corporate Blvd., N.W., Boca Raton, Florida 33431.

International Standard Book Number 0-8493-4953-2
Library of Congress Card Number 96-3959
Printed in the United States of America 1 2 3 4 5 6 7 8 9 0
Printed on acid-free paper

PREFACE

Like probably every other chemist, for many years I have been interested in learning how nature manages to make reactions from such "difficult" substrates as molecular nitrogen, to produce ammonia in biological nitrogen fixation, methane to form methanol in biological oxidation, and water to produce molecular oxygen in photosynthesis. The behavior of these stable molecules in chemical reactions described in textbooks did not give clear answers to my questions. In some cases (dinitrogen) it was difficult to understand how they were involved in any chemical reaction at such low temperature and pressure. In others (methane), it was also not clear how to stop the reaction at a necessary intermediate and not let it oxidize further to carbon monoxide or carbon dioxide. My questions, which were addressed initially to my friends and colleagues engaged in biochemical problems and later to the published literature, showed that my lack of understanding was due not entirely to my ignorance but also to the objective fact: virtually nothing was known about the mechanism of these highly important enzymatic reactions not only in the 1950s but in the 1960s and even later. Meanwhile it was becoming clearer with time that there is always a chemical analog for any enzymatic process and, therefore, there are probably simple analogs also for those reactions which have just been mentioned. Moreover, the search for these chemical analogs could help elucidate the mechanisms of their biological counterparts. It was interesting to try to find catalytic processes for the "difficult" substrates using transition metal complexes as catalysts: they had already demonstrated that they were a powerful tool for the activation of molecules of various types. I persuaded my colleagues to take part in this research and now I believe we all do not feel disappointed. Naturally, it would be too difficult to try to solve all these problems simultaneously, but as soon as we achieved some success in one direction we could explore the second problem and then the third.

My teacher, Professor N. N. Semenov, supported these endeavors and in the case of photosynthesis even persuaded me to start this field of research. Thereafter he followed the development of our work with keen interest and thus was an inspiring influence. I feel immensely grateful for this inspiration and decided to dedicate this book to his memory: this is but a weak gesture to express my gratitude.

Now we are living in a happy time. Of course I don't mean the state of the Russian economy or financial support for our science, which is completely inadequate. Nor do I have in mind the many gloomy prospects regarding various situations in the world. Among hopeful developments I see the successes of science that are proceeding in an accelerating rate; in particular, it is during our time that the problems of the structures of various enzymes and their active centers, including those of global significance, are being successfully solved. The investigation into chemical mechanisms of enzyme action by means of a search of their chemical analogs becomes more and more interesting and develops as an exciting detective story. We know now the principles of enzyme functioning and can formulate a working hypothesis how this process could be discovered from general considerations, taking into account the thermodynamics of the process, the properties of concrete molecules, and thus suggesting possible mechanisms. Then we search for a chemical model to check the hypothesis. Our theory gives us only general ideas about the areas where we may search. At present there is no way to predict precisely and we have to use not only a strictly scientific approach, but also our intuition coupled with some hope for luck. Indeed, sometimes we are lucky and discover a new interesting reaction. This is rewarding in itself, but our aim is to find a model for an enzyme, and we start to compare our reaction with the enzymatic process. The comparison shows that our hypothesis can be fully confirmed or completely rejected; in fact, usually it must be at least modified and we must again try to search for new experimental confirmations of the modified version. Of course, we are not the only people working along these lines; biochemists find new important details revealing the structure of enzymes and the mechanism of enzymatic catalysis; other chemists make important discoveries and slowly, collective efforts help us approach the truth from different directions. This is a long process, but the very existence of the corresponding enzymatic reaction gives us the hope that the answer will be eventually found.

At present the picture of remarkable processes of global importance, although perhaps not yet complete, becomes clearer and clearer. This book describes our understanding of certain enzymes from a chemical point of view and the development of chemical catalysts able to perform similar reactions. I have to acknowledge the direct and indirect involvement of many people who helped me in the preparation of the book. I have stressed already the inspirational role of my teacher, N. N. Semenov. Among my other teachers, I would like to mention my father, E. A. Shilov, who helped to develop my taste for the search of mechanisms of chemical reactions and also for poetry, which was of importance since we always needed much imagination in our research. I would like also to mention the late Bianca Tchoubar (France) who was my close friend for many years and whose illustrious personality influenced me and, therefore, this work a great deal.

My own colleagues made a formidable contribution, particularly: A. F. Shestakov, N. T. Denisov, A. K. Shilova, T. A. Bazhenova, A. A. Shteinman, E. I. Karasevich, V. Ya. Shafirovich, N. P. Luneva and many others. Some of them helped me write this book: E. Batova, E. I. Karasevich, N. P. Luneva, A. K. Shilova, T. A. Bazhenova, V. A. Nadtochenko.

I would like to thank the Russian Academy of Sciences, the Russian Foundation of Basic Research, the International Science Foundation and also the AMOCO company for the financial and other help that made our research possible.

I am also grateful to my distinguished friends and colleagues who, even though not directly working with me, helped to develop my understanding in the fields of chemistry and enzymology: Professors D. Barton, I. I. Moiseev, R. Hoffmann, J.-M. Lehn, S. Lippard, J. T. Groves, R. Crabtree, J. Chatt, G. J. Leigh.

Needless to say, none of the persons mentioned is responsible for the mistakes and shortcomings which are undoubtedly present in the book. I hope to be able to improve it in the future and will be grateful for critical remarks and comments.

One of the probable dangers in writing this book lies in the fact that the substrates of the catalytic reactions described have been studied in the laboratory of the author: on the one hand, it is always more pleasant to write about the things which are known not only from other books or articles but from one's own experience and which are understood in the process of continuous change from hopes to disappointments, with several successes and many more failures. On the other hand, this way of choosing the content may lead to a certain subjectivity. I have always tried to be objective and to present quite balanced views, but probably did not always succeed in my intentions. However, even if this is true, I would like it to be subjective in the sense of the words written in the book by Sir Cyril Hinshelwood* in whose laboratory at Oxford I had the pleasure to work for nine unforgettable months: "If the treatment is in places impressionistic, I hope it is so in the better sense which would allow a painting to be not less true than a photograph. At any rate I believe that the short-comings are in the execution rather than in the method."

* Hinshelwood, C. N., *The Kinetics of Chemical Change*, Oxford University Press, Oxford, 1940.

I believe that the most important field of research for the chemistry of the future is what we would call chemical bionics; now it is still at its infancy. It includes the use of the principles of chemical reactions in living organisms for development of new catalytic processes.

N. N. Semenov, "Science and Society," 1981

INTRODUCTION

The chemistry of the reactions occurring in living organisms is striking to a chemist's imagination. It is so different from "usual" chemistry that the term itself seems to be unsuitable. When the word "chemistry" is used, it immediately reminds us of industrial apparatus or laboratory vessels sometimes functioning at high temperatures and pressures, often using or producing very unpleasant chemicals. These chemicals are sometimes dangerous and, despite all precautions, pollute the environment, to say nothing of causing innumerable accidents leading to sad consequences and even to the death of people. Chemical processes waste much energy and expensive raw materials. When we want to stress the high quality of certain fruits or vegetables which are bought in a shop or are grown in our garden we say (at least in Russia) that they have been produced "without any chemistry".

Meanwhile, all of nature is filled with an immense number of various chemical reactions constantly proceeding in all living organisms and constituting the basis of life. Molecular biology is actually a chemistry of proteins, carbohydrates, fats, nucleic acids, etc. with different transient species such as free radicals, ions, and molecular complexes participating.

However, as distinct from the chemistry we are accustomed to in our laboratories or in chemical industry, the natural chemical processes are mostly ecologically safe. Moreover environmental pollution, which is produced by man, is efficiently removed by living organisms. Unfortunately, it is more and more difficult for nature to cope with the consequences of irrational human activity and we can no longer behave with impunity in recognizing our reckless large-scale projects.

In nature the processes are very economical from the viewpoint of energy waste; moreover, over the time span that life has existed on our planet, energy has been accumulating in the forms of coal, oil, and natural gas as the products of photosynthesis. Now mankind is carelessly wasting these energy deposits using very simple chemistry which mostly consists of burning organic materials in furnaces, car engines, etc. to form carbon dioxide and water with additions of such harmful compounds as carbon oxide and nitric oxides.

In spite of the strict limitation of conditions required for natural chemical reactions (usually atmospheric pressure, low temperature, the presence of water and in most cases oxygen), a variety of very complex chemical substances are produced which are often difficult to synthesize in the laboratory despite all our methodic achievements (which are great) and our freedom to choose the conditions. High selectivity, which is now required in all modern industrial chemical processes, has already been successfully achieved in reactions of living nature. With enzymes which are natural catalysts, the compounds needed for an organism are produced selectively and stereospecifically, and the compounds considered waste matter for one kind of organism are fully used by others.

Some basic chemical reactions realized in living nature are very simple from the viewpoint of their stoichiometry, but until recently they were completely unknown in the laboratory and, thus, have excited the chemist's imagination. The reactions which will be described in this book: molecular nitrogen reduction in solution, selective hydrocarbon oxidation (including that of methane), photodecomposition of water, belong to this category.

Molecular nitrogen (dinitrogen) is known to be chemically very inert. Until very recently, every chemical textbook stated that at low temperatures it was unable to react with anything except lithium metal. Industrial ammonia synthesis is performed at high pressures (500 to 600 atm.) and temperatures (ca. 500°C). The reacting gases must be kept very pure since various impurities may easily poison the catalyst. At the same time, it has been known since the last century that a number of microorganisms can fix dinitrogen, naturally, at ambient pressures and temperatures in the presence of water and dioxygen. The primary product of dinitrogen reduction is ammonia, which is subsequently used in protein synthesis. Even now, when industrial ammonia synthesis has been developed at a very large scale, biological nitrogen fixation remains the main route of dinitrogen circulation between the free atmospheric dinitrogen and nitrogen in organonitrogen compounds, e.g., proteins.

Methane can be of course easily burned to produce carbon dioxide, but to produce efficiently and selectively methanol or formaldehyde is very difficult, since usually the oxidation products formed from methane (except the final product, carbon dioxide) are more chemically reactive than methane. Meanwhile, there are enzymes in living organisms which selectively hydroxylate hydrocarbons to alcohols, the C–H bond in the terminal CH_3 group which is the strongest thermodynamically being often hydroxylated preferentially. Methane monooxygenase oxidizes methane to methanol more easily than any other alkane, although methane is known to be the least chemically reactive among other hydrocarbons.

Transformation of carbon dioxide and water into carbohydrate and dioxygen under solar light is known to be the primary reaction of photosynthesis, which is the fundamental process, providing energy to all other living organisms, i.e., responsible for the very existence of life on the earth. It looks chemically very simple but requires a complicated mechanism, only now becoming understood in its fundamental details. Photodecomposition of water, which is even simpler as to its stoichiometry, accumulating almost the same amount of light energy in reaction products and forming the basis of photosynthesis in plants, is still very difficult to realize in the laboratory at least with solar light and with sufficiently high efficiency.

Naturally, chemists were fascinated by the existence of these and other natural processes and wanted to reproduce them in the laboratory. It must be stressed that at present this is not only a dream but has become an urgent necessity: we must learn from living nature how to perform chemical reactions without harmful consequences, otherwise the very existence of the mankind will be threatened. Now it is becoming more evident that we shall not survive the pollution of the environment, even if it remains at the present level.

The task of mimicking the enzymatic reactions in the laboratory looked very difficult, if at all possible, until even comparatively recently. The main reason was the lack of knowledge of enzymatic mechanisms and the singularity of chemical reactions in biological systems. Sometimes this made the chemists suggest fantastic hypotheses to explain the existence of such reactions and their high rate. Now there exist a number of various methods to determine the structure of enzymes, and the mechanism of their functioning is becoming ever more clear. In many cases the enzymes can be obtained in crystalline form; the development of X-ray analysis of single crystals has revolutionalized our knowledge of their structure and the mechanism of enzymatic catalysis. Modern computer analysis allows us to solve the problem of enzyme structure comparatively quickly. Of course, it is far from being a simple task (one must realize that the molecules of enzymes have molecular weights reaching tens and hundreds of thousands), but we obtain the coordinates of all the atoms including those constituting an enzyme active center and its closest surrounding. The general conclusion that may be made from this analysis is that the mechanism of enzymatic catalysis is, on the whole, analogous to the mechanisms of "usual" chemistry. The main difference is the much more complex and perfect *organization* of the interacting molecules in the enzyme active center. It is this organization which leads to the qualitative differences in the chemistry of living nature as compared with ordinary chemistry. The situation may be compared with what we know about computers which are based on simple elements and simple principles. Their complex organization, however, leads to remarkable consequences and their functioning reminds us of the behavior of living creatures.

This result might be expected if one were to proceed from the assumption that life originated from inanimate nature as a result of the combination of certain processes of ordinary prebiological chemistry and the subsequent evolution of organized systems which came into existence. It is improbable that each separate process, such as those mentioned above, could appear in the living organism in its modern well-organized system. It is more likely that the processes were originated accidentally in imperfect, poorly organized form and then became more complicated and efficient over the process of evolution. It is of importance that already in an initially poorly organized form, the catalytic reaction had to be sufficiently efficient to promote the existence of prebiotic systems, otherwise it would have been unutilized and would have disappeared.

This allows us to consider optimistically the possibility of performing reactions which are known only in living nature in purely chemical conditions, particularly for small and simple substrates. We may count on the discovery of these reactions, at least in not very efficient systems, and then on their development to more perfect states proceeding from the knowledge of their mechanisms and using chemical means as well as principles of biological systems with their complex organization.

This approach, which was coined recently "biomimetics",[1] is more and more frequently used by chemists. Forming the corresponding organization of the active center (i.e., applying "molecular design" or "molecular engineering"), we can achieve a great acceleration of the reaction and make it considerably more selective as compared with the usual "unorganized" systems, although the mechanism remains essentially the same. The biomimetic approach turned out to be very effective in solving the problem of activation of small molecules. Recently, a number of new reactions were discovered which are essentially filling the gap which had existed between enzymatic reactions of such substrates as dinitrogen, dioxygen, dihydrogen, methane, carbon dioxide, and their reactions in purely chemical conditions.

The reason for the choice of the three types of the reactions mentioned above for this book, i.e., dinitrogen reduction, activation and oxidation of alkanes, including methane, chemical models of the photosynthesis (apart from that already mentioned: they all have been studied in the laboratory of the author), was first of all connected with their importance: in living nature they constitute the base of life on earth. At the same time, the reactions chosen for the book have something in common: the substrates are particularly inert molecules and to activate them transition metal complexes have to be used as catalysts both in living nature and in the laboratory.

In all three cases the role of di- and polynuclear complexes was found to be essential. Non-radical mechanisms which are thermodynamically favorable are demonstrated for the most difficult cases. At the same time, for different substrates optimal catalytic properties require different metals and different oxidation states of the metals. Consideration of the complexes used as catalysts helps us understand why certain metals have been chosen by nature to play a central part in the enzymatic processes.

As will be seen in this book, in order to overcome the inertness of the chosen molecules, somewhat general approaches were used (although the molecules are representative of very different classes of chemical entities). It may be hoped that the principles applied for their activation may be also used for other types of perhaps less inert molecules.

THE AUTHOR

A. E. Shilov, Ph.D., D.Sc., is presently director-organizer of the Institute of Biochemical Physics of the Russian Academy of Science (Moscow) and head of the laboratory of complex catalysts at the Institute of Chemical Physics (Chernogolovka, Moscow region).

Born in 1930, Dr. Shilov was graduated from Kiev University in 1952, then entered the Moscow Institute of Chemical Physics as a postgraduate student where he started work under the supervision of Professor N. N. Semenov. He obtained his Ph.D. in 1955 and his D.Sc. in 1967.

Although he initially worked primarily on problems related to gas kinetics and mechanisms, he later shifted to reactions in solution, homogeneous catalysis by metal complexes, and chemical models of metalloenzymes.

His achievements include: discovery of carbene mechanisms in thermal decomposition (1955–1957); discovery of branching chain reactions with branching due to excited molecules produced in exothermic chain propagation stages (1963); formulation of the free ionic mechanism of olefin polymerization in homogeneous Ziegler-Natta systems (1961–1963); discovery of new reactions of dinitrogen in solution of metal compounds, including catalytic N_2 reduction to hydrazine and ammonia in protic media; formulation of four-electron mechanism of N_2 reduction in polynuclear systems (1965–1985); discovery of activation of alkanes, including methane, in solution of metal complexes, particularly by Pt (II) complexes; demonstrations of metal-alkyl intermediates in reactions with alkanes; formulation of five-coordinate carbon mechanism in hydroxylation of alkanes by high-valent metal complexes in enzymatic as well as in model chemical systems (1969–1985); discovery of photocatalytic systems evolving dihydrogen and dioxygen from water; and development of chemical models of photosynthesis (1975–1985).

In 1981 he was elected corresponding member and in 1990 full member of the Russian Academy of Sciences. Since 1992, he has been a member of Academia Europaea.

To the Memory of My Teacher,
Professor N. N. Semenov

CONTENTS

Chapter I

NITROGEN FIXATION IN SOLUTION

It may be that this solution [catalytic synthesis of ammonia] *is not the final one. Nitrogen bacteria teach us that Nature, with her sophisticated forms of the chemistry of living matter, still understands and utilizes methods which we do not as yet know how to imitate.*

Fritz Haber, "The synthesis of ammonia from its elements, "
Nobel Lecture, June 2, 1920.

I.1 INTRODUCTION

The existence of the biological nitrogen fixation by certain bacteria, which has been known since the last century, has inspired chemists to search for purely chemical systems capable of fixing dinitrogen catalytically under mild conditions. Until the 1960s, however, all the attempts to observe dinitrogen reactions at low temperatures and pressures failed and the reaction with some metals, mainly with lithium to form nitrides, remained the only example of N_2 reaction at room temperature. Some publications appeared in the literature before the 1960s claiming that small amounts of ammonia can be observed in reactions of dinitrogen with some reducing agents. Later, however, it was realized that the ammonia was the product of some impurities present in reagents, e.g., nitrides, or nitrogen oxides in dinitrogen. At the beginning of our work in this field we often obtained "positive" results of ammonia formation, e.g., passing dinitrogen through the solution of molybdate in aqueous HCl in the presence of zinc metal, which was dissolving in it (hoping to use hydrogen *in statu nascendi*). Sensitive analysis always shows the presence of some ammonia. Blank experiments (e.g., without zinc or without dinitrogen) may demonstrate some decrease of ammonia yield. It gives hope to the researchers, but further attempts to increase the yield fail in spite of all efforts and eventually one suspects that ammonia originates from something other than dinitrogen. Special experiments with dinitrogen enriched with the isotope ^{15}N usually help to realize the mistake but must be performed with care because the mass-spectroscopic analysis that is generally applied to find out whether ammonia in solution is enriched by ^{15}N may itself be the source of an erroneous conclusion: the mass number 29, which corresponds to dinitrogen-29 produced at the oxidation of ammonia containing some amounts of ^{15}N, may be also due to small amounts of organic materials present, e.g., ether, which produces $C_2H_5^+$ in the ionization chamber of the mass-spectrometer with the same mass number (29).

Thus disappointments may happen with both usual chemical and mass-spectrometrometrical analysis for ammonia. Nobody knows how many investigators experienced these disappointments in their works because apparently only few of them were published (before the authors realized that they had committed mistakes...).

Biological nitrogen fixation was studied in parallel with these attempts by the chemists but until the beginning of the 1960s the process was observed only in living bacteria. For a long time the attempts to isolate functioning enzymes or at least to observe the reaction of N_2 *in vitro* in the decomposed cells failed. This of course hampered the attempts to investigate the chemistry of N_2 reduction. Even the fact that the primary product is ammonia was frequently put in doubt.

In 1960 the dinitrogen reduction in supernatant solution produced from the cells was observed for the first time in the presence of the ATP-generating system and this work has started a new development in biochemical investigation of nitrogenase chemistry. The enzyme was isolated rather soon after that and a number of physical methods were applied for investigations by many authors. Since then the progress in understanding of the chemistry of dinitrogen enzymatic fixation has greatly accelerated.

Almost at the same time in the early 1960s, first reactions of dinitrogen in solution were discovered in the presence of transition metal complexes first in aprotic media with strong reducing agents and then in water and water–alcohol mixtures with less stronger reductants. Almost simultaneously first dinitrogen complexes with transition metal compounds were discovered and turned out unexpectedly to be very stable. At the beginning of the development of this new field of coordination chemistry first metal dinitrogen complexes revealed no chemical reactivity of coordinated dinitrogen besides the ability of N_2 to liberate itself from the coordination sphere. Soon, however, first intermediate complexes with dinitrogen capable of further reduction in the metal coordination sphere was observed and isolated, and subsequently many dinitrogen complexes were discovered, both mononuclear and dinuclear, with N_2 capable of forming hydrazine and/or ammonia at protonation. These three fields, dinitrogen biological fixation, dinitrogen reduction in solution, and dinitrogen complexes with transition metal compounds, are clearly closely related and developed in parallel influencing each other, but for a long time the structure of nitrogenase and the mechanism of its functioning remained obscure, allowing different hypotheses to be tested. Perhaps it was not so bad because new systems were discovered and compared with nitrogenase. Some generalizations can be made now and it is clear that the enzymatic system is one of the number of systems reducing dinitrogen in protic media, all having something in common but each having its own peculiarities.

Meanwhile, the so-called iron–molybdenum cofactor was isolated from one of the proteins involved in nitrogenase, apparently playing a central role in the activation and reduction of dinitrogen, and finally X-ray structures of both proteins, Fe protein and MoFe protein, that form the nitrogenase complex were recently determined. Although the exact mechanism of dinitrogen reduction is not yet completely clear there is much less freedom of choice between possible schemes now, and with low-temperature dinitrogen chemistry already in existence the next steps in the development of chemical models of the enzymatic nitrogen fixation are clearer than ever before.

I.2 GENERAL CONSIDERATIONS

In this section properties of the N_2 molecule and the main causes of its chemical inertness will be considered, as well as the possible ways of overcoming it. These considerations were developed in parallel with experimental work on biological dinitrogen reduction, dinitrogen complexes, and new reactions of dinitrogen in solution. Some of the ideas were suggested before experimental evidence was obtained, while the others were the result of experimental findings. Not all of the conclusions are equally certain and perhaps some of them are still somewhat speculative.

Inertness of dinitrogen is very well known and reflected in its characteristics[2] N_2 has very high bond dissociation energy, high ionization potential, and negative value of electron affinity (Table I.1).

The dissociation energy of the N≡N bond is very large, the ionization potential is high, the electron affinity has a large negative value, and the proton affinity though positive is smaller than, for example, for methane (5.3 ± 0.3 eV).[5] Dinitrogen is a very weak base and does not interact even with the strongest acids.

It should be noted, however, that the strength of thc triple bond does not by itself explain the inertness of dinitrogen. Triple bond dissociation energy in acetylene (230 kcal/mol) is approximately the same as that for dinitrogen and carbon monoxide bond energy is even higher (256 kcal/mol), but both the latter molecules undergo multiple chemical reactions unknown for dinitrogen.

I.2.1 MOLECULAR ORBITALS OF DINITROGEN[2]

The electronic configuration of dinitrogen may be represented as $(1\sigma_g)^2(1\sigma_u)^2(2\sigma_g)^2(2\sigma_u)^2(1\pi_u)^4(3\sigma_g)^2$. Therefore the highest occupied orbital $3\sigma_g$ is higher than $1\pi_u$ due to the mixing of *s* and *p* orbitals. Figure I.1 represents the approximate boundary surfaces of the molecular orbitals and their energies. It is seen that the lowest unoccupied molecular orbital $1\pi_g$ is situated rather high and two occupied orbitals $3\sigma_g$ and $1\pi_u$ are strongly bonding, their energy differing by 1.5 eV.

TABLE I.1
Some Physical Constants and Characteristics of Dinitrogen

Interatomic distance	1.095 Å
Ionization potential	15.58 eV
NN bond dissociation energy	225 kcal/mol
Vibration frequency (gas)	2231 cm^{-1}
Electron affinity	–1.8 eV[3]
Proton affinity	5.12 eV[4]
Solubility	
In water	1.7×10^{-2} cm^3/cm^3
In benzene	1.11×10^{-1} cm^3/cm^3

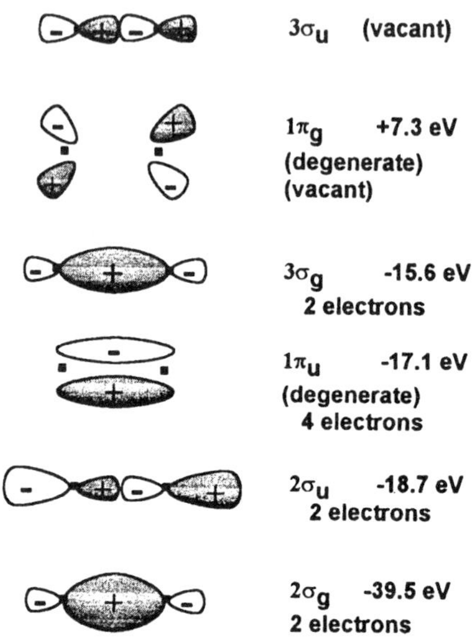

FIGURE I.1 Cross section of boundary surfaces of the molecular orbitals of dinitrogen and their energies.

I.2.2 THERMODYNAMICS OF CONSECUTIVE BOND CLEAVAGE IN DINITROGEN[5-7]

Important conclusions about the reasons of dinitrogen inertness as well as the clue to the overcoming it may be achieved when considering the energies required for consecutive cleavage of the three bonds in the N_2 molecule. Dissociation of the first of the three bonds (requiring more than 100 kcal/mol) corresponds to about half of the total triple bond energy. It distinguishes this molecule sharply from other molecules with multiple bonds, e.g., acetylene (Figure I.2) whose first splitting bond is, on the contrary, the weakest one (53 kcal/mol). The sharp difference between these isoelectronic molecules is evidently the consequence of the repulsion of the two unshared electron pairs and the electron pair of the remaining π-bond after the cleavage of one of the π-bonds of dinitrogen, whereas the electron repulsion is less strongly manifested in ethylene because four electrons are engaged in two σ-bonds with H atoms in the latter molecule.

This peculiarity of the N_2 molecule is its intrinsic characteristic rather than a reflection of the nature of the species reacting with it. Thus the energy of the dinitrogen triplet state (which may be considered as an N_2 molecule with one of the π-bonds cleaved) corresponds to a much higher level relative to the ground state than the corresponding triplet state level of acetylene.[3]

The difference in the energy distribution between the triple bonds in acetylene and dinitrogen leads to drastic differences in the reactivities of these molecules (although the total triple bond energies in both molecules are almost the same): one- and two-electron reactions that are thermodynamically favorable for acetylenes (and also olefins) are often forbidden for dinitrogen. Thus, hydrogenation of acetylene

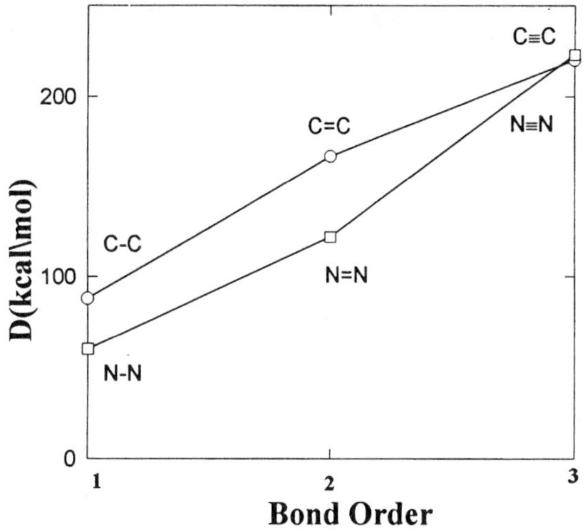

FIGURE I.2 Bond energies of triple bonds in dinitrogen and acetylene.

(as well as ethylene) is strongly exothermic (ΔH_r^o for the reaction $C_2H_2 + H_2 \rightarrow C_2H_4$ is 42 kcal/mol) whereas for dinitrogen the corresponding reaction $N_2 + H_2 \rightarrow N_2H_2$ is strongly *endothermic* and ΔH_r^o is 51 and 56 kcal/mol for *trans*- and *cis*-diazene, respectively.[8] The addition of the H atom to acetylene is 41 kcal/mol exothermic while the reaction: $N_2 + H^\bullet \rightarrow {}^\bullet N_2H$ is about 9 kcal/mol *endothermic*.[9] Therefore, neither a radical chain reaction of dihydrogen with dinitrogen, which could be initiated by the addition of a hydrogen atom to N_2, is possible nor the catalytic hydrogenation via diazene as an intermediate (and this is probably the main reason for the absence of activity of typical hydrogenation catalysts towards dinitrogen).

Formation of ammonia, which is a key process in the enzymatic dinitrogen fixation, is via N_2 reduction. The strength of the first bond of the N≡N bond that is being cleaved is reflected in the values of redox potentials E^o (Figure I.3).[6]

One- and two-electron transfer with simultaneous addition of protons:

$$N_2 + e + H^+ \ 1 \ N_2H \qquad (1)$$

$$N_2 + 2e + 2H^+ \ 1 \ N_2H_2 \qquad (2)$$

correspond to the negative values of E^o, i.e., requires stronger reducing agents than dihydrogen. (All redox potentials in this book are given vs. the normal hydrogen electrode.)

In aprotic media strong reducing agents, like lithium metal, are capable of reacting with dinitrogen to form nitrides. These reactions seem to be very unlikely in protic media because the conditions become very favorable to a simpler reaction of dihydrogen formation from two protons:

$$2H^+ + 2e \rightarrow H_2 \qquad (3)$$

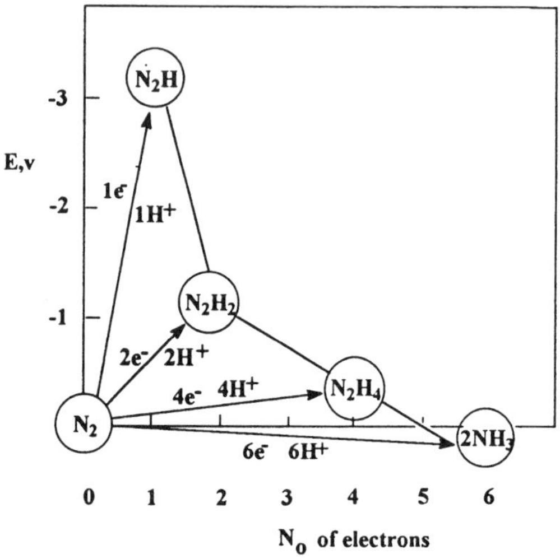

FIGURE I.3 Redox potentials of one-, two-, four-, and six-electron reduction of dinitrogen.

At the same time, thermodynamic considerations of consecutive bonds cleavage in dinitrogen demonstrate its "weak points" relative to reducing agents. The second and the third bonds that are being cleaved in the molecule are, on the contrary, very weak: about 60 kcal/mol each.[7] Therefore the addition of two dihydrogen molecules to dinitrogen forming hydrazine:

$$N_2 + 2H_2 \rightarrow N_2H_4 \qquad (4)$$

is endothermic to only 20.7 kcal/mol in the gas phase and 8.2 kcal/mol in water solution while the addition of three dihydrogen molecules forming ammonia is already exothermic (22 and 38 kcal/mol [NH_3] in the gas and in water, respectively). Thermodynamically, dinitrogen reduction in protic media is facilitated because the polar molecules formed, i.e., hydrazine and ammonia are solvated via hydrogen bonds formation. Reduction of dinitrogen to hydrazine and ammonia in water solution corresponds to redox potentials of –0.36 and +0.55 V, respectively, and thus to considerably weaker reductants than for one- and two-electron reduction.

Therefore, four- and six-electron reduction of dinitrogen opens the opportunity of the use of comparatively weak reducing agents. Evidently, simultaneous addition of three or even two hydrogen molecules is impossible (moreover, these reactions are symmetry forbidden and the second reaction is impossible also for thermodynamic reasons). Synchronous transfer of four or six electrons (and four or six protons) to dinitrogen from four of six molecules of a one-electron reducing agent not bound to N_2 is clearly impossible too; however, dinitrogen thermodynamic peculiarities make it possible to reduce N_2 by polynuclear transition metal complexes and to use such complexes as catalysts of dinitrogen reduction by mild reducing agents.

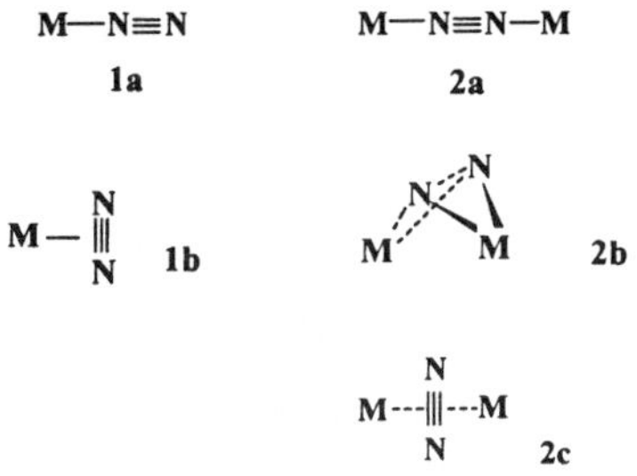

FIGURE I.4 Typical structures of dinitrogen complexes with transition metal compounds (M = metal + other ligands).

Let us suppose that we have a catalyst capable of accepting four electrons one by one and then transferring them to dinitrogen in one step. The catalyst may consist of a cluster including several metal ions capable of accepting and transferring electrons. If the redox potential of each one-electron step does not differ much from the others and is close to the redox potential of the hydrazine molecule, the thermodynamic restrictions for dinitrogen reduction will be fulfilled. This general principle (originally formulated in 1969 to 1970)[10] may be applied to all the molecules capable of multielectron reduction or oxidation. As a rule one of the one-electron steps in consecutive multielectron processes requires more energy than the others and a reductant (or an oxidant) capable of performing the most difficult step in the noncatalytic stepwise process is needed; in other steps the surplus energy will be wasted, being transformed into heat. In the case of a catalyst, capable of switching over to a multielectron process from the one-electron transfer, a reductant (or an oxidant) that corresponds to an average energy will be needed, i.e., definitely smaller than that required for the most difficult step in the multistep process. This is particularly important for the catalytic reduction of dinitrogen with its extremely uneven distribution of energy in the stepwise process.

I.2.3 TRANSITION METAL COMPLEXES: CATALYSTS FOR FOUR-ELECTRON DINITROGEN REDUCTION

As was already mentioned, dinitrogen, similarly to other unsaturated molecules, turned out to be able to form complexes with transition metal compounds in which N_2 reveals itself as both electron donor and electron acceptor, the latter properties being much more pronounced. Different kinds of the complexes already observed experimentally (*vide infra*) and widely considered theoretically (e.g., see Reference 11) are presented in Figure I.4.

In aprotic media these complexes may react further with a reducing agent to form hydrazides and nitrides, producing hydrazine and ammonia under the subsequent action of an acid. It is of importance that dinitrogen becomes a much stronger base in these complexes than in the free state and may be protonated, which in turn strengthens its acceptor properties and facilitates the reduction in protic media. Linear mononuclear and dinuclear complexes M–N≡N and M–N≡N–M are as a rule more stable and more thoroughly studied than other types of complexes. Their

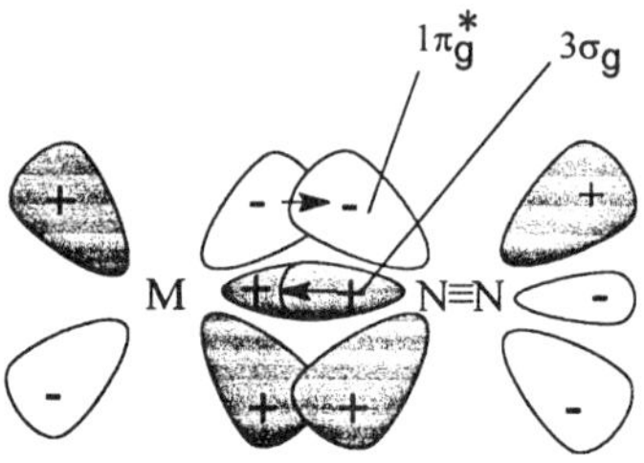

FIGURE I.5 Interaction of metal and dinitrogen orbitals in linear mononuclear complex M–N≡N.

relative stability is due apparently to the fact that the highest occupied molecular orbital ($3\sigma_g$) is a σ orbital. Complexes of other types may turn out, however, to be important as the intermediates in a reduction process. $d_\pi \rightarrow p_\pi$ interaction is the most essential for the reduction of dinitrogen in the coordination sphere of the complexes.

Now let use consider metal–dinitrogen complexes from the viewpoint of their participation in N_2 reduction.

I.2.3.1 Mononuclear Complexes

Linear end-on complexes M–N≡N were already mentioned to be more stable than side-on complexes $M-\overset{N}{\underset{N}{|||}}$, dinitrogen being more activated in linear complexes because metal *d* orbitals can interact with both vacant π_g^* orbitals of dinitrogen. Therefore protonation of dinitrogen and its further reduction in side-on complexes of the type 1b (Figure I.4) is less likely than in the linear end-on complexes, though the former may be intermediates in the formation of more complex di- and polynuclear complexes in the process of the reduction.

Figure I.5 shows the scheme of the interaction of a metal and dinitrogen orbitals in a linear complex of type 1a (Figure I.4). For the maximum weakening of the π-bonds in dinitrogen at least four electrons are needed and, taking into account that an electron may be shifted to the nonbonding d_{xz} orbital of M (according to the Hund rule), the number of *d* electrons at M should be increased to at least five. The reduction of dinitrogen may proceed at protonation provided it is thermodynamically allowed:

$$M^n - N \equiv N \xrightarrow{+4H^+} M^{(n+4)+} + N_2H_4$$
$$M^n - N \equiv N \xrightarrow{+6H^+} M^{(n+6)+} + 2NH_3 \qquad (5)$$

To avoid a thermodynamically difficult formation of diazene, N_2H_2, and to form directly hydrazine or ammonia, metal ion M has to increase its oxidation state by four or six units, respectively. Evidently this corresponds to an initially very low oxidation state of M and apparently M must be a sufficiently strong reducing agent. For a catalytic cycle even stronger reductant will be needed to return M to its initial reduced state active towards dinitrogen. Possibly the situation will improve for the feasibility of catalysis if M were to be attached to a cluster of other metals capable of transferring their electrons to replace those shifted to dinitrogen; however, it is not clear whether this will be sufficient for an effective catalytic process in protic media if dinitrogen remains bound to only one atom of the cluster.

I.2.3.2 Linear Dinuclear Complexes M–N≡N–M

The complexes of the type 2a (Figure I.4) are the most widespread among the dinuclear dinitrogen complexes (*vide infra*). The mechanism of dinitrogen reduction at protonation may be presented in the following way:

$$M-N\equiv N \xrightarrow{H^+} M{=}N-\overset{H}{\underset{+}{N}}{=}M - \begin{cases} \xrightarrow{3H^+} 2M^{(n+2)+} + N_2H_4 \\ \xrightarrow{5H^+} 2M^{(n+3)+} + 2NH_3 \end{cases} \quad (6)$$

Two electrons from each M are interacting in this mechanism with two π-orbitals of dinitrogen and therefore the cleavage of two π-bonds in N_2 proceeds simultaneously. This mechanism, which allows one to use weaker reducing agents than with mononuclear complexes, is a good concrete illustration of the possibility of the four-electron mechanism of N_2 reduction considered above. The complex M–N≡N–M may be represented by the two extreme structures:

$$M{-}N{\equiv}N{-}M \leftrightarrow M{=}N{-}N{=}M \quad (7)$$

with different weights of each for various M. The structure, which might be considered as the product of two-electron reduction of dinitrogen, M–N=N–M, does not have any physical sense in this scheme and definitely is not intermediate between the two structures in Equation 7. It should be mentioned here that the reduction of dinitrogen is traditionally considered in the literature as a sequence of two-electron processes: the formation of diazene is believed to be facilitated by its coordination on metal complexes. It is clear that in the proposed mechanism, with both π-bonds affected and four electrons from two M involved, the approach with coordinated diazene or its derivative is an unnecessary formalism: the cleavage of two π-bonds in the dinuclear dinitrogen complexes occurs simultaneously.

In aproptic media when M is a sufficiently strong reducing agent full cleavage of the NN bond may be expected to form metal nitride:

$$M{-}N{\equiv}N{-}M \rightarrow 2M{\equiv}N \quad (8)$$

Protonation of the dinitrogen ligand will shift the structure approaching the hydrazine derivative and can lead to hydrazine if this is a thermodynamically favorable reaction. Direct formation of ammonia may be also visualized if each M is capable of providing three electrons:

$$M-N\equiv N-M \xrightarrow{H^+} M{=}N-\overset{H}{\underset{+}{N}}{=}M \xrightarrow{5H^+} 2M^{3+} + 2NH_3 \quad (9)$$

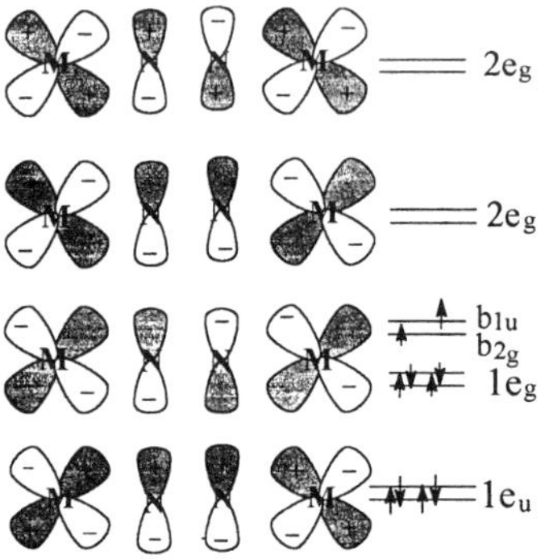

FIGURE I.6 Diagram of molecular orbitals for linear dinuclear complex M–N≡N–M with octahedral surrounding of M.

If in the dinuclear dinitrogen complex each M has two electrons available but two Ms are not sufficiently strong reductants to produce hydrazine at protonation then participation of an additional external reducing agent providing two electrons (simultaneously or one by one) may lead to the cleavage of the N–N bond and formation of the ammonia derivative, precursor of ammonia, which will be formed in hydrolysis, e.g.,

$$M-N\equiv N-M \xrightarrow{2e,\,2H^+} 2M=NH \xrightarrow{4H^+} 2M^{2+}+2NH_3 \qquad (10)$$

This stepwise overall 6-electron process will require a less strong reducing agent than for hydrazine formation, the external reducing agent being needed to provide what could be called *reductive coaction.*

For a linear dinuclear complex M–N≡N–M there is an optimum electronic configuration of M for the most effective activation of dinitrogen to the reduction through protonation.[6] Figure I.6 shows a simplified diagram of molecular orbitals for such a complex with the octahedral surrounding of M atoms.[12]

Doubly degenerated Ie_g orbital must be populated fully for maximum antibonding influence on the NN bond and maximum negative charge on N atoms for a given redox potential of M. This corresponds to the presence of four *d* electrons on two Ms, thus, two electrons on each. Also, it is desirable that each nonbonding b_{2_g} and b_{1_u} be populated by one electron to prevent electron transfer from the $1e_g$ orbital to *b* orbitals. The further increase in the number of *d* electrons for a given redox potential of M will not increase its reducing ability. When both b_{2_g} and b_{1_u} orbitals are fully populated by two pairs of electrons they become the highest occupied orbitals and the reducing power of M is used more for them than for the $1e_u$ orbital. When more electrons are added to occupy the $2e_u$ orbitals the effect is opposite to that desired for the activation of N_2 because NN bonding is increased. Therefore it might be expected that electronic configurations d^2 to d^4 will be suitable for the catalytic activity of M towards dinitrogen, the d^3 configuration presumably being the optimum one. The increase of the number of the *d* electrons to more than four would lead to the loss of the catalytic activity towards dinitrogen.

To improve the catalytic properties of the intermediate complex M–N≡N–M (with an appropriate electronic configuration of M) it is important to have a flexible bidentate ligand unifying both Ms and forming a cycle when dinitrogen is coordinated (Scheme I.1). In this case the coordination of N_2 will be accompanied by smaller loss in entropy than for two M complexes existing separately in solution, and this will increase the equilibrium constant of the intermediate complex formation.

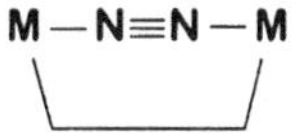

SCHEME I.1 Cyclic dinuclear dinitrogen complex.

I.2.3.3 Dinuclear Complexes of Other Types

Complexes 2b and 2c (Figure I.4) are apparently as a rule less stable than 2a because the highest occupied molecular orbital $3\sigma_g$ of dinitrogen is not effectively used. At the same time they are theoretically feasible[13] and have been observed experimentally. They might be expected to form particularly in two cases:

1. When both M are especially strong electron donors and their acceptor properties are not of particular importance. Other factors may play a decisive role, and 2b and 2c complexes may become stable. The complexes of this kind may be expected for aprotic media.
2. When both Ms are anchored to a common ligand and are situated at a distance insufficient to form a linear complex (the distance between the two metals for such a complex must be about 5 Å) but suitable for 2b or 2c (3.5 to 4 Å). There is no reason these complexes could not be intermediates in dinitrogen reduction in protic media.

For complexes 2b and 2c the optimum number of *d* electrons at M for the activation of dinitrogen may be larger than for the linear complexes; in any case, the limitation of the number of electrons for these complexes is not evident.

I.2.3.4 Polynuclear Complexes M_nN_2 and an Optimal Catalyst for Dinitrogen Reduction

When forming mono- and dinuclear complexes with transition metal compounds dinitrogen donor properties are strengthened and this facilitates the addition of an acceptor, e.g., proton. Other metal atoms may play the role of the acceptors, therefore three- and four-nuclear complexes can be formed, e.g., as shown in Scheme I.2.

This will make easier the subsequent reduction of dinitrogen, because the N–N bond will be loosened to a maximum extent and thus four- and six-electron reduction will be facilitated. Protonation of coordinated dinitrogen to form hydrazine or ammonia will lead to a minimum change of the oxidation state of metal atoms forming the complex and thus will require minimum energy.

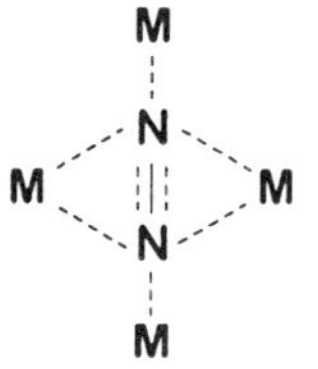

SCHEME 1.2 Tetranuclear η^2-dinitrogen complex.

We can try to visualize an optimal catalyst for dinitrogen reduction in protic media taking into account the above considerations. An optimal catalyst may be defined as the one that functions with the weakest, yet thermodynamically possible reducing agent at pH close to 7. The enzymatic dinitrogen reduction has to be presumably close to this definition.

The catalyst is expected to be a polynuclear cluster with a maximum number of contacts (four, five, or even six) with dinitrogen. In its reduced form the catalyst should acquire an appropriate redox potential, not too negative (which would be difficult to achieve) but sufficient to coordinate N_2 and induce its further protonation. Four electrons of the cluster should be available to enter π^* orbitals of dinitrogen. Metal atoms should be bound by a flexible common support to adjust to changes in distances in the process of dinitrogen reduction.

Electrons from an external reducing agent should move to the catalyst together with protons, each proton and electron simultaneously or immediately one after the other to avoid the accumulation of negative or positive charges, which would destabilize the catalyst. Protons can go directly to the dinitrogen ligand, if it is sufficiently basic to accept a proton from the existing donor, or temporarily may form hydride by addition to a metal. Each electron transferred will shift electrons from the cluster to coordinated dinitrogen. At some stage the protonated dinitrogen ligand will be transformed to a hydrazine or ammonia derivative. This step is likely to be particularly difficult because the oxidation state of several surrounding metal ions must be suddenly increased. It may require additional electron transfer from the external reducing agent (for *reductive coaction*) and this must be situated closely enough to the catalytic site to perform the effective electron transfer but not to strongly influence the catalytic center redox potential. It is clear that the optimal catalyst must be a highly organized polynuclear molecular system functioning in a concerted way. We shall see that enzymatic dinitrogen reduction fully corresponds to this picture.

I.3 BIOLOGICAL NITROGEN FIXATION

I.3.1 HISTORICAL REVIEW

Biological nitrogen fixation was discovered at the end of the 1880s when Hellriegel and Wilfarth provided clear-cut proofs of its existence.[14a] Later it was found that ability to fix nitrogen is very widespread among procaryots and archebacteria. It is observed for microorganisms of both anaerobic and aerobic bacteria, free-living and in symbiosis with plants, for example, with leguminous plants. The list of reliable diazotrophs includes not less than 200 microorganisms that live almost in any media where life exists.[14b]

The total amount of nitrogen fixed by microorganisms per year is estimated to be 150 to 200 million tons. For billions of years this was the main process in nitrogen circulation in nature. Even presently biological nitrogen fixation provides more bound nitrogen than industrially created by man, although industrial synthesis of ammonia from dinitrogen and dihydrogen is already approaching in volume that produced by microorganisms.

For a long time since the discovery of biological nitrogen fixation the chemistry of this process was completely unknown. This was mainly due to the difficulties in studies: until the 1960s nitrogen fixation was observed only on living cells of bacteria.

In the 1930s Bortels[15a] observed that metals (molybdenum and iron) essentially affect the process of nitrogen assimilation from molecular nitrogen. It was also shown that nitrogen assimilation was improved when vanadate was added instead of molybdate.[15b] Thus, already in the 1930s it could be concluded that nitrogen-fixing enzymes could contain molybdenum or vanadium, the metals that could participate in nitrogen binding.

Even the nature of the first product of the process long remained unclear. Winogradsky expressed the viewpoint that dinitrogen enters into the reduction reaction to produce initially ammonia.[15c] However, this conclusion did not seem unambiguous at that time and other hypotheses were advanced: it was suggested, for example, that hydroxylamine was formed from dinitrogen as an intermediate as well as (in another hypothesis) some organonitrogen compounds (see Reference 16).

From the beginning of the 1940s, labeled dinitrogen $^{15}N_2$ was introduced into the investigation practice.[17] This helped to intensify studies and make more unambiguous conclusions. Using this approach Burris and Wilson confirmed ammonia to be the first "key" intermediate.[16]

The principal advance in the study of biological nitrogen fixation happened in 1960, when Mortenson et al.[18] became the first to show that nitrogen can be reproducibly reduced not only by intact cells, but also by the extracts from specially destroyed cells of a nitrogen-fixing microorganism *Clostridium pasteurianum.* It was elucidated that ATP and magnesium salts must be introduced into the system, because dinitrogen reduction occurs as a process coupled with ATP hydrolysis, which is catalyzed by magnesium salts.

After this work, which made possible the use of biochemical and physical methods, the development of studies of enzymatic nitrogen fixation was greatly accelerated. It turned out that in experiments *in vitro* a simple inorganic reductant (sodium dithionite, $Na_2S_2O_4$) could be used,[19] which is no less efficient than natural reducing agents (ferredoxin or flavodoxin).

Soon the enzyme nitrogenase was isolated in a pure state. It was shown that it consisted of two proteins, Fe protein and MoFe protein; both were subsequently isolated and purified. The studies of these proteins have been made since then by available physical and chemical methods. Fe protein is a dimer of identical subunits with molecular mass of 60 kDa. It contains one $[Fe_4S_4]$ cluster and is a specific one-electron donor for MoFe protein. No other reductant of protein or nonprotein nature can replace Fe protein in this reaction. MoFe protein is a tetramer of two types of subunits $\alpha_2\beta_2$ with a total molecular mass of 220 kDa. It contains two atoms of Mo, 32 atoms of Fe and the same number of S^{2-} ions per molecule. These atoms are grouped together in two types of clusters: so-called P clusters and iron–molybdenum cofactors, FeMoco. Only in the presence of a combination of these two proteins does dinitrogen reduction proceed in the reaction whose stoichiometry was shown, under optimum conditions, to be represented by the equation:

$$N_2 + 8H^+ + 8e^- \text{ [for example, } S_2O_4^{2-}] + 16 \text{ MgATP} \xrightarrow{N_2ase} 2NH_3 + H_2 + 16 \text{ MgADP } 16 \text{ HPO}_4^{2-} \quad (11)$$

Two ATP molecules are hydrolyzed at the transfer of each electron. One mole of H_2 is yielded per mole of fixed dinitrogen even at saturating N_2 pressures.

In 1977 the isolation of the iron–molybdenum cofactor from protein was performed for the first time by Shah and Brill.[20] It attracted much attention and was extensively studied by various physical methods, because it was almost certainly the active center on which binding, activation, and reduction of dinitrogen occurred. The solution of the nitrogenase problem as a whole was connected with the elucidation of the cofactor structure. The results of studies of FeMoco are presented below.

Until very recently molybdenum was considered by many, including the author of this book, to play a decisive role in the action of the enzyme. As mentioned above, vanadium can also exert a similar stimulating influence on the growth of bacteria in dinitrogen atmosphere and for a long time was suspected to be able to replace molybdenum in activation of dinitrogen. Biochemical genetics of nitrogen fixation and elucidation of structures and functions of the major part of genes responsible for the nitrogen fixation ability of microorganisms was developed recently, and it was unambiguously shown that Mo is not a unique and obligatory metal for biological nitrogen fixation: there are molybdenum-independent "alternative" nitrogenases, which genetically differ and are encoded by different structural genes. One of them expectedly contains vanadium instead of molybdenum, but unexpectedly a third nitrogenase was detected and this turned out to contain only iron as transition metal. Presently nitrogenases are classified according to the metal content and are called iron–molybdenum, iron–vanadium, or iron-only nitrogenases.

A comparative study of the structures of isolated protein components and functions of all three nitrogenases show that they all are variations on the same theme and are of great structural and functional similarity.

A recent outstanding advance in studies of biological nitrogen fixation was the elucidation of molecular structures of both proteins of nitrogenase (Fe protein and MoFe protein) obtained by X-ray analysis of single crystals and determination of the molecular structure of metal-containing centers of these proteins (P clusters and cofactor).[21] The knowledge of three-dimensional structures of proteins allows studies of the mechanism of the nitrogenase reaction to be raised to a new level, because it generates new opportunities to introduce a real chemical content into considerations for a real chemical mechanism of dinitrogen activation and reduction.

I.3.2 COMPOSITION AND STRUCTURE OF PROTEIN COMPONENTS OF NITROGENASES

I.3.2.1 Fe Proteins

Fe proteins from nitrogenases of different types of microorganisms as well as from molybdenum-containing and alternative nitrogenases have been isolated, purified, and thoroughly studied. (For additional references and more detailed information concerning nitrogenases see References 22 to 28.) Proteins have much in common in sizes, structures, and properties. The proteins are dimers of identical subunits with the total molecular mass about 60 kDa. They contain one $[Fe_4S_4]$ cluster that, as shown by recent X-ray structural analysis of single crystals of Fe protein, binds its subunits using SH groups of cysteins from different subunits as ligands.

Amino acid sequences of Fe proteins of different nitrogenases from more than 20 types of bacteria have been determined and their identity was found to be within 45 to 90%. Fe proteins of molybdenum and vanadium nitrogenases of the same microorganism differ the least; for example, for *Azotobacter vinelandii* the identity is 90%, while for Fe protein of iron nitrogenase the identity is 60%. These data agree well with the data on cross-reactions of proteins: Fe proteins from Mo and V nitrogenases give efficient systems with MoFe and VFe proteins, whereas the Fe protein of Fe nitrogenase is not efficient in combination with either MoFe or VFe proteins.

The spectra and redox properties of all studied Fe proteins are almost the same, because they are due to the existence of an $[Fe_4S_4]$ cluster, which functions as a one-electron donor changing oxidation state from +2 to +1. The method of coordination of the cluster with the protein matrix is identical in all known Fe proteins, and it had been determined even before the three-dimensional structure was obtained by X-ray structural analysis. The parts of the amino acid sequence of enzymes that enter the composition of the active center and the part of the protein molecule that performs the protein–protein subunit interactions are found to be the most conservative. In primary structures of all studied Fe proteins four strictly retained residues of cystein per protein are found. These residues are the most suitable as ligands for binding the iron–sulfur cluster. Consecutive substitution of cystein residues by serine (CH_2–OH) by means of the insertional mutagenesis helps to establish that only two residues of cystein, Cys 98 and Cys 133 (numbering of residues is taken from the Fe protein of *A. vinelandii* sequence), are essential for the manifestation of the activity of the Fe protein. It is these residues of cystein belonging to different subunits that coordinate the cluster and make it the binding bridge between them. This structural peculiarity (coordination of metal centers by amino acid residues from different subunits) was observed previously in the photosynthetic reaction centers and in nitrate reductase; it was also found in MoFe protein of nitrogenases and, probably, it is a common feature of multisubunit metalloproteins.[27,28]

In Fe protein the cluster is located virtually on the surface of the protein globules and is easily accessible to solvent. This is likely to be the reason for strong sensitivity of protein to oxidation and seems to be necessary for its catalytic action.

Another common feature of all primary structures of Fe proteins studied is the existence of two nucleotide-binding parts Glu-X-Glu-X-X-Glu-Lys-Ser (Glu is the residue of glutamic acid, $-(CH_2)_2-COOH$, Lys is that of lysine, $-(CH_2)_3-NH_2$, Ser is serine residue, $-CH_2-OH$, and X is an arbitrary amino acid). The substitution of every one of glutamic acid residues in this fragment, for example, by proline results in the loss of activity of Fe protein, which is due to the blocking of the MgATP binding site on Fe protein. X-ray analysis shows that these parts are located in the slit on the surface of subunits and are situated approximately 20 Å from the iron–sulfur cluster, i.e., their direct interaction is impossible.

The addition of both MgATP and MgADP to Fe protein causes several effects. The midpoint potential is approximately shifted by 100 mV (from about –350 to –450 mV), and EPR and CMD spectra are considerably changed. These effects are usually explained by conformational changes occurring in the protein; the changes are different for the addition of MgATP or MgADP: thus the first facilitates chelation of iron in the cluster by α,α'-dipyridyl and *o*-phenanthroline, while the second does not reveal this influence.

The main function of Fe protein in the nitrogenase reaction consists of MgATP-activated one-electron transfer to the large MoFe protein, which contains centers of the substrate binding and reduction.

The method of insertional mutagenesis allows determination of the amino acid residues that play the principal role in the protein–protein interaction between MoFe protein and Fe protein. They turned out to be arginine and glutamic acid. They are found in all primary structures of Fe proteins. X-ray structural analysis of single crystals of Fe protein shows both of these amino acids to be located on the surface of subunits near the iron–sulfur cluster, where the contact of proteins must take place for electron transfer and ATP hydrolysis. When arginine is substituted, for example, by histidine, Fe protein loses its ability to transfer electrons to MoFe protein, although it continues to produce the complex with MoFe protein and catalyzes the ATP hydrolysis.

I.3.2.2 MoFe Proteins

The large protein components of nitrogenase (MoFe protein for Mo nitrogenase and VFe protein and FeFe protein for the alternative nitrogenases) have molecular mass of about 220 kDa. MoFe protein is a tetramer of two types of subunits, $\alpha_2\beta_2$. VFe and FeFe proteins are hexamers and have two additional subunits with mass of about 13 kDa. Each protein molecule (tetramer or hexamer) contains two Mo (or V) atoms, 32 Fe atoms, and the same number of S^{2-} ions. The studies of the structures of these proteins by EPR, ENDOR, Mössbauer spectroscopy, MCD, EXAFS, extrusion, and chemical modification provided important information about the structure of metal-containing centers in proteins. All the transition metals atoms in MoFe proteins were found to be organized into two types of redox centers. About half of iron is in the composition of $[Fe_4S_4]$ clusters, named P clusters, and the rest, together with molybdenum (or vanadium), is in M centers, also named iron–molybdenum (or iron–vanadium or iron-only) cofactors. P clusters differ somewhat from the usual ferredoxin-like complexes in the number of physical parameters (*vide infra*).

Primary structures were determined for more than ten of MoFe proteins from free-living (aerobic, anaerobic, and photosynthetic) and symbiotic microorganisms for classical and alternative nitrogenases. Considerable similarity was found in sequences of amino acids, primary structures, and subunits from VFe and FeFe proteins, showing a greater similarity to each other (55% of identity) than to MoFe protein (about 32% of identity). In all MoFe proteins isolated from different microorganisms and in analogous proteins of alternative nitrogenases strictly retained structurally similar domains were detected around the cystein residues. These were thought to be the ligands of redox centers in these proteins. This was later confirmed upon the determination of the molecular structure of MoFe protein.

Single crystals of MoFe proteins from three microorganisms had been obtained, but their molecular structures could not be elucidated by X-ray structural analysis due to difficulties in preparation of isomorphically substituted heavy-atomic derivatives of protein. This problem was first overcome by Bolin, who used a synchronous source of X-ray irradiation with several wavelengths. This allowed him to solve the phase problem and determine the three-dimensional structure of protein at a resolution of 5 Å using the method of anomalous scattering on intrinsic heavy atoms of nitrogenase. Results of the X-ray structural analysis of crystalline MoFe protein were published by Bolin et al. in 1990.[29] Two cofactors have been shown to be located in two different subunits of proteins about 70 Å apart. One large (double) P cluster containing eight Fe and eight S^{2-} groups is located at a distance of about 19 Å from each FeMoco. Apparently, the unusual physical parameters of P clusters observed are explained by this close spatial arrangement of two Fe_4S_4 clusters. The electron density pattern indicates the ellipsoidal shape of the FeMoco, but a resolution of 5 Å did not allow any conclusions about a mutual arrangement of iron, molybdenum, and sulfur atoms in the cofactor.

In 1992 Kim and Rees published the results of an X-ray structural analysis of single crystals in MoFe proteins from microorganisms of *A. vinelandii* and *C. pasterianum* at a resolution of 2.7 Å.[21] The results confirmed the data of Bolin about a mutual arrangement of metal-containing clusters in proteins. Additionally it was found that cofactors are located in α subunits (one per each of two subunits) and deeply immersed at a distance at least of 10 Å from the surface of the protein globule. The cofactor is surrounded mainly by hydrophilic residues, although there are also some hydrophobic residues.

I.3.2.3 P Cluster

The total form of the $\alpha_2\beta_2$ tetramer is such that it can be regarded as an αβ dimer with the rotation axis of the second order, which passes through the P cluster pair located between the α and β subunits. The protein surrounding a P cluster consists of hydrophobic residues. In agreement with Bolin's results, P clusters turned out to unite two Fe_4S_4 clusters. The bridges between the two are all S atoms, two S atoms from cysteine residues (Cys β95 and Cys α88) and one S–S bridge formed by sulfurs involved in both united Fe_4S_4 clusters. This structure of the P cluster and its hydrophobic surrounding suggests that its role is not only to transfer electrons to the FeMo cofactor, but also to create (in the fully reduced state) a redox potential more negative than that of a reduced *single* Fe_4S_4 cluster and to keep the electrons without interacting with protons of the media. This is an example of a general principle that is demonstrated in later sections of the book for other enzymatic and model chemical systems: when the cluster increases its size (in this case uniting with another cluster) one electron (or hole) will be stabilized by increasing the size of its orbital, but the second electron (or hole) placed on the same orbital will be destabilized because of the repulsion with the first electron (or the hole). *Hence the fully reduced large cluster will be a stronger reducing agent than the small one and the fully oxidized large cluster will be a stronger oxidant than the small one.* Accordingly, the semireduced cluster (in this case with one electron) is a less strong reductant than the reduced small cluster (also with one electron). Thus, two Fe_4S_4 clusters are united in the P cluster apparently to assist in the most difficult step of dinitrogen reduction. It can also serve as the pool of two electrons, the second one being transferred in a less difficult step. Hydrolysis of ATP is necessary to transfer electrons from the Fe_4S_4 cluster of the Fe protein to the P cluster against the thermodynamic potential, but the mechanism of the coupling that makes this transfer thermodynamically possible is still not clear.

I.3.2.4 FeMo Cofactor

In 1977 Shah and Brill reported about the isolation of the iron–molybdenum cofactor (FeMoco) of MoFe protein.[20] For this purpose they used a procedure that had been successfully applied for separation of other Mo cofactors from Mo-containing enzymes: acid–base protein denaturation. The distinction for MoFe protein of nitrogenase is that FeMoco remains bound to denaturated protein, and it is separated by extraction with some suitable organic solvent.

The state of the cofactor after its separation from protein was estimated by its ability to recover the activity of extracts from cells of a mutant microorganisms (with lost ability to synthesize FeMoco) towards acetylene reduction.

Naturally, the cofactor has attracted considerable interest because there is no doubt that it plays a central role in dinitrogen activation and reduction by the enzyme. Besides molybdenum the cofactor contains iron and sulfur. According to the data of various studies, the ratio Mo:Fe:S in the cofactor is 1:6–8:8–9. All sulfur exists as sulfide S^{2-}. The molecular mass of the cofactor is about 1500 kDa. It has been shown that it is not associated with any amino acid or peptide, does not contain sugars, lipoic acid, or coenzyme A. Reagents added at isolation were considered as potential ligands of the cofactor. However, it turned out that none of them was absolutely necessary: a number of acids, bases, and solvents in various combinations gave good results at isolation.

The search for the organic part of the cofactor took almost 10 years. Only in 1988, as the result of development of works on biochemical genetics of nitrogenase, was it found that homocitrate is an endogenous ligand of FeMoco and is necessary for its biosynthesis.[30]

The structure of the isolated cofactor remains unknown. Despite many attempts, it has not been obtained in the crystalline form needed to perform X-ray structural analysis.

Sufficiently good resolution obtained by Rees et al. upon studying single crystals of MoFe protein allowed them to suggest structural models of the cofactor and P clusters involved in this protein.[21,31,32a,b]

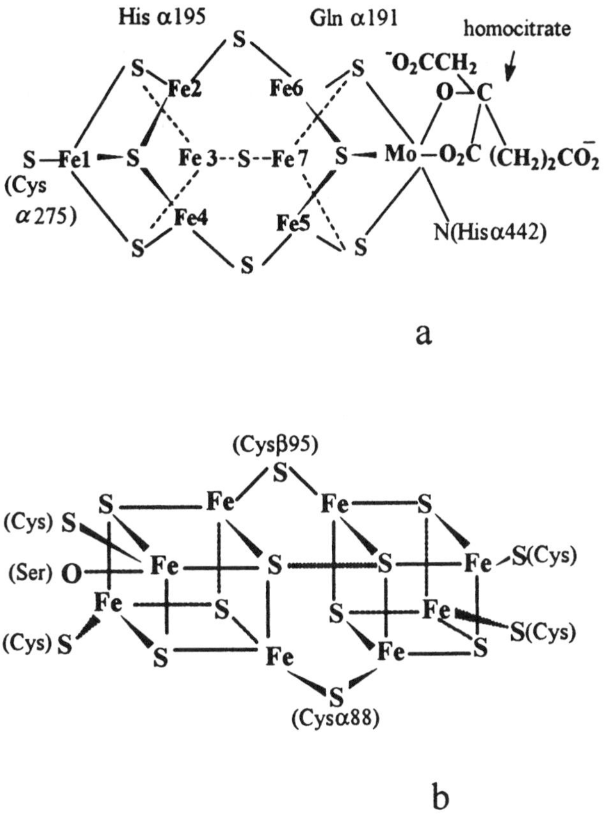

FIGURE I.7 The structure of FeMo cofactor (a) and P cluster (b). (Adapted from Kim, J. and Rees, D. C., *Science,* 257, 1677, 1992.)

A resolution of 2.7 Å (and later even 2.2 Å) did not allow the separate atoms in the cluster to be distinguished, therefore the authors used all available large information on FeMoco and P clusters to obtain molecular models: biochemical results on amino acid sequences, substitution of separate amino acids in proteins by insertional mutagenesis, EXAFS and other spectroscopic results, the data on chemical composition, and the structural data on model complexes. As a result, a convincing picture of the structure of the cofactor and P clusters was obtained. The structures suggested are presented in Figure I.7.

FeMo cofactor (Figure I.7a) consists of two cubane fragments [Fe_4S_3] and [Fe_3MoS_3] connected by three μ bridges, which are most likely all S^{2-}, although one of them was originally considered to be a different ligand and was designated Y (at first suggested to be NH or O).[32b] The three bridging ligands have comparable electron density in MoFe protein from *C. pasteurianum.* The somewhat lower density obtained earlier for MoFe protein crystals from *A. vinelandii* may reflect dynamical (disorder) properties or compositional heterogeneity of this site.

It may be noted that the ratio Mo:Fe:S = 1:7:9 in the proposed structure agrees quite well with the analytical data for FeMoco (Mo:Fe:S = 1:6–8:8–9) obtained earlier and presented above.

The cluster is connected with the protein matrix through side groups of cysteine (Cys 275), a ligand of Fe1, and histidine (His 442), which is a ligand of Mo. These atoms of Fe and Mo are on the third-order axis at a distance of 7.5 Å from each other. In this structure Mo has octahedral coordination and is connected, in addition to three sulfur atoms, with an N of histidine and two O atoms of hydroxyl and carboxyl groups of homocitrate. Two more side groups of residues (histidine 195 near Fe2 and glutamine 191 near Fe6) are located in the immediate vicinity of the cluster but it is unlikely that they are directly connected with metals of the cofactor. The side group of glutamine 191 ($H_2N–OC–(CH_2)_2$) interacts with one of carboxyl groups of homocitrate. All these amino acid residues (Cys 275, His 442, His 195, and Gln 191) are strictly retained in all primary structures of known MoFe proteins. In addition, studies by insertional mutagenesis show that Cys 275 and His 195 are involved in the binding of FeMoco with MoFe protein *in vivo*. The substitution of glutamine (Gln 191) by lysine causes the change in the substrate-reducing properties of nitrogenase, and these observations also confirm the assumption that Gln 191 interacts with homocitrate.

The tetrahedral coordination found for Fe1 and octahedral for Mo are typical of the coordination of these metals both in model compounds and iron–sulfur proteins. The main peculiarity of the suggested model is the fact that six of the seven iron atoms in the cofactor have an unusual triangular-pyramidal coordination and are coordinatively unsaturated. The solvent molecules, which could be coordinated near iron, have not been found, while the presence of hydride ions in diffraction experiments with such resolution was impossible to determine.

The electron density pattern obtained recently by the same authors with a resolution of 2.2 Å for single crystals of MoFe protein from *C. pasteurianum* confirms the model suggested previously. The Fe–Fe distances in the cluster are found to be equal to 2.4 to 2.6 Å, which indicates the existence of some bonding between iron atoms apparently additionally stabilizing the cluster. No other atoms are present inside the cluster cavity, whose diameter is approximately 4 Å.

The absence of a large number of bridges binding FeMoco and protein explains why the cofactor can be easily extracted from protein without loss of its catalytic activity when it is reset again on apoprotein (protein matrix deprived of the cofactor).

The immersion of the cofactor inside the protein molecule and the absence of water molecules near it, when it is bound to protein, explain its stability to water action in the protein-bound state and its rapid decomposition by water in the isolated state. This can be due to the ease with which coordinatively unsaturated iron atoms undergo the attack of water molecules with subsequent hydrolytic cleavage of Fe–S–Mo bridges.

EPR studies of the isolated cofactor show that after separation of protein it retains the characteristic signal with spin $^3/_2$ observed for the M center of MoFe protein. The broadening of the EPR signal of the isolated FeMoco, as compared with the signal of the M center, is attributed to a small change in the ligand sphere of the cofactor upon isolation. In MoFe protein the EPR signal of the M center has g factors about 4.3, 3.7, and 2.01, and in the isolated FeMoco the appearance of three different types of lines with g factors of 4.6, 3.4, and 2.0; 4.5, 3.6, and 2.0; and 4.9, 3.1, and 1.9 is observed in addition to the broadening of the signal. These particles have different pulse voltammetric redox waves at −0.37, −0.32, and 0.43 V, respectively. This indicates some uncertainty of the structure of FeMoco in the solution and may explain why the cofactor is not obtained in the crystalline form.

The M center in MoFe protein was shown to be able to exist in at least three oxidation states: dithionite-reduced or semireduced (s-r), oxidized (ox), and reduced (red). The s-r state of the M center is obtained as the result of the standard procedure of the protein purification in the presence of excess dithionite. Such an M center can be oxidized and reduced. M (s-r) gives the EPR spectrum with spin $^3/_2$. Its oxidation occurs at the potential of −0.05 V in the one-electron process to the state M(ox) by dyes, for example, thionine. The state M(ox) is diamagnctic and EPR-silent. The M(red) center is formed at the reduction of M(s-r) in the one-electron process at −0.47 V. This is a substrate-reducing state inside protein and it is achieved only under conditions of the nitrogenase reaction, i.e., in the presence of the complete nitrogenase system: Fe protein, MgATP, and dithionite. This state has a whole number spin and produces no EPR spectrum.

Redox processes typical of the protein-bound M center could be expected for the isolated FeMoco as well. The EPR signal of the cofactor is convenient for use as a control of its oxidation state. It has been shown that the same redox-active dyes efficient in the oxidation of M centers also oxidize the isolated dithionite-reduced FeMoco, which exhibits an EPR spectrum, to the EPR-silent state. The reaction is entirely reversible in an excess of dithionite. Dioxygen also oxidizes FeMoco, but an irreversible oxidation of the cofactor is observed even when stoichiometrical amounts of oxygen are used, because the addition of an excess of dithionite can only partially return the initial EPR spectrum with spin $^3/_2$.

Electrochemical studies of the cofactor solution in NMF (2.8×10^{-3} *M*) performed by Burgess et al. have shown that it undergoes a redox transformation (ox) → (s-r) at the formal potential $E_0 = -0.32$ V.[32c] The process is only quasi-reversible, anodic and cathodic peaks differing by about 100 mV. EPR spectra of the cofactor were observed in parallel with electrochemical experiments and its solutions were titrated with dithionite. The results of this work show that the EPR signal is completely restored after addition of 0.8 electron per Mo atom; it follows that one-electron reduction occurs in the process (ox) → (s-r). Additional amounts of dithionite do not affect the EPR signal. The process (ox) → (s-r) can be observed by UV-VIS spectroscopy as well. It is indicated that these spectra are not characteristic. After gel-filtration on G-25 to remove dithionite, the range of 400 to 270 nm is opened in NMF, and FeMoco(s-r) demonstrates the spectrum characteristic of the metal–sulfur charge transfer band. The oxidized FeMoco has a greater total absorption in the range of 600 to 270 nm and structural peculiarities near 300 nm. Owing to these FeMoco(s-r) is green-brown and FeMoco(ox) is red-brown. The spectral changes correlate with electrochemical studies. The spectrum of FeMoco(red) is unknown.

The location of the added reducing electron on the cofactor is still not quite clear. Early EPR studies of MoFe protein with ^{95}Mo and ^{57}Fe showed that the signal with spin $S = {}^3/_2$ belonged to iron only. These data firmly proved that Mo did not enter the EPR chromophore. Because it is known now that FeMoco is the EPR chromophore of protein and a particle, which undergoes redox transformations, one can suggest that the iron atom is an electron acceptor. However, more recent results on EPR of ^{95,96}Mo of MoFe protein and ENDOR results show that the molybdenum atom is integrated into the spin-system $^3/_2$ and, hence, it must be involved in the redox process.

Direct information about an electron acceptor was obtained from Mössbauer spectroscopy and X-ray absorption spectroscopy (XAS) on sulfur (K edge) and molybdenum (L edge). The observed change in the value of the isomeric shift on ^{57}Fe from 0.32 to 0.37 mm/s that accompanies the reduction FeMoco(ox) → FeMoco(s-r) indicates that the reducing electron is located (at least partially) on iron. Nevertheless, the observed shift is somewhat lower in value than that in the similar redox process on model iron–sulfur clusters (0.08 mm/s). Iron is known to be an electron acceptor in these clusters even if a cluster contains molybdenum. Therefore it is clear that the reducing electron on FeMoco(s-r) must be partially shared by other atoms. The study of the isolated cofactor in two states (oxidized and semireduced) by X-ray absorption near edge structure (XANES) showed that sulfur plays an active part in this redox process, because it is sulfur lines that undergo considerable changes, and Mo almost does not participate in this process.

The cyclic voltammogram of the cofactor on the glass carbon electrode shows, in addition to the process FeMoco(ox) → FeMoco(s-r), the reduction at –1.1 V with the corresponding oxidation at –0.9 V. These waves are ascribed by the authors to the pair FeMoco(s-r)/FeMoco(red).[32c] This is also a one-electron process, because at a relatively fast scanning rate (about 0.4 V/s) both reductions ([ox] → [s-r] and [s-r] → [red]) are almost equal in altitude. XANES spectra on Mo (L edge) of such samples showed that in this case as well the Mo atom is not involved in the redox process. The degree of involvement of iron or sulfur in this process is unknown: there has been no opportunity so far to study FeMoco(red) by XANES, Mössbauer, or any other method, because nobody has yet succeeded in isolation of the cofactor reduced to the substrate-binding state under the nonenzymatic conditions. The EPR spectrum is restored in the back transition from the reduced to the semireduced state.

Thus, one can conclude that redox properties of FeMoco are quite similar to those of M centers.

The following observation can serve as one of the most convincing proofs that the cofactor is a substrate-binding and a reducing center of the enzyme. The substitution of a single amino acid (glutamine, α-Glu 191) by lysine, or histidine (α-His 195) by asparagine, causes two effects simultaneously: a change in the substrate specificity (both ethylene and ethane are formed from acetylene instead of only ethylene in the native nitrogenase) and a change in the EPR spectrum of FeMoco.[33]

To be the catalyst the cofactor must be able to bind various small molecules. The addition of a particle must affect its EPR spectrum. EPR spectroscopy was used to demonstrate the reactivity of the cofactor towards different ligands, including substrates of nitrogenase. In this way it was shown that some thiols, selenols, EDTA, *ortho*-phenanthroline, and dipyridyl reacted with the cofactor.

Binding of EDTA to FeMoco causes a complete loss of the EPR signal. The reaction is nonstoichiometric, an approximately 40-fold excess of EDTA is needed. The process is reversible: the addition of stoichiometric amounts of Zn^{2+} completely restores the EPR signal and the activity of the cofactor. *ortho*-Phenanthroline acts in a similar way, and addition of Fe^{2+} helps to restore the cofactor activity. It can be concluded that the EPR chromophore is accessible to an exogenic ligand. It is surprising that unlike synthetic analogs that are rapidly destroyed by chelating agents, activity of FeMoco is not affected by such reactants.

Thiophenol interacts with FeMoco; in this process the EPR signal does not disappear but changes: all three lines narrow and g factors shift a little, so the resulting spectrum looks like the spectrum of FeMoco in protein. The stoichiometry of binding is 1:1, iron being the likely binding center. Selenophenol behaves in a similar way. Citrate binds to FeMoco without a change in the ESR signal somewhere in the external labile coordination sphere. Thus, FeMoco distinguishes between citrate and homocitrate, the latter not binding to FeMoco under similar conditions.

It has been shown that among the nitrogenase substrates only cyanide reacts with the cofactor in the dithionite-reduced state. It is directly attached to the cofactor, as indicated by the change in the EPR spectrum. The complete binding is observed at the ratio of 2.4 CN^- per Mo atom. Values of g factors change from 4.6, 3.3, and 2.0 to 4.3, 3.8, and 2.0. The data on titration indicate that each FeMoco binds two CN^- ions with different constants. The study of cyanide and methylisocyanide binding with the cofactor was also performed by NMR on ^{19}F nuclei using p-$CF_3C_6H_4S^-$ bound to the cofactor as a control ligand. These data obtained at room temperature do not fully agree with the results of experiments at the temperature of liquid helium. Although the latter confirm the fact of the interaction between CN^- (or CH_3NC) with the cofactor, they show that there is likely to be one binding center and dynamic equilibrium between bound and free CN^- (or CH_3NC). It has been shown that molybdenum does not participate in binding of CN^- (or CH_3NC). The same experiments (when the change was observed in the NMR spectra of a ^{19}F nucleus bound to the isolated FeMoco by p-$CF_3C_6H_4S^-$) show that neither azide ion nor CO interact with FeMoco in the dithionite-reduced state.

The attempts to observe the catalytic activity of the isolated cofactor using chemical reductants and in the electrochemical cell at the cathode have so far failed, and the problem of reducing the cofactor under nonenzymatic conditions and using it as a catalyst for the substrate reduction remains unsolved. However, there are some promising observations. First, the isolated cofactor possesses redox properties of the M center of MoFe protein, and, second, it reacts directly with an electrode. However, the formation of the catalytically active cofactor may not be the result of a simple electron transfer. Perhaps, the isolated FeMoco must be reduced in the presence of some additional ligands, because it is obvious that the protein surrounding the cofactor plays an important role in the catalytic activity of nitrogenase. For example, it has been found that the change even of a single amino acid in the region of the strictly retained residue of histidine in the α subunit of MoFe protein (His 195) located in close proximity to the cofactor but not directly connected with it (according to the results of X-ray structural analysis) gives mutants, which manifest different properties in reduction of acetylene and dinitrogen, and also in dihydrogen evolution: some of them demonstrate reduction of acetylene to ethane while others lose the ability to reduce N_2 but continue to reduce acetylene and yield dihydrogen.

I.3.2.5 "Alternative" Nitrogenases

From the study of biochemical genetics of nitrogen fixation it follows that molybdenum is not unique in the enzymatic reduction of N_2.

Mutant microorganisms, which are diazotrophs only in the absence of molybdenum in the medium, have been obtained by the change in genes encoding nitrogenase proteins of *A. vinelandii.* The ability of these microorganisms to fix nitrogen increased with the addition of vanadium. Subsequent isolation of vanadium-containing nitrogenase confirmed the existence of molybdenum-independent nitrogenases.

Later a third nitrogenase was shown to exist as well, and the latter contains neither molybdenum nor vanadium. It is synthesized by some nitrogen-fixing microorganisms in the absence of molybdenum and vanadium in the growth medium; moreover, the addition of molybdenum or vanadium suppresses the nitrogen fixation ability of strains of microorganisms that have this iron-only nitrogenase.

The existence of molybdenum-independent nitrogenases has been established among various nitrogen-fixing microorganisms (see References 22 and 23). It turned out that all microorganisms-diazotrophs studied by now had a molybdenum-dependent nitrogenase system. It is unclear why some of them possess only molybdenum nitrogenase, while other microorganisms have two or three nitrogenases, some having molybdenum and vanadium nitrogenases and others having molybdenum and iron ones.

The data on compositions and properties of three nitrogenases for one bacterial source, i.e., *A. vinelandii,* are summarized in Table I.2.[23] All three nitrogenases are composed of two proteins. Each of them has a specific homodimeric Fe protein of about 60 kDa with one iron–sulfur cluster. The second protein component has a different subunit structure in different nitrogenases: MoFe protein is a tetramer, and VFe and FeFe proteins are hexamers. Large proteins of all three nitrogenases contain two types of prosthetic groups, cofactors and P clusters. The FeV cofactor looks like FeMoco in composition and a number of its physical parameters. In VFe protein the chemical surrounding of vanadium (from EXAFS data) is very similar to that of molybdenum in MoFe protein; e.g., similarly to molybdenum, no noticeable changes in the ligand surrounding and the oxidation level of vanadium are observed in the course of the catalytic cycle. FeV cofactor can be separated from the protein in a similar way as for FeMo cofactor. It also contains homocitrate as an endogenous ligand. If FeV cofactor extracted from VFe protein is used for the reconstruction of activity of "apo-MoFe protein", the hybrid protein formed, unlike both "parents", cannot fix nitrogen but retains the ability of vanadium nitrogenase to produce C_2H_6 from C_2H_2. These results indicate that interaction of the cofactor with the nearest surrounding protein seems to be specific in each type of nitrogenase and affects the character of reactions with substrates. All these enzymes have similar requirements for manifestation of their catalytic activity.

Comparison of chemical behavior of the three nitrogenases shows the following similarities and differences:

TABLE I.2
Comparison of Averaged Composition and Activity Data for *Azotobacter* Nitrogenases

	MoFe protein	VFe protein	FeFe protein
Native M_r	230 kDa	210 kDa	216 kDa
Subunits	$\alpha_2\beta_2$	$\alpha_2\beta_2\gamma_2$	$\alpha_2\beta_2(\gamma_2)$
Subunits M_r	56 kDa, 59 kDa	50 kDa, 55 kDa, 13 kDa	50 kDa, 58 kDa, nd[a]
Metals			
Mo	2	<0.1	<0.1
V	nd[a]	2	<0.1
Fe	30	21	24
Activities			
Specific activity in nmol H_2 $(min \cdot mg)^{-1}$ under argon	2000	1400	250
Electron flux			
Under N_2	70% NH_3; 30% H_2;	40% NH_3; 60% H_2;	20% NH_3; 80% H_2;
N_2H_4	No	Yes	?
Under C_2H_2	90% C_2H_4; 10% H_2;	40% C_2H_4; 60% H_2;	15% C_2H_4; 85% H_2;
C_2H_6	No	Yes	Yes

[a] nd = not determined.

1. Mo nitrogenase can use up to 75% of electrons per target reaction: ammonia production from nitrogen while V nitrogenase uses 40%, and Fe nitrogenase uses only 20% of electrons.
2. A similar order is observed for acetylene reduction in competition with the reaction of hydrogen release.
3. Unlike MoFe protein, VFe protein and FeFe protein give some amount of ethane, the product of the four-electron acetylene reduction, in addition to ethylene, and VFe protein also produces a small amount of hydrazine (about 1%) at dinitrogen reduction. Similar data for FeFe protein are unknown. Apparently, all these data lead to the conclusion that molybdenum is an optimum heteroatom of the iron–sulfur cofactor for the main function of nitrogenase, while less perfect vanadium and particularly iron nitrogenases could be used by microorganisms to fix nitrogen when the concentration of molybdenum in the environment is not sufficient.

It can be noted here that the low activity towards reduction of substrates, particularly dinitrogen, of isolated preparations of the iron nitrogenase can simply mean that either optimum conditions of experiments have not been found, or something essential is lost in the course of isolation and purification of the enzyme from microorganisms, because experiments with mutant strains of *A. vinelandii* containing only the iron nitrogenase show that they grow well on N_2.

I.3.3 FUNCTIONING OF NITROGENASE

The minimum conditions for the performance of the cycle of nitrogenase reaction includes the presence of MoFe protein, Fe protein, ATP, Mg^{2+}, and dithionite and the absence of oxygen. Under these conditions dihydrogen evolution occurs with simultaneous hydrolysis of ATP to ADP and phosphate. The introduction of dinitrogen into the atmosphere causes the decrease of the amount of dihydrogen produced and formation of ammonia, while the rates of the ATP hydrolysis and the reductant oxidation do not change. Under optimum conditions the main nitrogenase reaction can be represented by Eq. 11:

$$N_2 + 8H^+ + 8e^- \text{ [for example, } S_2O_4^{2-}] + 16\text{ MgATP} \xrightarrow{N_2ase} 2NH_3 + H_2 + 16\text{ MgADP} + 16\text{ HPO}_4^{2-} \quad (11)$$

where the transfer of each electron is coupled with hydrolysis of two ATP molecules.

In addition to dinitrogen reduction, nitrogenase catalyzes the reduction of many other substrates of small size with multiple bonds. These compounds and the products formed are presented in Table I.3.[33a]

TABLE I.3
Nitrogenase Substrates and Reactions

Two-Electron Reductions

$2e^- + 2H^+ \rightarrow H_2$
$C_2H_2 + 2e^- + 2H^+ \rightarrow C_2H_4$
$N_3^- + 2e^- + 3H^+ \rightarrow NH_3 + N_2$
$N_2O + 2e^- + 2H^+ \rightarrow H_2O + N_2$

$$\begin{array}{c} CH_2 \\ / \ \backslash \\ HC{=}CH \end{array} + 2e^- + 2H^+ \rightarrow CH_2{=}CH-CH_3 + \begin{array}{c} CH_2 \\ / \ \backslash \\ H_2C-CH_2 \end{array}$$

Four-Electron Reductions

$HCN + 4e^- + 4H^+ \rightarrow CH_3NH_2$
$RNC + 4e^- + 4H^+ \rightarrow RNHCH_3$

Six-Electron Reductions

$N_2 + 6e^- + 6H^+ \rightarrow 2NH_3$
$HCN + 6e^- + 6H^+ \rightarrow CH_4 + NH_3$
$HN_3 + 6e^- + 6H^+ \rightarrow NH_3 + N_2H_4$
$RNC + 6e^- + 6H^+ \rightarrow RNH_2 + CH_4$
$RCN + 6e^- + 6H^+ \rightarrow RCH_3 + NH_3$
$NCNH_2 + 6e^- + 6H^+ \rightarrow CH_3NH_2 + NH_3$
$NO_2^- + 6e^- + 7H^+ \rightarrow NH_3 + 2H_2O$

Multielectron Reductions

$RNC \rightarrow (C_2H_6, C_3H_6, C_3H_8) + RNH_2$
$NCNH_2 + 8e^- + 8H^+ \rightarrow CH_4 + 2NH_3$

In the reaction catalyzed by nitrogenase, hydroxonium ions are reduced to dihydrogen:

$$2H_3O^+ + 2e^- \rightarrow H_2 + 2H_2O \tag{12}$$

H_3O^+ is considered an endogenous reducible substrate (because nitrogenase functions in aqueous media).

Dihydrogen evolution by nitrogenase is observed for all microorganisms studied. The conditions for H_2 evolution are the same as for dinitrogen reduction: the presence of ATP, corresponding reductant, and nitrogenase. Inhibitors for the reduction of hydroxonium ions are the same as for the nitrogenase reaction. As a rule, these are various chelate-forming compounds. No specific inhibitor of the reduction of H_3O^+ has been found. Neither CO nor H_2 inhibit the reduction of H_3O^+. Other reducible substrates can be considered inhibitors competing for electrons.

The term *ATP-dependent* is used to describe the reduction of hydroxonium ions by nitrogenase. This is done to distinguish this reaction from other reactions of dihydrogen formation, for example, from the reaction catalyzed by hydrogenase. One of the characteristic features of the nitrogenase reaction is the ability to catalyze the formation of HD from H_2O and D_2 only in the presence of N_2.

Of all substrates of nitrogenase only acetylene seems to be able to competitively suppress dihydrogen evolution so that at saturating concentrations of acetylene almost the total flow of reducing equivalents produced by nitrogenase is used for the reduction of C_2H_2. It was mentioned already that the dinitrogen-reducing reaction consumes about 75% of all electrons at saturating dinitrogen pressures (about 0.2 atm), and the remaining 25% of electrons are consumed in dihydrogen evolution.

For highly purified nitrogenase the dinitrogen reduction rate is 75 to 100 molecules/min.

In the cases of Mo-independent nitrogenases the ratio of the electron flow for dihydrogen evolution to that for ammonia production under dinitrogen is considerably higher than that for Mo nitrogenase: 50 and 75% of electrons are used to yield H_2 for V and Fe nitrogenases, respectively. Because the efficiency of coupling of the MgATP hydrolysis with electron transfer is the same for Mo and V nitrogenases, this difference is not the consequence of the limitation of electron transfer. This also does not relate to the use of dithionite as an electron donor for purified proteins, because cultures of *Azotobacter* growing as Mo-independent nitrogenases also release considerably more H_2 than in the case when their growth depends on Mo.

The efficiency of the competition of dinitrogen with H^+ for reducing equivalents for V nitrogenase increases with the decrease of the temperature. In the range from 30 to 5°C activity of Mo nitrogenase towards N_2 decreases much more as compared with corresponding decrease for the V nitrogenase. For both Mo-independent nitrogenases, acetylene as a substrate competes with H_3O^+ for electrons less efficiently than for Mo nitrogenase.

There are two important reactions connected with the functioning of nitrogenase that have something to do with the intermediate steps in dinitrogen reduction. One is formation of small amounts of hydrazine when nitrogenase is decomposed by an acid or an alkali while it is reducing dinitrogen.[34a] Hydrazine itself is a poor substrate of nitrogenase and is not formed as an intermediate in free state in the case of molybdenum nitrogenase (although it was mentioned above that about 1% of hydrazine was found to be formed for the case of V nitrogenase). This result indicates that a hydrazine *derivative* (the product of four-electron reduction of dinitrogen) is formed intermediately and normally is reduced further to the ammonia level. Another reaction is the already mentioned catalytic formation of HD in the presence of D_2. It was found to be not the simple exchange $D_2 + H_2O = HD + HDO$, but rather "reductive hydrolysis":

$$D_2 + 2e + 2H^+ \rightarrow 2HD \tag{13}$$

because the process requires electrons of the reducing agent and proceeds competitively with dinitrogen reduction. At the same time reaction 13 *requires dinitrogen* and therefore indicates the presence of an intermediate in the process of N_2 reduction capable of reacting with D_2. It was regarded initially as the evidence for the intermediate formation of diazene (N_2H_2) bound to the enzyme, although its stoichiometry does not provide an easy explanation from the viewpoint of N_2H_2 as the intermediate.[16] Both the reactions (hydrazine formation at the decomposition of nitrogenase in the process of N_2 reduction and HD production according to 13) now have analogues among nonenzymatic chemical processes (actually the first reaction indicating the presence of a hydrazine derivative was discovered in chemical dinitrogen reduction *earlier* than in the enzymatic process) and we shall see that the second reaction does not require the intermediate diazene formation (see Section I.4.2.).

I.3.3.1 Mechanism of Nitrogenase Functioning

In the processing of the nitrogenase catalytic cycle several reactions proceed consecutively and/or in parallel: formation of the MgATP complex with Fe protein, reduction of MoFe protein by this complex, ATP hydrolysis releasing ADP and phosphate, coordination and reduction of a substrate, and release of product molecules. Dinitrogen transformation to ammonia is the result of a series of electron and proton transfers to the FeMo cofactor of the enzyme. The sequence of electron transfers is virtually certain: an external reducing agent, flavodoxine or ferredoxine *in vivo* or dithionite *in vitro,* transfers its electron to Fe protein, then, one by one, from the latter to MoFe protein where N_2 is reduced to NH_3. As was already mentioned, presumably initially electrons move to a P cluster, than to FeMoco and a substrate. Kinetic and EPR data confirm the oxidation of P clusters resulting from the electron transfer to FeMoco.[34-36] The three-dimensional structural data show that Fe protein and MoFe protein have to interact in such a way that the distances between the centers transferring electron in succession (Fe protein → P cluster → FeMoco) have to be in the range of 12 to 14 Å. The rate of Fe protein oxidation (200 s^{-1}) is in agreement with the theory of electron transfer inside a protein matrix.[37]

Kinetic investigation of the nitrogenase reaction indicates the MgATP hydrolysis ($k = 9.4\ s^{-1}$) to precede the electron transfer inside the complex Fe protein–MoFe protein ($k = 3.0\ s^{-1}$).[34,35] As was already mentioned, MgATP binding provokes a conformation change in Fe protein involving an $[Fe_4S_4]$ cluster, and ATP hydrolysis starts only after the complex Fe protein–MoFe protein is formed. These observations together with the data of Fe protein molecular structure led to the hypothesis explaining the regulation by the ATP hydrolysis of one-way electron transfer to MoFe protein.[28] The $[Fe_4S_4]$ clusters and the nucleotide-binding centers are far apart, therefore their direction interaction is impossible, and ATP hydrolysis results in the Fe protein conformational changes increasing its affinity first to the electron donor (ferredoxin) and then to the acceptor (MoFe protein). These conformational changes allow the Fe protein to function as a "molecular clock" determining the time sequence during the reduction of dinitrogen by nitrogenase. Kinetic investigation of the nitrogenase reaction led Lowe and Thorneley[34-36] to the conclusion that the electron transfer is accompanied by an association–dissociation cycle between the Fe and MoFe proteins, the dissociation being the rate-determining step of the whole process ($k = 5\ s^{-1}$). However, according to one work[38a] the dissociation of the proteins *in vitro* is due to the presence of dithionite, which provokes a salt effect. The authors of the work[38a] found that photoelectron transfer can be used instead of dithionite with NADH as an electron donor and eosin as photosensitizer. Without dithionite the complex of Fe and MoFe protein seems to be stable, effecting, nevertheless, the electron transfer and the reduction of nitrogenase substrates including dinitrogen.

The detailed mechanism of the electron transfer and coupled process of ATP hydrolysis is still unclear: such problems as the interaction of the proteins, the electron and protons paths through the proteins, the exact nature of intermediates, etc. remain to be determined. The role of ATP is particularly obscure. Because presumably two ATP molecules are hydrolyzed per electron transferred to a substrate molecule, it is possible that there are two centers of ATP hydrolysis in the enzyme: e.g., one could be used for necessary conformational changes and the other could form an active proton to help the protonation of substrate to afford the final product. Hopefully new information will be obtained in the near future in experiments conducted with the knowledge of the three-dimensional structures of Fe and MoFe proteins and molecular models of the M centers.

Lowe and Thorneley proposed the full schematic description of the reactions in the cycle of the nitrogenase action based on the experimental kinetic results.[35] The scheme in Figure I.8 represents the changes in MoFe protein in the process of electron accumulation in its active center. Here E is MoFe protein; the subscript with E corresponds to the number of electrons transferred to it. Each step of the scheme is the result of the total cycle involving the dissociation of the complex of the Fe and MoFe proteins. As was mentioned above, this is probably not the case in the enzyme functioning *in vivo* without dithionite.

To be able to coordinate and activate dinitrogen, the FeMo cofactor of the MoFe protein must be sufficiently reduced: according to the scheme in Figure I.8 this corresponds to three electrons transferred to the MoFe protein from the Fe protein. At the step E_3 FeMoco can already reversibly coordinate N_2. Recently it was experimentally confirmed[36] that the transition from the E_3 state (the reversible binding of N_2) to the E_4 state is connected with the oxidation of the P centers and this is a critical point of the catalytic cycle of the MoFe protein when dinitrogen becomes irreversibly reduced to the intermediate state that corresponds to the hydrazine

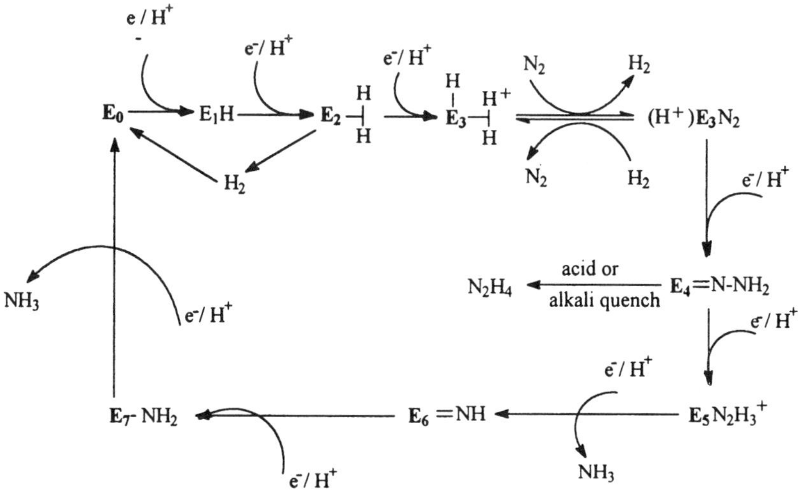

FIGURE I.8 Catalytic cycle for the reduction of N_2 by Mo nitrogenase of *K. pneumoniae*. E_0: resting state of MoFe protein; N_2 binds to species E_3, displacing H_2.

derivative producing hydrazine when decomposed by acid or by alkaline. Consecutive addition of two more electrons and two protons leads to the formation and liberation of two molecules of ammonia. The scheme, in particular, the structure of the intermediate $E_4 = N - NH_2$ reflects the views of the authors[34-36] who prefer a single molybdenum atom as the active center of dinitrogen activation. However, independently of the structure there is an intermediate producing hydrazine and this corresponds to the four-electron reduction mechanism; it may be concluded, therefore, that the mechanism does not involve a concerted six-electron reduction, although these data alone may be also treated as two consecutive steps of two-electron reductions.

A concrete chemical mechanism of the interaction of dinitrogen with FeMoco is the central point in the understanding of the catalytic action of nitrogenase. Before the molecular model of the M centers was reported almost all hypothetical mechanisms of a substrate reduction had been based on the participation of the molybdenum atom; the differences in opinions concerning the mechanism usually were connected with the number of molybdenum atoms (one or two) and with the participation of iron in the intermediate complex with dinitrogen formed in the process of the reduction (dinuclear molybdenum–iron complex was often proposed). Discovery of alternative molybdenum-independent nitrogenases has already led to some doubts concerning the direct participation of molybdenum: catalytic properties and physical characteristics of the M centers in these nitrogenases are very similar to each other and to molybdenum-containing nitrogenase, and if dinitrogen can be reduced on the iron center of the FeFe cofactor then why can the similar surrounding for dinitrogen not function in Mo and V nitrogenases?

The problem of the detailed mechanism and the site of dinitrogen activation and reduction has not yet been finally solved and likely will remain under discussion for some time; however, the molecular structure of the cofactor seems to be more suitable for coordination on the central coordinatively unsaturated iron atoms than on coordinatively saturated molybdenum, both by forming mononuclear complexes with molybdenum or via dinuclear complexes involving molybdenum and iron. The role of molybdenum (or vanadium in the corresponding V nitrogenase) might be in stabilizing the cluster and this is revealed in a shortening of the iron–iron distance at the side closer to molybdenum. Molybdenum in the state Mo^{IV} (d^2) decreases the number of electrons in the cluster without a drastic change of its reducing ability. One might suggest that in an iron-only cluster the larger number of electrons (if all the iron atoms are in Fe^{II} state there are four electrons more) is a destabilizing factor (presumably occupying some antibonding molecular orbitals). Therefore there seem to be more similarities between Mo and V nitrogenases (presumably vanadium is in the V^{III} state with the same number of electrons) than between both these nitrogenases and the iron-only nitrogenase. The differences between Mo and V nitrogenases are due to somewhat different reducing properties: reduced FeV cofactor could be a stronger reductant than its molybdenum analogue, therefore some hydrazine is formed in parallel with ammonia in the case of V nitrogenase, while in the case of molybdenum-containing enzyme hydrazine is observed only upon the addition of acid or alkali with proton or hydroxide ions coaction, and the only final product at neutral pH is ammonia. We shall see in the next section that similar behavior is observed in certain model chemical systems that are not sufficiently strong reductants to produce free hydrazine as an intermediate.

Since publication of the data on the three-dimensional structure of MoFe protein and possible structure of FeMoco, many different hypotheses have been suggested concerning the mode of dinitrogen activation including interaction with one, two, four, and even six atoms of iron and/or molybdenum.[32,38b,38c]

If the conclusion about the activation of N_2 at the central iron atoms is correct then the intermediate complex might involve iron atoms, and diagonal arrangement of dinitrogen may be suggested: it will be then coordinated side-on to two iron atoms and end-on to two others. Four contacts will strongly activate dinitrogen. No precisely identical complexes have yet been observed in model systems, but we shall see some similarity with certain model dinitrogen complexes in the following sections.

A concrete scheme then may be tentatively suggested for the activation and reduction of dinitrogen on the cofactor (Figure I.9). Reversible coordination of dinitrogen on complex **1** to form complex **2** increases its electron-accepting properties and induces the following electron transfer to form the more reduced center **3** (without dinitrogen H_2 is formed in the reaction of complex **1** with protons in the surrounding). Complex **3**, which is still in equilibrium with N_2, can react with D_2 to produce dideuteride and subsequently two molecules of HD in the reaction with protons. This sequence of events explains why HD is formed only in the presence of N_2, but at higher pressures of N_2 the rate of HD formation is again decreased because D_2 competes with N_2 for electrons; there is no necessity to suggest intermediate diazene formation (as some researchers believe).

In the scheme (Figure I.9) complex **3** corresponds to the state E_3 in the catalytic cycle of the Thornley–Lowe scheme (Figure I.8) in which N_2 is still reversibly bound

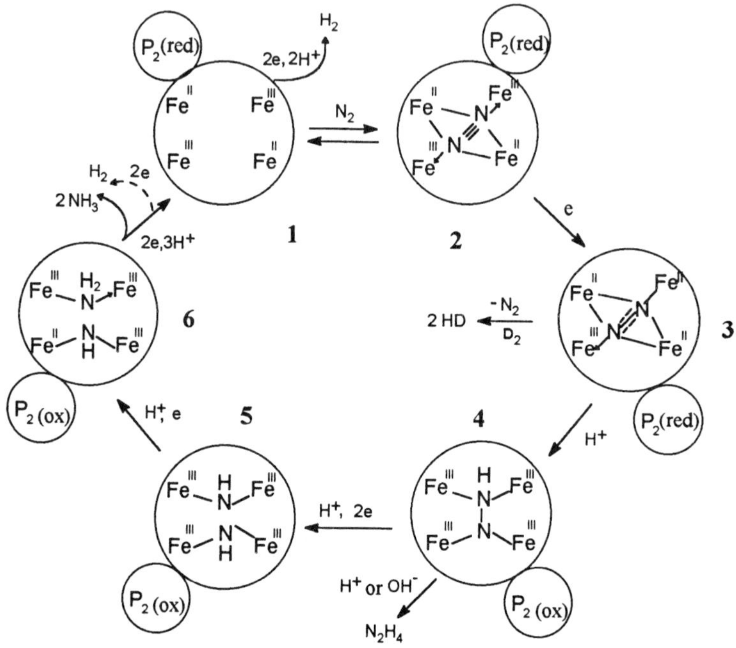

FIGURE I.9 Tentative scheme for nitrogenase catalytic cycle.

to the cofactor, and the next step is the irreversible four-electron reduction to the hydrazine derivative (state E_4 in the Thornley–Lowe scheme), which requires one more electron and is accompanied by an electron transfer from the P cluster (simultaneously or immediately after formation of the hydrazine derivative). It is presumably in this step that the P cluster is needed for *reductive coaction* to help to perform the crucial step of N_2 complex transformation into the hydrazine derivative. We shall see the analogues of this action among the model systems in the next sections. In the following steps a single N–N bond is cleaved and two ammonia molecules are formed. Some of the electrons may be used in formation of additional dihydrogen: one molecule for molybdenum nitrogenase and several molecules for alternative ones.

Although still speculative, this scheme seems to represent "optimum catalytic performance" according to the considerations presented in the previous section for such "difficult" substrates as dinitrogen and taking into account the data on dinitrogen reduction in model systems (*vide infra*). Of course, it is not the only possible mechanism. It should be emphasized that there is little hope that all the details of dinitrogen reduction mechanism in the enzymatic process will be elucidated from the investigation of the enzyme itself. Rather, it could be expected that more and more plausible schemes will be gradually developed in the interdisciplinary research, i.e., combination of experimental investigations of the enzyme and model systems together with theoretical studies based on quantum chemistry.

I.4 DINITROGEN REDUCTION IN SOLUTION

I.4.1 APROTIC MEDIA

I.4.1.1 The Systems

Enzymatic nitrogen fixation involves a donor of electrons (ferredoxin *in vivo* or dithionite *in vitro*), the system of electron transfer, and the site of dinitrogen activation where the N_2 molecule is bound to a transition metal complex and subsequently reduced to ammonia with participation of protons. As we have seen, this site, presumably FeMo cofactor, is a very complicated, well-organized molecular system necessary for dinitrogen reduction by comparatively weak reducing agents. If the reducing agent is strong the system for N_2 activation and reduction need not be so well organized. In aprotic media we can use very strong reducing agents, but there must be of course another source of protons (if we want to produce ammonia) besides water or acids because they would react with the reducing agent to oxidize it. However, at least stoichiometric reduction may be expected to form nitrides from intermediate dinitrogen complexes and then to produce ammonia upon the subsequent addition of an acid. These simple arguments are quite obvious now, but in the early 1960s dinitrogen was considered so chemically inert that no attempts were made by investigators working with transition metal compounds in the presence of strong reducing agents to verify the possibility of dinitrogen reduction: dinitrogen gas was considered a suitable inert atmosphere for conducting various reactions. Even when some indications of N_2 involvement were present no serious attention was paid by the investigators.

In 1964 Vol'pin and Shur published their pioneering paper[39] describing the results of dinitrogen reduction in aprotic media. The reaction turned out to be quite general, and a new stage in the study of dinitrogen reactions in solution, which had been virtually unexplored, was initiated.

In the first experiments[39] chlorides of transition metals were used ($CrCl_3$, $MoCl_5$, WCl_6, $FeCl_3$, and $TiCl_4$). When they reacted with such strong reducing agents as $LiAlH_4$, EtMgBr, or *i*-Bu_3Al, ammonia was produced in the hydrolysis of the reaction products. Initially, appreciable yields of the reaction product (up to *25%* of ammonia relative to the amount of transition metal employed) was observed at comparatively high dinitrogen pressure (100 to 150 atm). Soon Vol'pin and his co-workers succeeded in obtaining more active systems. Thus in the reaction of N_2 with the products of the reaction of Cp_2TiCl_2 (Cp = η^5-C_5H_5) with EtMgBr in ether the yield of ammonia (after hydrolysis) was 67% per mole of the titanium compound employed, even at atmospheric pressure.[40] On increasing the pressure, the yield of ammonia became nearly quantitative. In the search for new nitrogen-fixing systems the number of transition metal compounds and possible reducing agents was greatly enlarged by Vol'pin's team and by many other investigators who joined this field of research (see References 6 and 41 to 43).

Typical examples of nitrogen-fixing systems are presented in Table I.4.[41-43] The activity of the systems towards dinitrogen varies widely depending on the systems and reaction conditions. Some generalizations can be made, however. First, as already mentioned, the number of transition metals compounds capable of taking

TABLE I.4
Typical Examples of Dinitrogen-Fixing Systems in Aprotic Media

System (and the ratio of reagents)	Solvent	P{N_2}, atm	Time, h	NH_3 yield per mole of M
$TiCl_4$ + *i*-Bu_3Al (1:3)	*n*-Heptane	150	11	0.25
$TiCl_4$ + $Ph_2^-Na^+$	THF	80	6	1.2
Cp_2TiCl_2 + EtMgBr (1:9)	Et_2O	1	9	0.67
Cp_2TiCl_2 + Mg + $MgBr_2$ (1:14:5)	Et_2O/C_6H_6	100	7	1.3
VO(acac) + *n*-BuLi (1:9)	*n*-Heptane	75	10	0.32
$FeCl_3$ + EtMgBr (1:9)	Et_2O	150	18	0.09
$FeCl_3$ + $Napht^{2-}$ $2Li^+$ (1:8)	THF	110	4	1
$ZrCl_4$ + Ph_2-Na^+ (1:6)	THF	80	6	0.5
$MoCl_5$ + Mg + MgI_2 (1:14:5)	Et_2O/C_6H_6	100	7	0.36
Cp_2Yb + $Napht^-Na^+$ (1:10)	THF	1	2	0.23
Cp_3Sm + $Napht^-Na^+$ (1:10)	THF	1	24	0.14

part in the nitrogen-fixing systems is very large. Apart from those already mentioned, vanadium, manganese, and cobalt chlorides, tetraalkoxy derivatives of titanium, acetylacetonates of vanadium, chromium, molybdenum, manganese and nickel, cyclopentadienyl derivatives of zirconium and niobium, and triphenyl phosphine complexes of titanium and iron proved to be active. Later lantanide compounds were included in the list of nitrogen-fixing systems, the most effective being the compounds of samarium and ytterbium.[44] The strongest N_2-reducing capacity is displayed by transition metals of Groups IV, V, and VI of the periodic table namely Ti, V, Cr, Mo, and W. Particularly active are titanium compounds. In the first row transition metals, the ammonia yields decrease generally from left to right, in line with decreasing stability of nitrides. However, iron compounds are as a rule more efficient than manganese compounds. Cobalt and nickel compounds are usually of low or no activity. Palladium, copper, and platinum complexes apparently demonstrate no activity in any of the system tested.

Yields of ammonia depend strongly on the reducing agent. The reductants include organometallic compounds (RMgX, RLi, and R_3Al), metal hydrides ($LiAlH_4$ and LiH), free metals (alkali, alkaline earth, and rare earth), aromatic radical anions and dianions, and dihydrogen (in combination with other reductants). The function of a reducing agent is to produce a low-valent transition metal complex activating dinitrogen. Usually for stronger reducing agents the yields of ammonia are higher. For example, $FeCl_3$ produces only about 9% yield of ammonia per iron in the presence of EtMgBr, whereas under the action of such a powerful reducing agent as dilithium naphthalene the yield of ammonia (after hydrolysis) reaches 100%.[45] Cp_2TiCl_2 forms extremely active systems with EtMgBr but does not reduce N_2 in the presence of *i*-Bu_3Al. The reason is apparently the inability of the latter reductant to essentially reduce titanocene to a lower oxidation state than Ti^{III}, while at least bivalent titanium is needed to activate and reduce dinitrogen. Usually the yields of ammonia (after hydrolysis) vary from <0.01 to 1 mol/mol of transition metal complex, indicating that two transition metal atoms are likely to participate in the reduction of one dinitrogen molecule, though in some of the systems (particularly with very strong reducing agents) up to 2 mol of ammonia are produced per mole of the complex.

The Vol'pin and Shur systems are, of course, very far from the biological dinitrogen reduction, primarily being much stronger reductants. Nevertheless, there are some similarities, e.g., carbon monoxide and acetylenes are strong inhibitors of N_2 reduction in the chemical systems, similarly to their behavior in the enzymatic process. It is significant that although water cannot be used as a solvent in the presence of such strong reducing agents, aprotic Lewis bases (ether or tetrahydrofurane) do not inhibit dinitrogen reduction, i.e., dinitrogen can successfully compete with the bases for a coordination site in low-valent transition metal complexes.

Attempts have been made to design systems for catalytic hydrogenation of N_2. However, dihydrogen usually inhibits the reaction, decreasing the yield of NH_3 even in systems that normally catalyze hydrogenation of unsaturated compounds. In some cases H_2 does not affect the NH_3 yield, and in a few cases, for example, $Ti(OEt)_4$ + (*i*-Bu)$_3$Al + N_2 an appreciable increase of the yield of ammonia (after hydrolysis) was observed following the introduction of H_2. However, with increase of dihydrogen pressure, the yield of ammonia passes through a maximum, which is still less than 100% with respect to the titanium compound chosen, and then falls. Nevertheless, these systems demonstrate the possibility of participation of dihydrogen (perhaps forming hydrides and subsequently some N–H-containing compounds in the process of the reduction). Indeed there is some evidence that N–D bonds are present in the products in the systems when deuterium is substituted for dihydrogen. Though in none of the instances was catalytic hydrogenation observed, Vol'pin and Shur and their co-workers achieved the catalytic reduction of dinitrogen under the influence of a more powerful reducing agent. They found that aluminum metal reduces dinitrogen at 130°C and a pressure of 100 atm in the presence of titanium tetrachloride and aluminum bromide (Table I.5).[41-43]

Aluminum is known to react with dinitrogen (actually it can burn in dinitrogen) at much higher temperatures. Both the rate and yield of ammonia essentially depend on the amount of aluminum bromide present. Almost full conversion of the initial aluminum can be achieved in an excess of $AlBr_3$ when its mole ratio to aluminum reaches about 3:1. The role of $AlBr_3$ is the cleavage of the strong Ti–N bonds in the nitride complexes formed. However, without $AlBr_3$ the system not only stops being catalytic but ceases to reduce dinitrogen altogether, indicating that aluminum bromide is involved in the catalytically active particle reacting with dinitrogen. Accordingly, the authors have found that a complex $(C_6H_6){\cdot}TiX_2{\cdot}2AlBr_3$ (X = Cl, Br) is catalytically active in the system $TiCl_4 + Al + AlBr_3$ when the reaction is undertaken in benzene. The complex $(C_6H_6){\cdot}TiCl_2{\cdot}2AlCl_3$ itself turned out to reduce N_2 at 130°C, producing a nitrido complex forming ammonia at the hydrolysis.

TABLE I. 5
Catalytic Reduction of N_2 by Aluminum and Lithium Aluminum Hydride[a]

System	Molar ratio of reagents	Reaction time (h)	Ammonia yield: Mol/g-at. of Ti	Ammonia yield: Referred to reducing agent (%)
$TiCl_4 + Al + AlBr_3$	1/6/0	8	0.02	0.4
	1/6/2	8	1.3	23
	1/12/33	14	10.7	92
	1/150/200	30	95	64
	1/600/1000	30	286	48
$TiBr_4 + Al + AlBr_3$	1/116/370	45	74	64
$Ti(OBu)_4 + Al + AlBr_3$	1/150/250	36	105	70
$C_6H_6 \cdot TiCl_2 \cdot 2 \cdot AlCl_3 + Al + AlBr_3$	1/150/250	30	115	77
$TiCl_4 + LiAlH_4 + AlBr_3$	1/4/44	14	3.7	74
	1/8/44	24	9.8	95
	1/16/90	14	17.8	85
	1/16/90	40	19.2	92
	1/16/120	40	19.8	94
	1/300/1500	40	125	31
$Ti(OBu)_4 + LiAlH_4 + AlBr_3$	1/16/90	40	17.6	84
$TiCl_4 + LiAlH_4 + AlBr_3 + AlCl_3$	1/300/1500/690	35	100	25

[a] $P(N_2) \sim 100$ atm; with $TiCl_4 + LiAlH_4 + AlBr_3 + AlCl_3$ the reaction was carried out at 60 to 70°C; in the other cases at 130°C.

Lithium aluminum hydride can be used as a reducing agent instead of aluminum in catalytic systems. In the case of the catalytic system $TiCl_4 + LiAlH_4 + AlBr_3$ more than 100 mol of ammonia (after hydrolysis) per mole of initial titanium tetrachloride can be obtained even at 60 to 70°C when the reaction is performed in a molten mixture of aluminum chloride and aluminum bromide. When the triple mixture $AlCl_3$–$AlBr_3$–C_6H_6 is used as a solvent, some catalytic reduction of ammonia can be observed (NH_3/Ti is about 5) already at 30°C.

Here we may note that although catalytic hydrogenation of dinitrogen has not so far been observed in aprotic solutions, the introduction of alkali metals was shown to strongly activate heterogeneous catalysts of the gas phase synthesis of ammonia. Thus, the reaction of iron, cobalt, titanium, and molybdenum complexes of phtalocyanines with metallic sodium results in the formation of a donor–acceptor complex that can catalyze the synthesis of ammonia from dinitrogen and dihydrogen starting from a temperature as low as 110°C. Alkali metals sharply increase the catalytic activity of ruthenium, osmium, iron, molybdenum, and other metals in the synthesis of ammonia; ruthenium activated by potassium is capable of catalyzing the reaction at a temperature as low as 146°C and at atmospheric pressure (see Reference 46).

Another approach to the preparation of the catalysts for ammonia synthesis from dinitrogen and dihydrogen was the use of the systems on the basis of lamellar graphite compounds (LGC) with transition metals and their chlorides as well as the systems prepared from the potassium salts of transition metals carbonyl hydrides supported on the activated carbon.[43,47] In both cases potassium metal was used to activate the systems to the catalysis of ammonia production (in the latter system after thermal decomposition of the carbonyl hydrides). LGC with iron forms active catalysts only in the presence of potassium metal. The presence of potassium is essential also for the catalysis of the isotope exchange $^{28}N_2 + {}^{30}N_2 = 2{}^{29}N_2$. The most active systems both for isotope exchange and ammonia synthesis are based on $OsCl_3$ and $RuCl_3$ (the reaction was observed already at 250°C). For $FeCl_3$ the exchange starts only at temperatures higher than 350°C. Carbonyl hydrides produce more active systems. The most active systems were obtained with $K_2Fe_2(CO)_8$ + K and with $K_2Ru_4(CO)_{13}$ + K at the activated carbon. The former system catalyzes the ammonia synthesis already at 150°C and atmospheric pressure and at 250°C its activity is 2.5 times higher than that of the classical catalyst of industrial ammonia synthesis. The catalyst can also work in a two-step manner. If the catalyst prepared from $K_2Fe_2(CO)_8$ + K at the activated carbon is heated first in the N_2 atmosphere at 250°C and then after removal of dinitrogen it is replaced by dihydrogen and the catalyst is kept at the same temperature, ammonia is produced at 0.4 to 0.5 mol/mol of iron. The hydrolysis of the catalyst after its reaction with N_2 at 250°C also produces ammonia. These results indicate that surface nitrides are formed in the reaction of iron (produced in the decomposition of the carbonyl hydride) with dinitrogen at 250°C and potassium is essential in this reaction. Therefore iron evidently forms a complex with N_2 that is reduced to a nitride with participation of potassium and this nitride can be transformed to ammonia in the reaction with dihydrogen or with protons in the hydrolysis. There are indications that dihydrogen takes part in the reaction already at the stage of dinitrogen reduction. In the presence of dihydrogen the formation of the surface intermediate, which can be subsequently transformed to ammonia upon reaction with dihydrogen, is faster than the reaction of pure dinitrogen to form nitrides. Therefore dihydrogen presumably takes part in the reduction together with potassium, perhaps producing surface amides or imides.

There is a clear similarity of these reactions with not only the classical ammonia synthesis on the industrial iron catalyst and at the same time with some dinitrogen reactions in aprotic media (particularly with those facilitated in the presence of dihydrogen), but also in some aspects with the enzymatic dinitrogen reduction: in all the cases polynuclear iron forms the active center, and protons can produce ammonia from the intermediates involving iron–nitrogen bonds. The main difference from the natural process is the requirement of a much stronger reducing agent (potassium) and/or much more difficult conditions (higher temperatures and pressures). Thus, acid that produces ammonia from surface nitrides (or N–H-containing intermediates: amides or imides) simultaneously decomposes the catalytic system. To bring the chemical systems nearer to the enzyme one needs to use less strong reductants stable enough in protic media to continue functioning after ammonia is produced. As indicated, better organized systems are necessary in order to fulfill this goal.

I.4.1.2 Reaction Products and the Mechanism[6,41-43]

I.4.1.2.1 Nitrides

As was mentioned above, complex nitrides are very likely products in many of the dinitrogen reducing systems in aprotic media, though their exact nature remains mostly unknown, because of difficulties in isolation. Nevertheless, certain conclusions can be made for some of the systems. For example, the already mentioned complex of titanium(II), $C_6H_6(TiCl_2{\cdot}2AlCl_3)$, obtained by the reduction of $TiCl_4$ with aluminum in benzene reacts with N_2 at 130°C to form the compound $[(TiCl_2{\cdot}2AlCl_3)_3N]{\cdot}C_6H_6$, which releases NH_3 on hydrolysis. Presumably the product is a nitride containing three titanium atoms per nitrogen atom. Yamamoto et al.[48] investigated the products of two similar N_2-fixing systems: $TiCl_3(THF)_3$–Mg and $VCl_3(THF)_3$–Mg. In both cases the reaction proceeds at room temperature and atmospheric pressure with yields of ammonia (after hydrolysis) reaching 1 mol/mol of transition metal compound. Black diamagnetic powders were isolated from both systems, having the formulae $MNMg_2Cl_2$ (THF) (M is Ti or V for titanium and vanadium systems, respectively). The products do not release N_2 on heating, produce NH_3 when hydrolyzed, and evidently are nitrides.

In a recent work[49] formation of a nitrido complex was observed in the reaction of dinitrogen with molybdenum(III) complex, $Mo(NRAr)_3$, where R is $C(CD_3)_2CH_3$ and Ar is 3,5-$C_6H_3(CH_3)_2$. The reaction occurs at low temperature (–35 to 30°C) and pressure (1 atm). At –35°C formation of a complex is observed in ether and toluene and upon warming to 30°C a conversion to nitrido molybdenum (VI), N≡Mo $(NRAr)_3$, follows. The authors believe that the intermediate complex contains a linear Mo···N$\equiv$N···Mo fragment with bridging dinitrogen. Another example of a nitride formation from a definite starting material was observed recently with monomeric vanadium(II) complex, $[(OEPG)VLi_4Cl_2(THF)_4]$ (OEPG is octaethylporphyrogen and THF is tetrahydrofurane). The treatment of this species with tetramethilethylenediamine (TMEDA) under dinitrogen results in formation of the nitrido bridged dimeric species $\{[(OEPG)V]_2(\mu\text{-}Li)_4\}\{Li(TMEDA)_2\}$.[50] In both latter systems dinitrogen is activated and reduced by metal complexes with d^3 electronic configuration of the metal. Their activity in dinuclear complexes with dinitrogen (apparently without too strong a reducing power of the complexes) is in agreement with the theoretical considerations presented in Section I.2.3. We shall see more examples of the activity of d^3 complexes towards dinitrogen in the systems reducing N_2 in protic media (*vide infra*).

In those Vol'pin systems that increase the yield of ammonia in the presence of dihydrogen, amide or imide complexes are likely products. Thus, there are indications that in the presence of D_2 deuterium is found in the ammonium salt produced by hydrolysis, confirming the presence of an N–D bond in the product obtained.

I.4.1.2.2 Organonitrogen Compounds

Some of the Vol'pin–Shur systems produce aromatic amines on hydrolysis of the reaction mixture. For example, aniline, as well as NH_3, is present in the hydrolysis products of Cp_2TiCl_2–LiPh–N_2 or Cp_2TiPh_2–LiPh–N_2 in ether. The yield of aniline can be increased from about 3 to about 17% with respect to titanium complex at the increase of dinitrogen pressure from 1 to 80–100 atm. Mainly *p*- and *m*-toluidines are formed with *p*- and *m*-tolyllithium, respectively, whereas a mixture of *o*-, *p*-, and *m*-toluidines is found using *o*-tolyllithium. Only traces of aliphatic amines are detected in similar systems when alkyllithium is used as a reducing agent.

Formation of N–C bonds in the reactions with aryllithiums occurs evidently in the nucleophilic attack on coordinated dinitrogen (as in the cases with *p*- and *m*-tolyllithium) or in the interaction of coordinated N_2 with benzyne, a triple-bond-containing cyclic molecule formed intermediately from *o*-tolyllithium. Some other interesting reactions forming N–C bonds are mentioned later when intermediate dinitrogen complexes are discussed.

Organonitrogen compounds were obtained when ketones or acid chlorides were added to nitrides formed in the systems Cp_2TiCl_2–Mg–N_2 or Cp_2TiCl_2–NaNp–N_2 in THF.[51] Ketones produce primary and secondary amines and acid chlorides produce nitriles. These results demonstrate higher reactivity of the nitrides formed in N_2-reducing systems compared with nitrides prepared from free metals.

I.4.1.2.3 Hydrazine

Hydrazine, the product of partial N_2 reduction, had been demonstrated in chemical dinitrogen-reducing systems before it was found in biological systems.

In 1969 two papers appeared simultaneously with reports that hydrazine was detected in dinitrogen reduction. In one of the papers[52] formation of unstable N_2 complex with low-valent Ti was reported at low temperature in the system Cp_2TiCl_2–*i*-PrMgCl–N_2 in equilibrium with free dinitrogen (*vide infra*). At –60°C dinitrogen in the complex is slowly reduced to a product that can be hydrolyzed to N_2H_4. The yields of hydrazine over several hours correspond to the amount of coordinated N_2 measured at lower temperature (–100°C) where the complexation is virtually complete. No nitride is formed at –60°C, but further reduction to nitride occurs with an excess of *i*-PrMgCl at temperatures above 0°C.[53]

In another system, Ti(O-*i*-Pr)$_4$–NaNp–N_2, hydrazine was also found in the hydrolysis products.[54] The hydrazine yield in the system increases with decrease of the initial Ti^{IV}:NaNp ratio, to a maximum of 15 to 19% based on Ti. Ammonia remains the main product. The hydrazine-forming compound can be reduced to nitride by excess NaNp.

Later hydrazine was detected in a number of other systems including the systems based on titanium, molybdenum, iron, etc. (*vide infra*).

Good yields of hydrazine of 80% per titanium were obtained, for example, after solvolysis in the system $TiCl_4(Ph_3P)$–LiNp–N_2 in THF at –80°C, whereas only ammonia and naphtylamine are reported for the same system without Ph_3P at room temperature.[55]

I.4.2 DINITROGEN REDUCTION IN PROTIC MEDIA[6,56-59]

To approach nitrogenase it was necessary to find the N_2 reduction with participation of solvent protons; in aprotic media with strong reducing agents, acid (to produce ammonia from nitrides) could be used only after the reduction was over. As was mentioned above, dinitrogen reduction in protic media is thermodynamically more favorable than in aprotic solvents because basic reaction products, hydrazine and ammonia, are stabilized by hydrogen bondings. Presumably the intermediates formed in the reduction have to be also more polar than the N_2 molecule, therefore kinetically the process may be facilitated and not so strong reducing agents can be used. Moreover, the achievement of catalytic dinitrogen reduction seems to be easier: formation of N–H bonds in the process of the reduction will be accompanied by the cleavage of M–N bonds, thus liberating the catalytic site to the new dinitrogen molecules.

However, an important obstacle is to be met in attempts to find systems reducing N_2 with the participation of protons. Strong reductants are unstable in protic media and react with solvent protons:

$$2H^+ + 2e \rightarrow H_2$$

This simple two-electron reaction will apparently always successfully compete with more complex dinitrogen reduction. One- and two-electron mechanisms for N_2 reduction require very strong reducing agents that are hardly able to exist for any appreciable time in water or alcohols. The four-electron mechanism requires milder reductants that can be reasonably stable in water solutions but even in this case the reduction requires a stronger reducing agent than H_2 and the reaction must be performed in nonequilibrium conditions because hydrazine is a stronger reductant than dihydrogen.

Of course, one may count on the possibility of six-electron reduction directly to ammonia level. However, whereas formation of a hydrazine derivative by the mechanism described above is a truly four-electron reduction, it is hardly possible to avoid this derivative and cleave the NN bond in one elementary step. The hydrazine derivative is likely to be more stabilized than hydrazine itself (because of π-bonds conjugation in a complex M=N–N=M) but this stabilization can hardly be sufficient to overcome the necessity of using a reducing agent stronger than dihydrogen. The example of nitrogenase confirms these expectations. Even long biological evolution has failed to create a catalytic system not producing H_2 in parallel with N_2 reduction. Although, as some suspected, H_2 evolution may play some physiological role (e.g., create reducing media to prevent interference of molecular oxygen) it is more likely that protons really compete with dinitrogen for electrons. The most perfect among the three known nitrogenases, molybdenum nitrogenase, which produces a minimum amount of H_2, still develops a sufficiently strong reductant to perform this reaction.

Consideration of the great number of unsuccessful attempts to find the reactions in which dinitrogen could be reduced in protic media confirms the view that it was not an easy task.

The first reproducible results demonstrating effective dinitrogen reduction with participation of protons were obtained in our laboratory in 1970.[60] We proceeded from the assumption that molybdenum (or vanadium) is essential for dinitrogen reduction by nitrogenase and we wanted to find hydrazine as the product of the four-electron reduction mechanism. Both assumptions turned out to be very important, although the first one, as we know now, was almost certainly not absolute. The

TABLE I.6
Systems Reducing Dinitrogen in Protic Media

M	Reducing agent	Products	Yield[a]	Conditions[b] (°C)
Ti^{II}	Na(Hg)	N_2H_4, NH_3	0.01	20
Mo^{III}	$Ti(OH)_3$	N_2H_4, NH_3	1	60
	$Ti(OH)_3 + Mg(OH)_2$	N_2H_4, NH_3	170	110
	(Without Mo)		0.005	180
	$Cr(OH)_2$	N_2H_4, NH_3	0.80	90
	(Without Mo)		0.015	90
	$Na(Hg) + Mg^{2+}$	N_2H_4	2.5	20
	$Na(Hg) + Mg^{2+} + PC^{c} + R_3P$	N_2H_4	10, 000	20
	Cathode + $Ti(OH)_3$	N_2H_4, NH_3	0.5	20
V^{II}	$V(OH)_2 + Mg(OH)_2$	N_2H_4, NH_3	0.65	20, water, pH 14.3
		NH_3	0.35	20, water, pH 12
	V^{II} + catechol	NH_3	0.75	20, water, pH 10.5
Nb^{III}	$Nb(OH)_3$	N_2H_4, NH_3	0.09	35
Ta^{III}	$Ta(OH)_3$	N_2H_4	0.02	35

[a] Moles per mole Mo and Ti(II) for systems with Mo and Ti(II); current yield for electrochemical reduction: for other systems according to the stoichiometry $4\ Red + 4H^+ + N_2 \rightarrow N_2H_4 + 4\ Ox$

[b] Medium: aq. methanol unless otherwise stated; pressure: ca. 100 atm N_2.

[c] PC: phosphatidylcholine.

search for hydrazine was a particularly happy choice: the analytical probe for N_2H_4 is very sensitive and when very tiny amounts of hydrazine were at last detected in one of the Mo-containing systems (as a result of certain luck)[6] we were sure that it was really the result of N_2 reduction because, unlike ammonia, hydrazine can hardly be the result of any impurity present. Further traditional chemical approaches (changing conditions, media, and additions) helped to find optimum systems for quite effective dinitrogen reductions.

Dinitrogen was reduced initially by freshly prepared hydroxides exerting strong reducing ability and containing molybdenum. The use of hydroxides was not planned but when they were found to be efficient in dinitrogen reduction it was promptly realized of course that this was in full agreement with theoretical expectations. In strongly alkaline media hydroxides can greatly increase their reducing properties and they naturally combine several metal ions to perform four- or six-electron dinitrogen reduction.

I.4.2.1 Dinitrogen Reduction by Metal Hydroxides

Table I.6 summarizes the systems reducing N_2 in protic media that include metal hydroxides.[6,56-59a] Molybdenum(III) and vanadium(II) hydroxides were found first and they turned out to be most active, molybdenum(III) in the presence of stronger reducing agents, vanadium(II) hydroxide with excess Mg $(OH)_2$. Later some other hydroxides were added to this list. They include $Ti(OH)_2$, $Nb(OH)_3$, and $Ta(OH)_3$, thus, they all combine negative redox potentials with d^2 or d^3 electronic configuration of the transition metal present. $Ti(OH)_3$ (d^1) is inactive, while $Cr(OH)_2$ (d^4) is only weakly active. In all cases they reduce dinitrogen to hydrazine and/or ammonia when freshly prepared under N_2 by additions of alkali to the corresponding salt in water or alcohol solution. They lose their activity at standing, some, e.g., $V(OH)_2$ (without $Mg(OH)_2$) and $Ta(OH)_3$, very quickly, the others, like $V(OH)_2$–$Mg(OH)_2$ and $Nb(OH)_3$ keep the activity much longer. Each of the systems has its own peculiarities, but they have much in common. Optimum reaction media for almost all the systems is aqueous methanol (2 to 10% H_20). It is more effective than water alone first because of higher solubility of dinitrogen and second perhaps because the hydroxide containing methanol and methylate as ligands preserve its amorphous character longer until undergoing transformation to stable and rigid structures of hydroxides, which were found to be inactive towards dinitrogen. In alcohols with longer alkyl chains the yields are generally decreased, probably because low dielectric constants disfavor the formation of polar products from dinitrogen.

High alkaline concentration (1 *M* and higher) leads first to N_2H_4 formation, which can be further reduced to ammonia. At lower alkaline concentrations (pH 8 to 12) the reaction is generally much slower, and the only product is ammonia, which is directly produced from dinitrogen without free hydrazine as an intermediate. This important occurrence, which resembles ammonia formation in the enzymatic reaction (where hydrazine is a poor substrate), was determined from kinetic measurements of dinitrogen and hydrazine reduction: hydrazine is reduced too slowly to be the intermediate and special experiments with $^{15}N_2$ confirmed this conclusion. In the intermediate range hydrazine and ammonia are produced in parallel.

Dihydrogen is formed in all dinitrogen-reducing systems and this reaction, competing with N_2 reduction, is one of the factors decreasing the reaction yields. Carbon monoxide inhibits dinitrogen reduction. Acetylene is reduced in all the systems that reduce N_2, but the range of C_2H_2 reduction is much broader: e.g., acetylene is reduced by both $Cr(OH)_2$ and $Ti(OH)_3$ formed without molybdenum as well as by Mo $(OH)_3$ alone, i.e., in the systems inactive or poorly active towards N_2. Moreover, it has been long established that Cr^{2+} ions reduce acetylene in homogeneous acid solutions in the reaction

$$2Cr^{2+} + C_2H_2 + 2H^+ \rightarrow 2Cr^{3+} + C_2H_4 \qquad (14)$$

An intermediate complex was detected upon cooling the solutions of Cr^{2+} with acetylene. The complex contains two chromium ions per acetylene molecule and its structure seems to be that shown at Scheme 1.3.

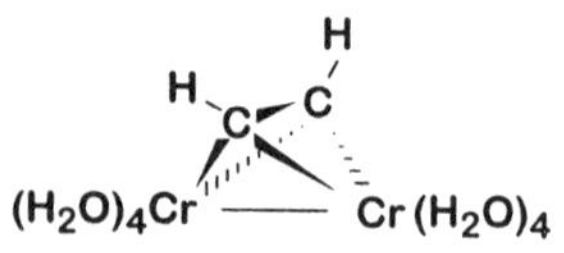

SCHEME 1.3 The structure of the intermediate acetylene complex with chromium(II).

In the case of the hydroxides ethane is formed together with ethylene, C_2H_6 being produced directly from acetylene and not from free ethylene formed intermediately. This four-electron reduction, found first in chemical nonenzymatic systems, is again analogous to enzymatic acetylene reduction in the presence of two alternative nitrogenases, vanadium- and iron-only enzymes.

I.4.2.2 Reductive Hydrolysis of Molecular Deuterium

Considering the action of different substrates competing with dinitrogen for electrons of a reducing agent it is worthwhile here to pay attention to the formation of HD from D_2 in the presence of the functioning nitrogenase (see Section I.3.3). It should be remembered that the reaction, which is in fact a "reductive hydrolysis" because its stoichiometry involves electrons of the reductant,

$$D_2 + 2e + 2H^+ \rightarrow 2HD \tag{13}$$

requires the presence of dinitrogen and was considered as the evidence for formation diazene, N_2H_2, as an intermediate. In one paper[59b] reaction 13 was observed in a model chemical system including rhodium complex with diphenyl-μ-sulfophenylphosphine and sulfate of vanadium(II) as the reducing agent. Formation of HD in this system proceeds along two paths. Together with the reaction 13 there is also the simple exchange $D_2 + H^+ \rightarrow HD + D^+$ but at high concentration of vanadium the stoichiometry of reaction 13 corresponds to the main path. Naturally this reaction does not require the presence of dinitrogen, but to efficiently catalyze the reaction the rhodium complex initially must undergo oxidation by air. The mechanism proposed to explain the observed data includes the following steps, well known in other systems:

$$M^{n+} + D_2 \rightarrow \left[M\langle{}^{D}_{D}\right]^{n+} \tag{15}$$

$$\left[M\langle{}^{D}_{D}\right]^{n+} + 2H^+ \rightarrow M^{(n+2)+} + 2HD \tag{16}$$

$$M^{(n+2)+} + 2e \rightarrow M^{n+} \tag{17}$$

The role of dioxygen may be explained by formation of oxidized rhodium complex, which is more easily reduced by V^{II} to the state forming an active catalyst than the initial complex. This model system provides the explanation for reaction 13 in the presence of nitrogenase. The role of dinitrogen is the formation of a more oxidized complex that accepts another electron from the reducing agent and induces the catalytic reaction of reductive hydrolysis. This mechanism, which suggests formation of deuterides, seems to explain the results more naturally than the hypothesis of intermediate diazene formation.

I.4.2.3 Mixed Vanadium(II)-Magnesium(II) Hydroxide

This system is a more detailed example of dinitrogen reduction. It remains one of the simplest and most active dinitrogen reducing systems. The reduction of N_2 occurs in aqueous or alcohol suspensions of freshly prepared hydroxide formed by adding excess alkali to the solution of a mixture of VCl_2 and $MgCl_2$. Apparently, vanadium(II) (d^3) in the mixed hydroxide combines the ability to both activate and reduce dinitrogen. The reaction produces hydrazine and ammonia at high rates at room and lower temperatures and at atmospheric N_2 pressure. Dinitrogen fixation takes place even from air, dioxygen only somewhat diminishing the yield. The easiest way to observe N_2H_4 formation from N_2 is to combine the solution of the double VSO_4–$MgSO_4$ salt, which is stable in air in neutral solutions, with 20% solution of KOH in methanol in a vessel open to air, shake for a few minutes, and then filter the solution to remove hydroxides. After acidification and subsequent addition of N,N-dimethylaminobenzaldehyde, a yellow azine is formed in the solution and can be identified and measured spectrophotometrically.

At high concentrations of alkali and high dinitrogen pressures the reaction ends up with hydrazine formation according to the stoichiometry

$$4V(OH)_2 + N_2 + 4H_2O \xrightarrow{Mg(OH)_2} 4V(OH)_3 + N_2H_4 \qquad (18)$$

Hydrazine formed competes with dinitrogen for electrons of vanadium(II), particularly at low N_2 pressures, and ammonia can be the final product. The highest yields of hydrazine + ammonia reach about 65% per vanadium electron.

At low alkali concentrations (pH 8 to 12), as was already mentioned, ammonia is directly produced from dinitrogen without intermediate formation of free hydrazine, thus the reaction

$$6V(OH)_2 + N_2 + 6H_2O \xrightarrow{Mg(OH)_2} 6V(OH)_3 + 2NH_3 \qquad (19)$$

takes place.

The role of excess $Mg(OH)_2$ appears to be prevention of formation of the rigid structure of $V(OH)_2$ that is inactive towards N_2. In the absence of $Mg(OH)_2$, vanadium(II) hydroxide is active only in the first few seconds after its formation and quickly loses its activity. In the presence of excess magnesium hydroxide ([Mg]:[V] >10) hydroxovanadium species form polynuclear clusters that are reasonably stable and only slowly lose their activity towards N_2. In their turn vanadium ions in the mixed hydroxide prevent the crystallization of Mg $(OH)_2$ and the system remains amorphous for a long time. Some indirect evidence indicates the number of vanadium ions in the clusters to be equal to four or six, though it is not excluded that there is a distribution of the cluster sizes with some variation of vanadium ions number from one cluster to another. For example, introduction of foreign ions such as V^{3+}, Co^{2+}, or Ni^{2+} in the initial solution inhibits N_2 reduction, presumably transforming active clusters into inactive ones by replacement of V^{2+}. The quantitative analysis of V^{3+} inhibition effect leads to the conclusion that tetramers are the likely species; tetramers are also suggested from the analysis of the results of ethane and ethylene formation in the reduction of acetylene. Maximum values of the hydrazine yields (65%) correspond to hexamers, because two V^{2+} ions out of six would remain nonoxidized after formation of hydrazine.

Some of the vanadium ions in the clusters are likely to be separated from each other by the bridges involving magnesium ions. Therefore linear dinuclear VNNV complexes can be suggested as the intermediates in N_2 reduction, which have to correspond to a high level of dinitrogen activation (see the previous section), and the intermediate complex shown in Scheme I.4 was suggested to explain the role of magnesium ions.

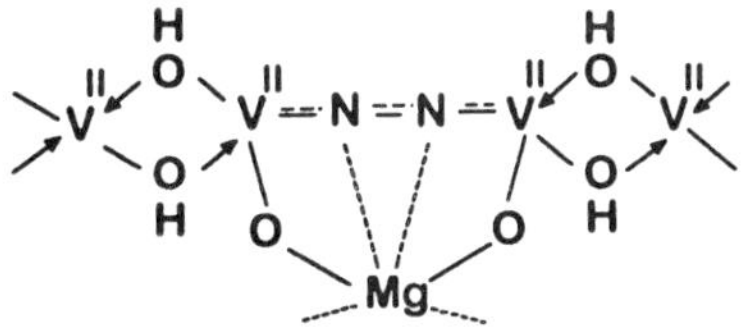

SCHEME 1.4 Proposed structure of the intermediate complex in N_2 reduction in the $V(OH)_2$–$Mg(OH)_2$ system.

The reaction mechanism involves transformation of dinitrogen in the complex into hydrazine derivative, each vanadium increasing its oxidation state to vanadium(III).

The system may be considered as a quasi-homogeneous solution of polynuclear vanadium(II) complexes and usual kinetic analysis may be applied.[61] The reaction is diffusion controlled at room temperature (the rate is determined by N_2 diffusion through magnesium hydroxide to the reaction center) but at 10°C and lower the reaction is sufficiently slow that the reduction process becomes the rate-determining step. The dependence of the rate constant on the dinitrogen pressure shows some curvature, indicating the approach to saturation. The following simple scheme may be suggested to interpret the kinetic results:

$$A^* + N_2 \underset{k_{-1}}{\overset{k_1}{=}} [A^* N_2] \xrightarrow{k} N_2H_4 \tag{20}$$

This scheme leads to the well-known Michaelis–Menten equation:

$$v = k[A^*]_0[N_2] / (K_m + [N_2]) \tag{21}$$

where $[A^*]_0$ is the initial concentration of active centers and K_m is the Michaelis constant equal to $K^{-1} + k/k_1$. (K is the equilibrium constant of the dinitrogen complex formation with A*.) If $K^{-1} \gg k/k_1$, which is likely the case, both the equilibrium constant K and the rate constant of the intermediate complex reaction k can be determined from the slope of the plots of v^{-1} vs. $[N_2]^{-1}$, which should be linear. Accordingly, the plots are linear (Figure I.10) and from their temperature dependences

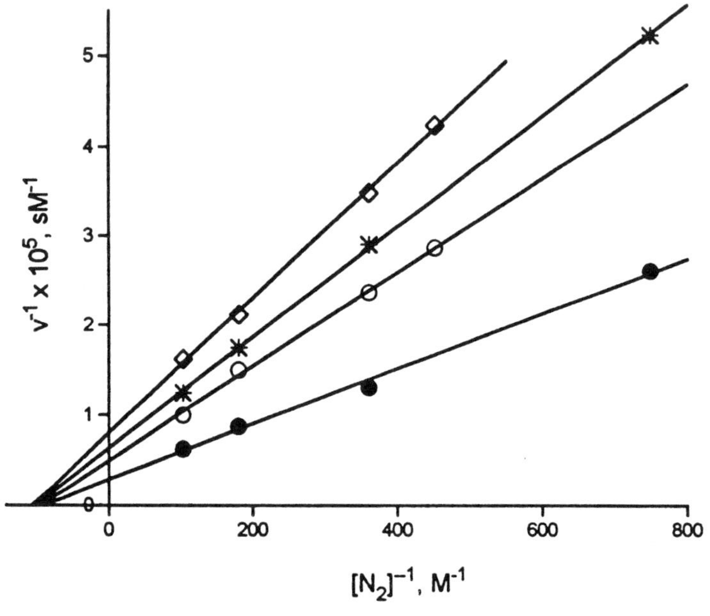

FIGURE I.10 The dependence of the rate of dinitrogen reduction on dinitrogen pressure in the system $V(OH)_2$–$Mg(OH)_2$.

the values of the enthalpy ΔH_r for complex formation and the activation energy E for the complex reaction to form hydrazine were evaluated. They are $\Delta H_r = -4$ kcal/mol and E = 8.4 kcal/mol. Both values are very low even among the systems reducing N_2 in protic media and correspond to the low stability and high reactivity of the intermediate complex.

Mixed V^{II}–Mg^{II} hydroxide is one of the systems (and we shall see several examples of such systems later) that disprove the generally accepted views on dinitrogen as very inert molecule. N_2 can be very active in the systems that are neither strong acids or bases nor too strong reductans or oxidants. The important condition for dinitrogen to reveal high chemical reactivity is the possibility of accepting at least four electrons in an intermediate polynuclear complex including several electron donors.

I.4.2.4 Soluble Complexes of Vanadium(II)

The exact structure of the intermediate polynuclear V^{II} complex in the mixed V^{II}–Mg^{II} hydroxide is, of course, difficult to elucidate and it was important to find a homogeneous system reducing N_2 in protic media. This was realized in our laboratory in 1972 for the complexes of V^{II} with catechol and substituted aromatic diols, which were found to be able to effectively reduce dinitrogen in homogeneous water and alcohol solutions.[62,63] In the case of catecholate complexes in water the reaction proceeds in the range of pH 8.5 to 13.5 with maximum yields at about pH 10. For substituted catechols the pH range corresponding to the most active complexes reacting with N_2 may differ noticeably (Table I.7).

TABLE I.7
Homogeneous Dinitrogen Reduction by V(II) Complexes with Diols at Room Temperature, $[V^{II}] = 0.05\ M$; $[Diol] = 0.5\ M$

Diol	Solvent	$P(N_2)$, atm	Time, h	NH_3 yield, %	pH
Catechol	CH_3OH	1	0.03	47	[a]
Catechol	CH_3OH	15	0.17	75	[a]
Catechol	H_2O	100	1.5	60	10.5
Gallic acid	H_2O	100	4.0	48	12.9
Pyrogallol	H_2O	100	4.0	60	13.5

[a] Catechol/NaOH = 1:1.14.

Later the number of catecholate V^{II} complexes active towards N_2 was greatly increased,[64] but this family of the complexes of vanadium(II) with aromatic diols and similar compounds remains unique: no other complexes have been found that would reduce dinitrogen in protic media homogeneously with reasonably high yields of products. Hydrazine formation was reported also in homogeneous water solutions (pH about 4.5) containing vanadium(II) and α,ω-dicarboxylates, $^-OOC–(CH_2)_n–COO^-$, with highest yields for n = 5 or 6.[65] The yields, though very small (about 0.01%), are reported to correspond to the equilibrium

$$N_2 + 4V^{2+}_{aq} + 5H^+ = 4V^{3+}_{aq} + N_2H^+_{5aq} \qquad (22)$$

The authors[65] explain these results by dinuclear complex VNNV formation with carboxylate ligands coordinating two vanadium ions in an appropriate way to form the cyclic complex with dinitrogen.

The reduction of dinitrogen by vanadium(II) catechol complexes in methanol occurs at room temperature and atmospheric N_2 pressure. The yield of NH_3, which is the only reaction product, reaches about 50% with respect to V^{II} as a one-electron reductant (Table I.7). In aqueous solutions such yields are observed only at elevated N_2 pressures. In the absence of dinitrogen a parallel reaction of V^{II} oxidation by the solvent protons proceeds with H_2 evolution. As N_2 pressure increases the yields of NH_3 increase and those of H_2 decrease. But at no pressure does the yield of NH_3 exceed 75%, 25% of electrons being used for H_2 formation. Thus, analogous to the enzymatic reduction in the presence of molybdenum nitrogenase, the stoichiometry of the reaction at saturating N_2 pressure correspond to the following equation:

$$8V^{2+} + N_2 + 8H_2O \rightarrow 8V^{3+} + 2NH_3 + H_2 + 8OH^- \qquad (23)$$

This equation including dihydrogen evolution *coupled* with the reduction of dinitrogen reflects the polynuclear character of the intermediate complex (*vide infra*).

Acetylene is quantitatively reduced by vanadium(II) catecholate to ethylene with *cis*-dideuteroethylene formed selectively from C_2D_2, also similarly to enzymatic reduction of acetylene by nitrogenase. Distinct from dinitrogen reduction, the pH range for the reduction of acetylene is much broader than for dinitrogen (from about pH 5 to concentrated alkali solution).

Carbon monoxide inhibits the formation of ammonia similarly to what is observed in the enzyme, but, apparently, unlike the enzyme, V^{II} catecholate reduces carbon monoxide mainly to methanol. Formaldehyde is also produced in low yields, but using ^{14}CO it was found that more than 90% of carbon monoxide reduction to methanol proceeds without intermediate formation of formaldehyde.[63]

Kinetic studies of the oxidation of the V^{II} catechol complex in the presence of dinitrogen produced the following equation for the reaction:

$$-d[V^{II}]/dt = k_1[V^{II}]^2[N_2] + k_2[V^{II}]^{0.5} \tag{24}$$

The first term corresponds to the reduction of dinitrogen to ammonia and the second to the parallel and *independent* reaction of H_2 evolution from the solvent protons.

Hydrazine which could be the intermediate in the reduction, is easily and quantitatively reduced to ammonia in the vanadium(II) catechol solutions. When the reaction of dinitrogen reduction is stopped at its initial stages by addition of an acid or an oxidant ($VOSO_4$) a small quantity of hydrazine is found. However, a kinetic study of hydrazine reduction has revealed that its rate constant is at least two orders of magnitude smaller than necessary for free hydrazine to be the only intermediate producing ammonia. The results show that hydrazine is formed at the decomposition of an intermediate that contains a single N–N bond and appears to be a vanadium hydrazido derivative. Later similar results indicating intermediate N_2^{4-} derivative formation were obtained in the enzymatic dinitrogen reduction (*vide supra*).

The second order of vanadium concentration in the rate equation for the N_2 reduction, and at the same time the one-half order for dihydrogen formation, are indicative of the complex polynuclear structure. It is possible, for example, to assume that the active complex contains four V atoms and two complexes are needed to reduce N_2 to ammonia, whereas a dinuclear complex produces dihydrogen and this dinuclear complex is formed in equilibrium dissociation of the tetramer into two monomers. In this case the H_2 formation coupled with N_2 reduction could be explained assuming that after ammonia formation by the octamer, a dimer of V^{II} is formed readily producing H_2.

Important information about the state of vanadium catechol complexes was obtained from their EPR spectra.[66] Addition of lithium methylate or sodium methylate to the solution of V^{2+} and catechol in methanol results in the vanishing of the spectrum of free V^{2+} and appearance of new signals that were attributed to three V^{II} catechol complexes replacing each other at different methylate concentrations. At low methylate concentrations a complex exists whose EPR signal has a g-factor of 2.006 and consists of 22 components. The increase of the methylate concentration leads to a drop in the intensity of this signal and increase of the intensity of another signal with g-factor of about 5.3 consisting of ten components. The intensity of the latter signal reaches maximum at $[LiOCH_3] = 0.57$ to 0.60 *M* and in the case of sodium methylate at $[NaOCH_3] = 0.7$ to 0.8 *M*. Further increase of methylate concentration

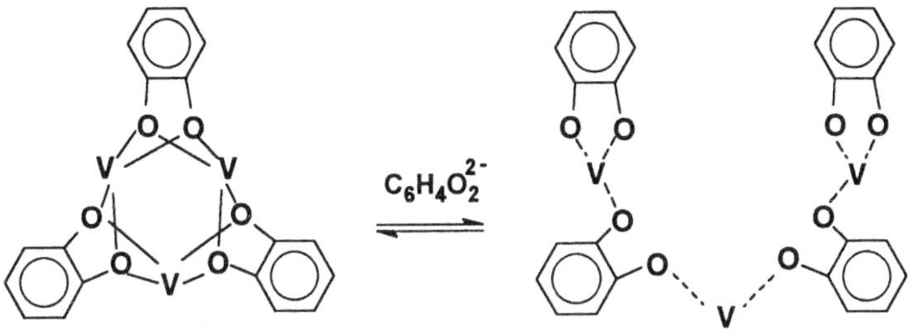

X Y

FIGURE I.11 Simplified structures of "rigid" and "flexible" trinuclear V^{II} catechol complexes. (Adapted from Luneva, N. P., Morwavsky, A. P., and Shilov, A. E., *Nouv. J. Chim.*, 6, 245, 1982.)

removes completely the 22-component signal and leads to a decrease of the intensity of the ten-component signal. Instead a new signal appears and increases its intensity. It consists of three lines with g-factors of 4.50, 3.41, and 2.000 that change their intensities in the same way with the increase of base concentration and evidently belongs to one complex.

The 22-component EPR spectrum of the first complex (designated X) was attributed to three equivalent vanadium atoms with strong spin exchange between them (number of components = $2 \times 3 \times 7/2 + 1 = 22$). (Nuclear spin of ^{51}V nucleus is $7/2$.) The simplified structure of the complex is presented in Figure I.11.

The spectrum of the second complex (Y) may be explained if one were to assume that it also includes three vanadium atoms but without noticeable spin exchange interaction. The observed ratio of intensities 1:1:3:3:3:3:3:3:2:2 corresponds to the superposition of eight lines corresponding to two equivalent vanadium atoms and eight lines corresponding to one vanadium different from the first two and therefore somewhat shifted in g-factor.

1	1	1	1	1	1	1	1		
		2	2	2	2	2	2	2	2
1	1	3	3	3	3	3	3	2	2

The simplified structure for this open trinuclear complex (Y) that is formed from the first one (X) under the action of the base and is apparently less rigid than X is shown in Figure I.11.

The third complex (Z) appears to be a familiar mononuclear octahedral V^{II} $(catechol)_3$. The structure of the similar but more stable V^{III} $(catechol)_3$ was determined by X-ray analysis.[67]

All three complexes are in equilibrium with each other, with the equilibrium shifting to the right upon an increase of the base concentration

$$X \underset{}{\overset{RO^-}{\rightleftharpoons}} Y \overset{RO^-}{\rightleftharpoons} Z \tag{25}$$

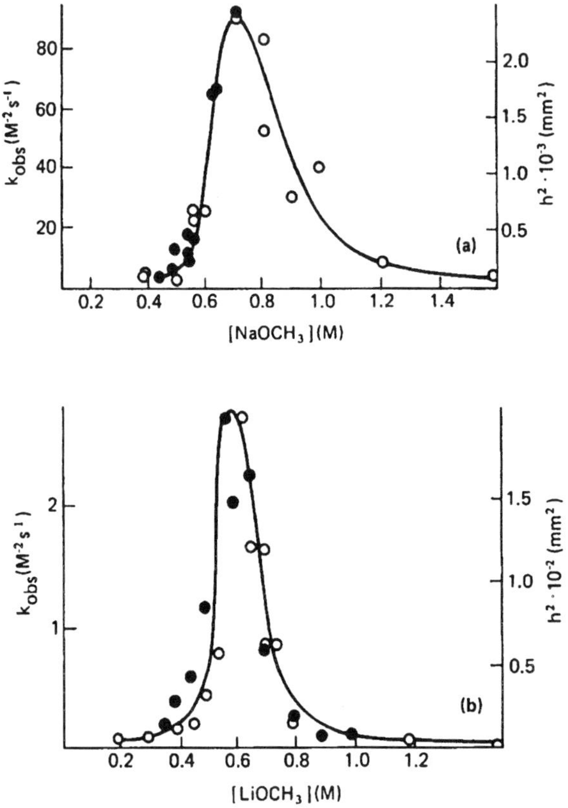

FIGURE I.12 Dependence of the N_2 reduction rate (●) and the intensity of the complex Y EPR signal (○) on the concentration of alkali. (Adapted from Luneva, N. P., Morwavsky, A. P., and Shilov, A. E., *Nouv. J. Chim.*, 6, 245, 1982.)

This is apparently due to the increase of a negative charge and therefore to destabilization of the trinuclear complex. Initially this leads to the weakening of the bridges and the cleavage of one of the bridges to produce the complex Y and at the higher base concentration to the decomposition of the trinuclear complex to the mononuclear species.

Considering the dependence of the reaction rate on the base concentration it is clear that both the "rigid" trinuclear complex X and mononuclear complex Z are not active towards dinitrogen, because the system is inactive in the ranges of the base concentration where these complexes are present as the main species. Concentration of the "flexible" trinuclear complex Y reaches maximum just in the region of the base concentrations where the system is the most active towards N_2, therefore it could be the active species. Comparison of the reduction rate with h^2 (h is the intensity of the EPR signal of Y) confirms this assumption quantitatively (Figure I.12). (It must be remembered that the rate is proportional to the second order of vanadium concentration.) This result also explains why $LiOCH_3$ and

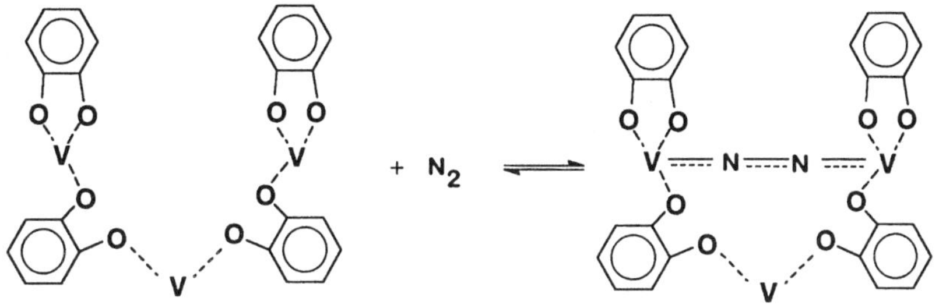

FIGURE I.13 Proposed scheme of dinitrogen activation by the "flexible" trinuclear complex Y. (Adapted from Luneva, N. P., Morwavsky, A. P., and Shilov, A. E., *Nouv. J. Chim.*, 6, 245, 1982.)

$NaOCH_3$ produce somewhat different effects on the reaction rate: $NaOCH_3$ is a stronger base and more effectively forms the active complex Y. Thus trinuclear complex activates dinitrogen and another similar complex is needed to perform the reduction. This may be explained by the equilibrium in Figure I.13 with dinitrogen forming a bridge between two vanadium atoms in the flexible complex Y.

The rate-determining step is proposed to be the interaction of the intermediate complex with another Y particle to form a hydrazine derivative of V^{III}, mononuclear V^{III} complex, and a dimer of V^{II}, which may react with protons to form H_2 molecule. Trinuclear hydrazine derivative can form N_2H_4 when it reacts with an acid, or in the absence of acid it can react with V^{II} to cleave N–N bond and produce ammonia (Scheme I.5).

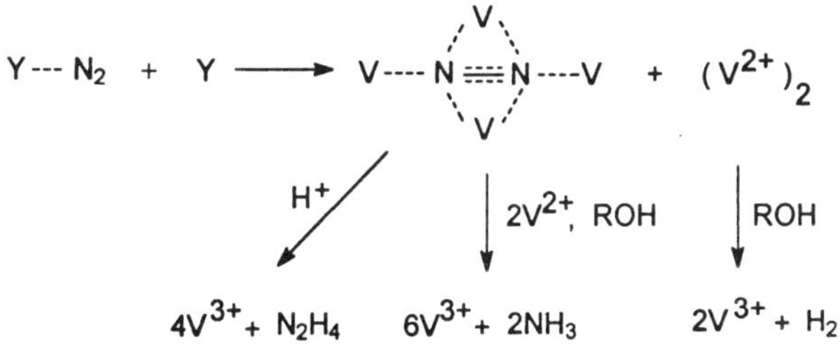

SCHEME 1.5 Mechanism of vanadium(II) catechol complex reaction with dinitrogen.

Scheme I.5 nicely explains all the facts concerning the dinitrogen reduction but fails to explain the one-half order for the parallel and independent dihydrogen evolution. There are some indications that, unlike dinitrogen reduction, H_2 formation is sensitive to the impurities present in the solution and presumably requires some additional catalyst. The main result of the studies of vanadium(II) catecholate reduction of dinitrogen, i.e., the evidence for polynuclear complex activating and reducing N_2, invokes no doubts. Other aromatic diols may form different complexes with bivalent vanadium. This is indicated in particular by the results of studies of a number of various ligands.[64] For example, hydrazine was found to be the final product of N_2 reduction in parallel with ammonia formation in some cases and thus some modification of the scheme is required for these ligands.

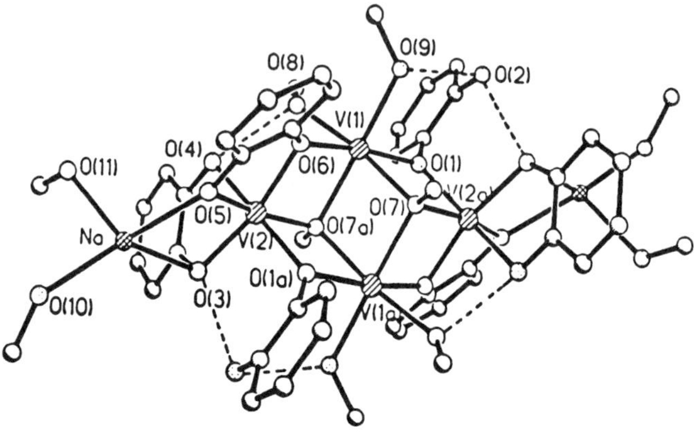

FIGURE I.14 The structure of vanadium di-*tert*-butyl catecholate complex. (From Luneva, N. P. et al., *Angew. Chem., Int. Ed. Engl.*, 32, 1178, 1993. With permission.)

I.4.2.5 Structure of the Vanadium Di-*tert*-butylcatecholate Complex

Other evidence for a polynuclear vanadium catecholate complex follows from the structure of the complex produced in alcohol solutions of V^{II} with *di-tert*-butylcatecholate.[68] The structure of the complex is presented in Figure I.14. The complex contains four vanadium ions, two of them being divalent and two trivalent, but they are indistinguishable, thus their oxidation state is 2.5, and the complex may be considered an intermediate in V^{II} oxidation to V^{III}. Each vanadium atom exhibits a distorted octahedral coordination involving oxygen atoms. Sodium atoms in the complex have coordination of distorted tetrahedron formed by oxygens. Each of the sodium atoms is coordinated by two oxygen atoms of methanol molecules and by two oxygen atoms of different catechol molecules. Due to this coordination, oxygen atoms are μ_2-bridging sodium and vanadium atoms. One of the three independent catechol molecules participates in the coordination of the vanadium atoms only by one oxygen atom while the second atom is protonated and does not participate in the metal coordination. Apparently, bulky *tert*-butyl substituents prevent the formation of mononuclear complex of V^{III} with three catecholate ligands as it occurs with nonsubstituted catechol, and the V^{III} formed remains bound to nonoxidized V^{II}. Accordingly, both dinitrogen reduction and dihydrogen evolution proceed with V^{II} *tert*-butylcatecholate only to 50% of the full extent of vanadium oxidation and the subsequent process develops very slowly. The structure shows the catechol molecule as a bidentate ligand at a single vanadium ion and a bridging ligand unifying two vanadium ions. The structure suggests a similar bonding in the active polynuclear complexes reducing dinitrogen.

I.4.3 CATALYTIC DINITROGEN REDUCTION

I.4.3.1 Molybdenum(III)-Titanium(III)-Magnesium(II) System

A catalytic system for dinitrogen reduction requires the presence of a sufficiently strong reducing agent that would reduce the oxidized form of the catalytic complex after or during the reaction with dinitrogen to the initial reduced state. The first candidate to be a catalyst from the data of Table I.6 is molybdenum(III), which activates N_2 only in the presence of a stronger reducing agent, such as $Ti(OH)_3$, $Cr(OH)_2$, or $Ta(OH)_3$. Presumably, these strong reductants take part in a process that may be called reductive coaction and was already described above. Molybdenum(III) activates dinitrogen in a di- or polynuclear complex, forming an intermediate, e.g., of the type Mo···N≡N···Mo, which in this case is unable to produce hydrazine or ammonia at the expense of molybdenum oxidation. The reaction may become thermodynamically favorable if a stronger reducing agent situated nearby simultaneously reduces molybdenum to its initial trivalent state. To perform many catalytic cycles the catalyst has to be readsorbed at a different site of the reductant or an electron flow has to be realized through the reductant. Therefore, for the latter case, the reducing agent must adsorb the catalytic complex at the surface and it must be electroconductive to use its electrons for the catalytic reduction. For electrochemical reduction electrons of a cathode can be used if the catalytic complex is adsorbed at the surface of the electrode.

First catalytic system for dinitrogen reduction in protic media was realized with molybdenum(III) as a catalyst and titanium(III) hydroxide as a reductant. With pure $Ti(OH)_3$ the yield of the products of N_2 reduction in the presence of molybdenum(III) formed in the process of coprecipitation of both metal hydroxides by addition of alkali reaches at best only equimolar amounts with respect to molybdenum even at elevated temperatures and pressures of dinitrogen. In the presence of salts of some other metals in solution before the addition of alkali, e.g., Mg^{2+}, Ca^{2+}, or Sr^{2+}, the yields increase and the system becomes catalytic. The effect of magnesium salts turned out to be particularly pronounced. The yields of hydrazine and ammonia are strongly increased and at high temperatures and pressures can reach several hundred turnovers per molybdenum present. The highest yields are obtained when the ratio of Mg^{2+}:Ti^{3+} is 1:2. At this ratio a compound of formula $MgTi_2O_2$ is produced and form fine crystals as was revealed by X-ray analysis and electron microscopy. The catalytic effect of molybdenum(III) is observed when its complex is adsorbed at the surface of $MgTi_2O_2$. Accordingly, whereas for mixed hydroxides it is essential to have both metals present in a homogeneous solution before the addition of alkali, in the case of the catalytic N_2 reduction $MgTi_2O_2$ can be formed initially without molybdenum and precipitated by the addition of alkali. After subsequent molybdenum catalyst addition catalytic activity towards dinitrogen is observed.

The experimental results show that the mechanism of the catalytic action involves reductive coaction of titanium ions surrounding the molybdenum catalyst, oxidation of these ions to titanium(IV), while coordinated dinitrogen is reduced to a hydrazine derivative, and subsequent electron transfer from the remote titanium ions to the catalytic center. The magnesium–titanium compound $MgTi_2O_2$ possesses semiconducting properties and evidently a molybdenum–dinitrogen complex on its surface forms an electron trap. Dihydrogen is evolved in fast reaction in the absence of dinitrogen, particularly at high temperatures. Introduction of the molybdenum catalyst even in very small amounts compared with titanium strongly inhibits dihydrogen evolution in *the presence of dinitrogen* and the oxidation of titanium to the fourvalent state proceeds much longer. It may be explained by the mechanism in which molybdenum complexes are adsorbed at the active sites of $MgTi_2O_2$, preventing electron flow to the hydrogen evolution centers. Dinitrogen forms intermediate complexes with molybdenum(III) and then is reduced to hydrazine or to ammonia with the participation of the solvent protons. The dependence of the reaction rate on dinitrogen pressure shows the Michaelis type form, which allows the equilibrium constant for dinitrogen complex formation and the rate constant for the reduction of activated dinitrogen in the complex to be estimated. Temperature dependences of the constants give: $\Delta H = -7$ kcal/mol; $\Delta S = 17$ e.u. for the enthalpy and the enthropy of the complex formation equilibrium and $E = 19.8$ kcal/mol for the activation energy to N_2 reduction in the complex.[69]

I.4.3.2 Catalytic Dinitrogen Reduction by Sodium Amalgam

The mechanism proposed for the catalytic N_2 reduction by electroconductive reductant has prompted the use of amalgams, particularly sodium amalgam, as a reducing agents.[57,58] Sodium amalgam is a strong reductant (redox potential for $Na/Hg \rightleftharpoons Na^+/Hg + e$ is -1.84 V), reasonably stable, however, in contact with water or alcohol. As a liquid it can easily be divided by shaking or agitating into small droplets to increase the surface-to-volume ratio and thus to make more effective its contact with solution. Finally, the amalgam can be prepared electrochemically using a mercury cathode and passing an electric current through the solution containing sodium ions; therefore it can be used in the electrochemical reduction of dinitrogen.

Introduction of the sodium amalgam has increased the yield of the reduction products, hydrazine and ammonia, in the catalytic system based on molybdenum(III) and $MgTi_2O_2$ described above. Apparently, the amalgam transfers its electrons through magnesium–titanium mixed oxide to the molybdenum catalytic center. Obviously the electron-transferring material is unnecessary and the amalgam can serve in principle as the only reducing agent. Accordingly, upon shaking the molybdenum(V) compound formed by dissolving $MoCl_5$ (without titanium–magnesium reductant) in methanol, under 100 atm N_2 with sodium amalgam, hydrazine can be detected in the solution. However, the yields are extremely small and irreproducible. This poor activity of the system reflects apparently inefficient contacts of the amalgam with the catalytic complex as well as instability of the complex in the solution. Nevertheless, the system was subsequently developed and transformed into *the most active among known nonbiological catalytic systems for the reduction of dinitrogen at ambient temperature and pressure.*

The first improvement of the system was achieved when magnesium salt was added to the solution of $MoCl_5$ in methanol before reducing the latter by sodium amalgam. The yields, albeit still low, became reproducible and so the reaction kinetics could be measured and conclusions made about the nature of the reaction rate-controlling step. Apparently, magnesium ions that are involved in the polynuclear structures (*vide infra*) stabilize the complexes and may make them more suitable for the activation of dinitrogen. The system's low activity was due most probably to inefficient contact of the complexes with the surface of the amalgam. Therefore it was necessary to find the means to bind the complex to the amalgam to improve the electron transfer from the reductant to the catalytic complex. After long search without much success, following the example of nature, we tried a

TABLE I.8
Development of the Catalytic Systems on Na(Hg) and Mo^{III} (Methanol, 20°C)

Catalytic system	Specific rate $[N_2]_{red}/[Mo]$, s^{-1}	Yield (N_2H_4 + ½NH_3) Per Mo (cycles)	Per red (%)
$MoCl_5$ + Na(Hg)			
70 atm N_2	<0.0001	0.5	0.004
+ Mg^{2+}	0.0003	2.5	0.02
+ Phospholipid (PL)		25	0.25
The complex $Mo_8Mg_2^{2-}$, PL, R_3P			
1 atm N_2	1.1	100–200	~10
70 atm N_2		10,000	30
Nitrogenase	~0.6		

phospholipid (phosphatidylcholine), which turned out to strongly increase the reaction rate and the product yield (Table I.8).[70]

With improved means of catalyst preparation (*vide infra*) it became possible to study the reaction at atmospheric pressure and room temperature in a very simple reaction vessel. This facilitated, in turn, the further efforts for system improvement. Finally phosphines were found to increase the yield of the dinitrogen reduction products, hydrazine and ammonia.[71]

Hydrazine and ammonia are produced in parallel reactions, their ratio depending mainly on the nature of the catalytic complex and also on the length of a phosphatidylcholine aliphatic chains. In the case of purified polynuclear molybdenum complexes (*vide infra*) hydrazine is the main product of dinitrogen reduction and the yield of ammonia amounts to only about 10% of total dinitrogen reduced.

After all the improvements the specific activity of the catalytic system in dinitrogen reduction (per the catalytic complex) reached or even exceeded that of nitrogenase. Up to 1000 turnovers can be observed with respect to molybdenum catalytic complex at atmospheric pressure and more than 10,000 turnovers at elevated pressure (see Table I.8).

The increase of the catalyst activity is reflected also in activation energy of the N_2 reduction. The activation energy drops from 18 kcal/mol without the phospholipid (but in the presence of Mg^{2+}) to 8.7 kcal/mol with the phospholipid and further to virtually zero in the full system with improved catalyst.[72]

The role of the phospholipid is to form a thin film on the surface of the amalgam with the catalyst incorporated and thus bound to the electron donor (Figure I.15). The film is kept at the surface apparently because of the attachment of the phosphatidylcholine positive heads to the negatively charged surface. Electrons move from the amalgam to the catalytic complex to perform the catalytic reduction. Proton-containing molecules (water and alcohol) diffuse through the hydrophobic part of the film to protonate the activated dinitrogen and to form the final products. The hydrophobic film helps to keep the amalgam droplets separated and prevents their fast agglutination. This leads to the increase of the surface-to-volume ratio and thus to the increase of the catalytic activity. Presumably the hydrophobic film also helps to control the proton flow to the catalyst from the solution. Indeed, the dihydrogen

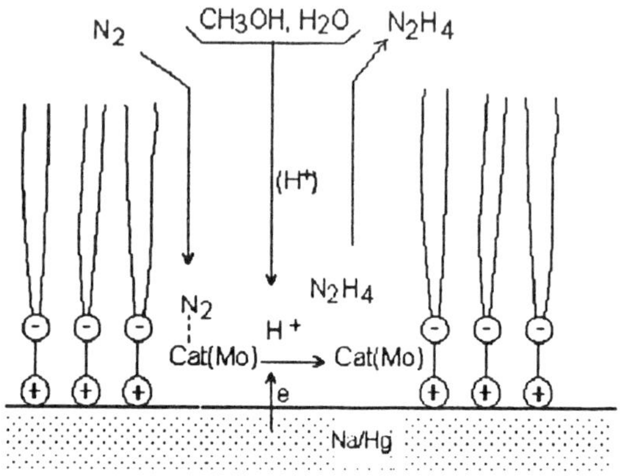

FIGURE I.15 Proposed mechanism of dinitrogen reduction catalyzed by molybdenum(III) complex incorporated in the phospholipid film on the amalgam surface.

evolution is also catalyzed by the molybdenum complexes and also facilitated by the presence of the phospholipid, but its rate increase is much less pronounced than that of dinitrogen reduction.

Apparently the chemical nature of the polar heads of the phospholipid is important for the incorporation of the catalyst because many other surface-active materials including phospholipids proved to be inactive. For example, tetraalkylammonium salt with long aliphatic chains, though it forms a film on the surface of the amalgam, failed to make a catalytic system with the molybdenum complexes. Nevertheless, polyvinylalcohol turned out to be another substance, of quite different nature, that acts as a cocatalyst in dinitrogen reduction by sodium amalgam, albeit less effectively than the phospholipid, indicating that neither the phospholipid nor the polyvinylalcohol drastically changes the nature of the catalytic complex when absorbing it at the amalgam surface.[73]

The role of phosphines increasing the yields of the reaction products is more difficult to elucidate. The effect of the phosphines depends on their basic strength and their size, increasing for stronger bases and molecules of smaller sizes. An evident suggestion should involve the formation of PR_3 complexes with molybdenum compounds. However, even the ratio of hydrazine and ammonia is not changed in the presence of phosphines; therefore it is unlikely that the nature of the catalyst is at all changed. Moreover, the initial rate of the reduction remains virtually unaffected in the presence of phosphines.[73] Therefore their most probable role is to increase the catalytic chain without affecting the catalyst itself. It may be suggested that at least one of the possible ways of catalyst destruction is the reduction of molybdenum by the amalgam to the low oxidation state, perhaps molybdenum(0). This reduced molybdenum may react with other species of the catalytic complex, reducing molybdenum and deactivating the complex. Formation of the phosphine complex may stabilize the low-valent molybdenum complex and prevent or slow down the reaction with molybdenum(III) in the catalyst, thus increasing the catalytic chain and the yield.

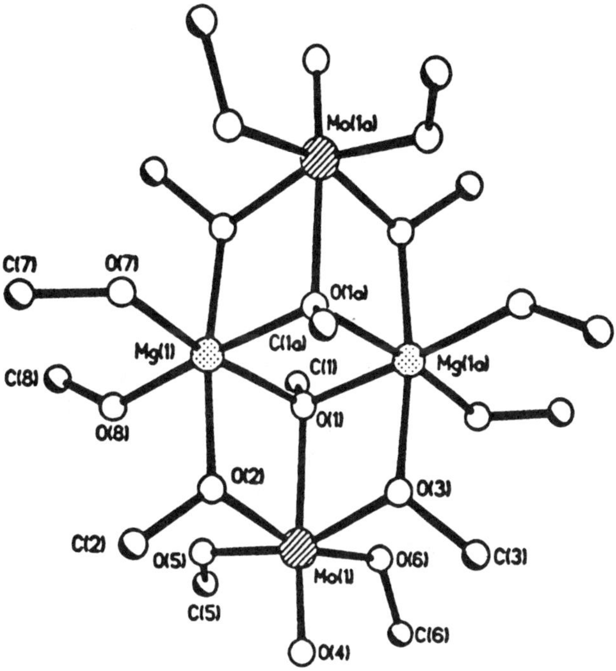

FIGURE I.16 The structure of dinuclear molybdenum(V) complex (I). (From Antipin, M.Yu. et al., *Gazz. Chim. Ital.*, 123, 265, 1993. With permission.)

This catalytic system provides an example of the development of a catalyst along the lines of a biological catalyst; that is, construction of an organized molecular assembly, approaching in some aspects enzymes, and thus drastically improving its catalytic performance.

I.4.3.3 Structure of Polynuclear Molybdenum Complexes

Bi- and polymolybdenum(V-VI) complexes containing –O–Mg–O– and μ-O bridges have been isolated from the solutions of molybdenum and magnesium compounds and their structures investigated by X-ray diffraction.[74] When reduced to the Mo^{III} state the complexes are active catalysts for dinitrogen reduction by sodium amalgam in the presence of surface-active phospholipids and organic phosphines. Crystals of all three molybdenum complexes isolated from the solution and suitable for X-ray analysis were extremely unstable out of the solution, easily collapsing to a powder. The single crystals used were therefore quickly transferred in a cold nitrogen gas stream (163 K) directly onto the diffractometer and kept at this temperature during data collection.

The structures of complexes I and II (Figures I.16 and I.17) are similar and differ only by the number of terminal methoxy (OMe) and oxide (O^{2-}) groups at molybdenum atoms. The central frame of I and II contains two molybdenum and two magnesium atoms (Mo^{V} and Mo^{VI} in the cases of I and II, respectively). They are linked by two μ_3-OMe ligands and four μ_2-OMe ligands. Each magnesium atom is

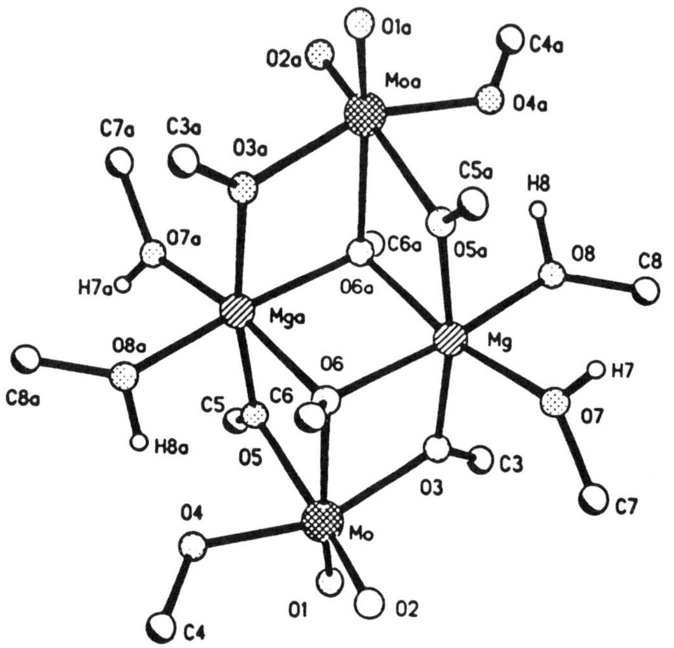

FIGURE I.17 The structure of dinuclear molybdenum(VI) complex (II). (From Antipin, M.Yu. et al., *Gazz. Chim. Ital.*, 123, 265, 1993. With permission.)

also coordinated by two methanol molecules and as a result the total magnesium coordination is a distorted octahedron.

The mixed valence Mo^{V}–Mo^{VI} complex III includes a number of weakly bonded methanol molecules and this probably explains its particular instability outside of the solution. The structure of the centrosymmetric dianion of III is shown in Figure I.18. The architecture of this complex is more complicated than that of I and II. Metal atoms of the cluster are linked by bridging oxygens of the μ_2, μ_3, and μ_4 types, bridging μ_2-OMe groups, and also Mo(1)–Mo(2a) and Mo(1a)–Mo(2) bonds between the Mo^{V} atoms with bond length 2.588(1). The atoms Mo(3) and Mo(4) are formally six-valent. Each Mo atom has one terminal oxide (O^{2-}) ligand.

Molybdenum in all the complexes can be reduced to the molybdenum(III) state when put in the contact with sodium amalgam in methanol. The completeness of the reduction was checked by the disappearance of the band at 310 nm which corresponds to Mo=O bonds, in the spectra of the complexes. According to the results obtained by titration with vanadium(IV) sulfate solution, molybdenum reaches the oxidation state III in the process of the reduction of all the complexes.

The complexes of molybdenum(III) formed can be isolated from the solution by additions of ether. Dark powders precipitated from the solutions were filtered off and washed by the mixture of ether and methanol. Analysis for molybdenum and magnesium have shown that the ratio [Mg]:[Mo] did not change during the reduction. All the complexes are diamagnetic, as are their precursors. The results obtained at Sephadex gel filtration of complexes I and III are shown in Figure I.19, as are the

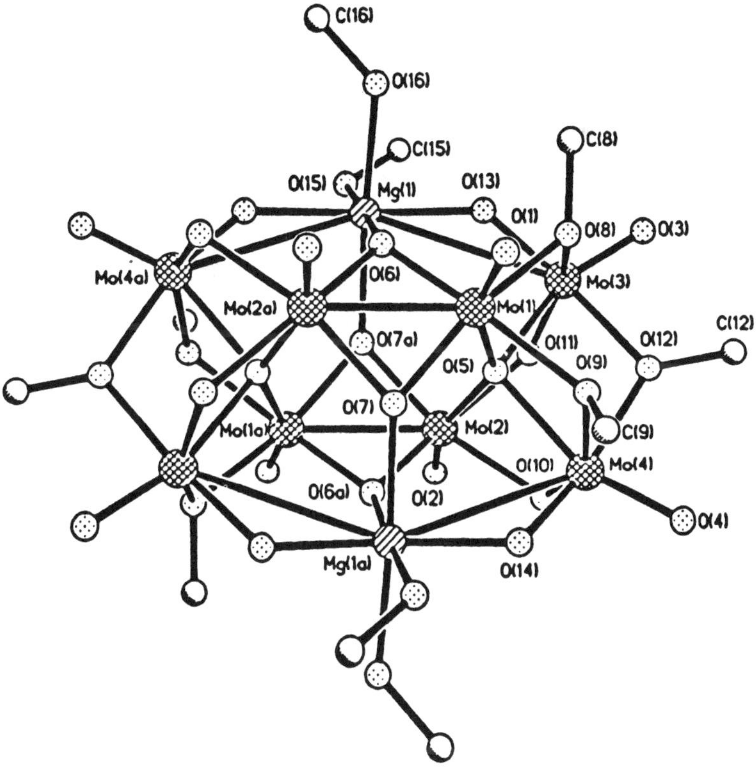

FIGURE I.18 The structure of the anionic part of the octamolybdenum complex (III). (From Antipin, M.Yu. et al., *Gazz. Chim. Ital.*, 123, 265, 1993. With permission.)

complexes produced by their reduction. It is seen that the lag time of the complexes remains virtually the same after the reduction, thus their molecular volumes and probably their nuclearity remain essentially unchanged during the reduction, although molybdenum(III) complexes are probably slightly more voluminous than their precursors.

In alkaline solutions molybdenum(III) complexes remain stable in the solution, but in acid media they lose their stability and apparently start to polymerize. The position of the peak of gel filtration of molybdenum(III) complex, which was produced by the reduction of III and was then kept in acid solution ($1 \cdot 10^{-2}$ *M* HCl) for several hours and then again made alkaline by sodium methylate, is shown in Figure I.19. It is seen that the molecular volume of the complex is considerably increased. The complex remains unchanged in alkaline media for a long time.

Figure I.20 shows kinetic curves of hydrazine formation in dinitrogen reduction by sodium amalgam in the presence of three complexes: Ia, formed upon the reduction of I; IIIa, formed upon the reduction of III; and IVa formed after polymerization of the complex IIIa in acid solution. In all the cases molybdenum concentrations were the same. It may be seen that under identical conditions the reaction rate and the yield of hydrazine are increased with the increase of the number of molybdenum

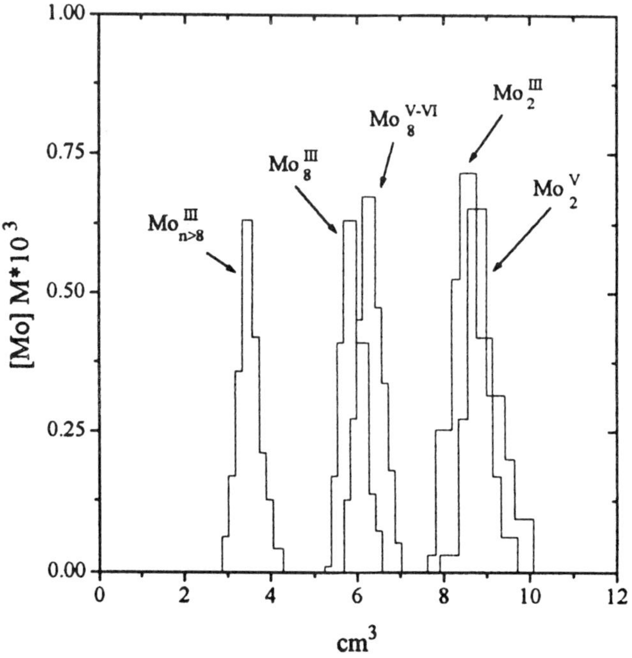

FIGURE I.19 Chromatogram of the gel filtration of the oxidized and reduced molybdenum complexes. (Adapted from Antipin, M.Yu. et al., *Gazz. Chim. Ital.*, 123, 265, 1993. With permission.)

atoms in the complexes. Complex IIa, produced upon the reduction of II, demonstrated the same activity as Ia. Taking into account that linear dependence of the reaction rate on complex concentration was observed in the range of molybdenum concentrations chosen for comparison, it may be concluded that catalytic rate constants are even more strongly dependent on the number of molybdenum atoms in the complexes than may appear in Figure I.20, because at equal concentrations of molybdenum the number of complexes will be correspondingly larger in the case of lower molecular weight complexes.

Although the structures of the Mo^{III} complexes, which are the catalysts for dinitrogen reduction, remain undetermined due to the difficulties in preparation of single crystals for X-ray study, some conclusions concerning their compositions and structures may be made on the basis of the structures of their precursors. Conservation volumes of the complexes and magnesium atoms during the reduction allows the suggestion that the structures do not change too much during the reduction. All the evidence shows that O^{2-} ligands disappear from molybdenum atoms at the reduction and are probably replaced by MeO^- ligands or coordinated methanol molecules. Other details of the structure could remain unchanged, though all the ligands, including bridging OMe groups, are probably more loosely bound to low-valent Mo^{III} than to Mo^V or to Mo^{VI}. Catalytic activity of the complexes seem to increase with the increase of the number of molybdenum atoms in the complexes. This finding supports the conclusion that dinitrogen becomes more activated in the complexes with the increase of their electron capacity, because this helps to decrease

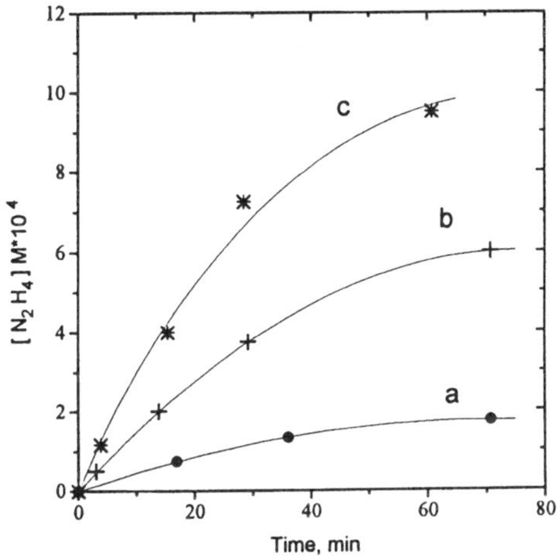

FIGURE I.20 Kinetic curves of dinitrogen reduction in the presence of molybdenum complexes of different nuclearity. a: dinuclear complexes I and II; b: octanuclear complex III; c: Mo^{III} complex of higher nuclearity (n > 8). P_{N_2}: 1 atm; temperature 20°C; solvent: methanol; [MO] = 5 × 10^{-6} M; [Phosphatidylcholine] = 4 × 10^{-4}; [n-Bu_3P] = 4 × 10^3 M.

NN bond order in the intermediate complex and to increase negative charge on nitrogen atoms of coordinated dinitrogen.

The structures found do not yet explain the exact means of dinitrogen activation in the catalytic complexes. Apparently, dinitrogen can replace two ligands on molybdenum to form an intermediate capable of producing hydrazine upon protonation. A dinuclear complex with bridging dinitrogen might be suggested. Such a structure with two Mo^{III} (d^3 electronic configuration) was already proposed for dinitrogen activation and reduction in the case of the systems based on V^{II}. Recently the dinuclear complex was proposed for Mo^{III} complex reaction with dinitrogen in aprotic media (see Section I.4.1).[49] The essential role of magnesium ions that strongly increase the rate of N_2 reduction supports this mechanism. Forming a bridge between two molybdenum atoms, magnesium atoms shift two molybdenums from each other to a distance suitable for forming the cyclic dinitrogen complex shown in Scheme I.6a. Another possibility is the formation of a bridging N_2 complex coordinated to molybdenum atoms by its π-bonds, i.e., of η^2 type (Scheme I.6b). Although linear MNNM complexes with d^3 electronic configuration of M are apparently more stable, two molybdenums may be situated in the complex at a distance suitable for the latter type of coordination. The structure of the FeMo cofactor of nitrogenase and also some results on the structures of bridging dinitrogen complexes (*vide infra*) support the suggestion that the energies of both types of structures are not too different and both are possible as intermediates in dinitrogen reduction.

I.4.4 DINITROGEN REDUCTION IN THE PRESENCE OF IRON COMPLEXES

In view of the finding of the alternative nitrogenase based entirely on iron as the only transition metal and of the structure of the FeMo cofactor, which suggests iron

SCHEME 1.6 Two possible structures of the intermediate dinuclear molybdenum(III) complex with dinitrogen.

as the likely core for dinitrogen activation, it is natural to expect the existence of iron systems reducing dinitrogen in protic media. Apparently, iron(II) hydroxide is too weak a reducing agent to reduce dinitrogen even in the presence of a stronger reductant, therefore $Fe(OH)_2$ does not reduce N_2 either alone or when formed together with $Ti(OH)_3$. In the enzymatic system, sulfide ligands, being stronger one-electron donors than hydroxo- or oxo-ligands, apparently help to activate and reduce dinitrogen, hence suitable polyiron sulfide complexes are obviously likely to be the catalysts in dinitrogen reduction.

There are two reports indicating the possibility of reducing N_2 to ammonia in methanol in the presence of iron salts by a very powerful reducing agent, lithium *p*-polyphenylene.[75a,b] The activity of the system was found to increase in the presence of some organic and inorganic sulfides.[75a] The same reducing agent was reported to reduce dinitrogen catalytically in the presence of polyiron–molybdenum carbonyl complex $Fe_7Mo_2S_8(CO)_{12}$.[75b] The yields, however, are low and the reductant is too strong to consider these systems as models for nitrogenase.

I.4.5 DIHYDROGEN AS A POSSIBLE REDUCING AGENT

In all the systems reducing dinitrogen in solution, including those in protic media, the reductants used are stronger than dihydrogen and in protic media H_2 is evolved in parallel with N_2 reduction. Therefore dihydrogen could not be used for catalytic ammonia synthesis in solution. However, it was shown in a comparatively early work[76] that H_2 can be used as an electron donor for stepwise dinitrogen reduction, if first the transition metal compound is reduced in an acid solution, then alkali is added and N_2 is in reduced alkaline solution by a hydroxide. Thus, whereas both $Ti(OH)_3$ and hydrazine are stronger reductants than H_2, reduction of Ti^{IV} to Ti^{III} can be accomplished with H_2 in an acid solution of Ti^{IV}–Mo^{V} on a Pt catalyst and then the mixed hydroxide can be made by addition of alkali and subsequently reduction of N_2 to N_2H_4 by $Ti(OH)_3$–Mo^{III} can be observed. The process can be rendered cyclic with H_2 as reductant and alternate additions of alkali and acid. This may have some principal similarity with ATP action in the natural system. The ATP hydrolysis may produce a localized proton that can be used to help to reduce a transition metal and then after the proton goes to the solution the reductant will be strong enough to transfer electrons to the activated dinitrogen.

I.4.6 ELECTROCHEMICAL REDUCTION

It is natural to expect that at sufficiently negative potentials of the cathode, electrochemical reduction of dinitrogen will take place provided the presence of a catalyst able to activate dinitrogen and to be adsorbed at the cathode surface. To prevent dihydrogen evolution at negative cathode potentials, it is desirable to use materials for the cathode that have a considerable overvoltage, such as mercury and lead. At a negative potential of mercury cathode, ions of alkali metals can be discharged to produce an amalgam, which will serve as a reducing agent for N_2 in the presence of the catalyst (*vide supra*).

Dinitrogen reduction was detected under the conditions of electrolysis for the heterogeneous system containing a mercury cathode and a $Ti(OH)_3$–Mo^{III} system.[77] The system showed a very pronounced dependence of the products (hydrazine and ammonia) yield on current density. The maximum ammonia yield is observed at 0.5 A/cm^2 current density. This is explained by the change of the mercury electrode interface charge.[77] At a current density of 0.5 A/cm^2 there are optimum conditions for adsorption of the negatively charged hydroxide on the surface and the potential is sufficiently negative to reduce dinitrogen. At lesser current densities, the potential appears to be insufficient, while at larger ones the negative charge increases, impeding the hydroxide adsorption on the electrode. Under optimum conditions, the current yield reaches 43% with respect to ammonia and 4% with respect to hydrazine.

In the seemingly homogeneous $Ti(OH)_3$–Mo^{III} system with additions of guanidine the rate of hydrazine formation ($v_{N_2H_4}$) passes the maximum at the potential –1.9 V. In the range of –1.6 to –1.9 V, a linear dependence of log $v_{N_2H_4}$ on the potential is reported. Thus the system behaves according to the Tafel equation

$$\phi = a + b \log i \qquad (26)$$

where i is the partial current used to reduce dinitrogen to hydrazine. Hence, the rate-determining step of the electrode process is thought to be electron transfer from the cathode to the catalyst active center containing the coordinated dinitrogen. The reaction was found to be sensitive to the concentrations of the base ($LiOCH_3$) and water. The latter serves as the main donor of protons and additions of water are essential to keep the process going with constant rate for many hours, particularly at high base concentrations.[78]

In the system Na/Hg–Mo^{III} complex with the phospholipid coaction described above, hydrazine formation rate was measured as a function of the sodium amalgam potential.[71] The formation of hydrazine was found to start at the mercury cathode at potentials $E < -1.75$ V. At these potentials the generation of sodium amalgam begins to take place. In the range of the potentials from –1.75 to –2.0 V the reduction rate is fully determined by the concentration of sodium in the amalgam and does not depend on the polarization potential of the amalgam electrode. The dinitrogen reduction rate increases with the shift to more negative potentials up to $E = -2.15$ V and then stops depending on E at more negative potentials. It is explained by the presence of a purely electrochemical mechanism of N_2 reduction together with amalgam reduction in the range $-2.15 < E < 2.05$ V when the reaction rate is determined by the electron transfer to the reaction center. At $E < 2.15$ V all the reaction centers activating dinitrogen are reduced and the further shift of the potentials to more negative values does not influence the rate.

Electrochemical reduction of dinitrogen in the coordination sphere of a mononuclear tungsten dinitrogen complex is described below.

I.4.7 OTHER SYSTEMS

As was mentioned in the introduction, quite a number of articles have been published claiming that dinitrogen could be reduced in various systems mainly to ammonia. In some cases subsequent attempts to reproduce the results failed. In other cases the authors continued their research for many years and it is difficult to deny or to confirm their findings because the yields were small and evidently difficult to reproduce.

A series of articles was published by Schrauzer et al. beginning with Reference 79. The authors reported the reduction of dinitrogen by sodium borohydride in the presence of complexes that include molybdenum and SH-containing compounds such as cysteine, glutatione, and thioglycerol. Schrauzer considered these systems as models of nitrogenase (e.g., see References 80 and 81). The systems definitely reduce nitrogenase substrates other than dinitrogen including acetylene, nitrous oxide, cyanide, azide, nitriles, and *iso*-nitriles. As to dinitrogen, the situation remains unclear because the attempts to reproduce the results on N_2 reduction in some laboratories, including the laboratory of the author of this book, were unsuccessful, but some important details could have been missed in performing the experiments.

Schrauzer proposed the mechanism with mononuclear molybdenum(IV) complexes as the active species reducing N_2 initially to N_2H_2. Apparently, it is difficult to reconcile the proposed mechanism with the reaction thermodynamics, and it is very different from what we know now about the nitrogenase and its chemical analogues.

I.5 DINITROGEN COMPLEXES WITH TRANSITION METAL COMPOUNDS AND THE MECHANISM OF N_2 REDUCTION IN THE METAL COORDINATION SPHERE

I.5.1 INTRODUCTION

The reactions of dinitrogen described in the previous section clearly involve intermediate dinitrogen complexes: this follows from all the experience of coordination catalysis where such complexes have been detected in reactions of olefins, acetylenes, carbon monoxide, dioxygen, etc. Although discovery and isolation of the intermediate complexes in catalytic reactions is not an easy task, investigation of similar but more stable model complexes contributed much to the real understanding of the catalytic mechanisms.

Until the 1960s dinitrogen complexes remained unknown. This could be attributed to their instability, and the results presented in the previous section seem to support this conclusion: all the indications are that the *intermediate* complexes are indeed very unstable.

The first dinitrogen complex that was isolated from solution was reported by Allen and Senoff in 1965.[82] They found it as the product of the reaction of $RuCl_3$ with hydrazine and the complex was found to be $[Ru(N_2)(NH_3)_5]I_2$. It is of interest that a book with the characteristic title *Nonexistent Compounds* was published the same year,[83] containing a section discussing possible dinitrogen complexes. Comparing dinitrogen with the isoelectronic carbon monoxide known to produce stable carbonyls the author, W. E. Dasent, wrote in 1965: "Superficially the nitrogen molecule is electronically equipped to form similar compounds, and it is therefore surprising that none has ever been isolated." The author explained the fact that no complexes had been observed by their instability. Meanwhile, paradoxically, the Allen and Senoff complex turned out to be remarkably stable. It was isolated in crystalline form and there are strong indications that this and similar complexes had been prepared earlier by other investigators who had not identified them as dinitrogen complexes. It was so difficult to accept that "inert" dinitrogen can form such a stable complex that Allen and Senoff's paper describing their results with the N_2 complex was initially rejected for publication by a journal. The complex might be considered an inorganic diazocompound rather than a real N_2 complex, the more so as it was obtained not from N_2 but indirectly from N_2H_4. However, a few months after the Allen and Senoff publication we found in our laboratory[84] that similar complexes could be obtained directly from N_2 in the reaction:

$$2Ru^{III}Cl_3L_3 + Zn + 2N_2 \rightarrow 2L_3Ru^{II}Cl_2N_2 + ZnCl_2 \tag{27}$$

where L = H_2O or THF. Dinitrogen was found to easily react with coordinatively unsaturated L_5Ru^{II} formed intermediately.

Another important development was preparation by Taube et al.[85] of the first dinuclear dinitrogen complex with Ru^{II}:

$$\left[Ru(NH_3)_5(H_2O)\right]^{2+} + N_2 \rightarrow \left[Ru(NH_3)_5N_2\right]^{2+} + H_2O \tag{28}$$

$$\left[Ru(NH_3)_5N_2\right]^{2+} + \left[Ru(NH_3)_5(H_2O)\right]^{2+} = \left[(NH_3)_5RuN_2Ru(NH_3)_5\right]^{4+} + H_2O \tag{29}$$

Thus, dinitrogen was found to be able not only to add to coordinatively unsaturated species, but also to replace ligands in coordinatively saturated ones (probably by a dissociative mechanism again via a coordinatively unsaturated intermediate). Later, using these and other reactions, quite a number of various dinitrogen complexes were prepared, initiating a new branch of coordination chemistry.[86,87] The structure of the complexes revealed that they had mostly linear structure of the fragments M–N≡N and M–N≡N–M and an NN bond only slightly elongated as compared with the NN bond in the dinitrogen molecule. Moreover, to the great disappointment of the chemists, dinitrogen in the coordination sphere of the complexes turned out to be as inert as free dinitrogen and did not enter any reaction besides replacement by other ligands to form free N_2.

Now it is not difficult to explain this inertness. The reactions in the coordination sphere are mainly two-electron processes, e.g., the well-known insertion into M–H or M–C bond (Scheme I.7). As we have seen in Section I.1.2, two-electron reactions of this kind for dinitrogen are as a rule thermodynamically unfavorable. The thermodynamic stability of the N_2 complexes naturally only increases their chemical inertness. Four-electron reduction at protonation:

$$Ru^{II}N_2 + 2H^+ \rightarrow \left[Ru^{IV}{=}N-NH_2\right]^{2+} \xrightarrow{2H^+} Ru^{VI} + N_2H_4 \quad (30)$$

is apparently thermodynamically forbidden and strongly shifted to the left.

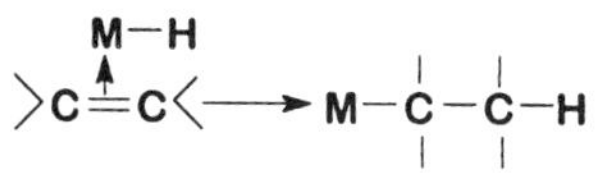

SCHEME 1.7 Insertion of olefin into the M–H bond.

A first *intermediate* complex of dinitrogen with a transition metal compound was detected in our laboratory in 1969 in one of the Vol'pin–Shur systems: Cp_2TiCl_2 + *i*-PrMgCl + N_2 in ether solution.[52,53] At low temperatures (below –60°C) an intensely blue complex (already mentioned above) was observed in solution under dinitrogen and later the equilibrium was established:

$$2CP_2TiR + N_2 = (Cp_2TiR)_2N_2 \quad (R = i-Pr) \quad (31)$$

At –60°C at standing in the presence of excess reducing agent the complex undergoes a slow transformation to another complex, $Cp_2Ti(MgCl)NNTiCp_2$, which produces hydrazine when decomposed by acid. Compared with other complexes known at that time, the complexes looked very unusual. However, later they were found to be the first representatives of important classes of dinitrogen dinuclear complexes (*vide infra*).

Meanwhile, new mononuclear dinitrogen complexes were synthesized, in particular, in 1969, the complexes of molybdenum(0), which played important role in future development.[88]

A few years later, Chatt and his colleagues in Brighton found that in mononuclear tungsten and molybdenum complexes dinitrogen could be reduced upon protonation to produce ammonia and hydrazine. Interesting new reactions of dinitrogen were observed in the coordination sphere of the metals. These discoveries were characterized by Chatt as "the beginning of the new era" in the chemistry of dinitrogen under ambient conditions.[89]

I.5.2 PROTONATION OF DINITROGEN IN MONONUCLEAR END-ON COMPLEXES

The first reaction of a coordinated dinitrogen in an end-on mononuclear complex was that observed with $W(N_2)(dppe)_2$ interacting with CH_3COCl:[94]

$$W(N_2)(dppe)_2 + CH_3COCl \rightarrow \left[WCl(N_2COCH_3)(dppe)_2 + N_2\right]$$
$$dppe = Ph_2PCH_2CH_2PPh_2 \quad (32)$$

TABLE I.9
X-Ray and IR Data on Some Mononuclear Dinitrogen Complexes

Complex	r(M–N), Å	r(N–N), Å	<M–N–N	ν(NN), cm^{-1}	Ref.
trans-$[V(N_2)_2(dmpe)_2]^-$ [a]	1.915	1.130	180.0	1793	96a
trans-$[Mo(N_2)_2(dppe)_2]$[a]	2.014	1.118	177.6	2020w, 1970s	96b
trans-$[Re(N_2)Cl(PMe_2Ph)_4]$	1.97	1.06(3)	177	1920	97
$Fe(N_2)(dmpe)_2$ [a]				1975	98
$FeH(N_2)(dmpe)_2$				2094	98
$[Ru(N_2)(NH_3)_5]Cl_2$	2.10	1.12	180	2105	99
$[Os(N_2)(NH_3)_5]Cl_2$	1.84	1.16	175	2088	100a
$K[Co(N_2)(PMe_3)_3]$	1.70–1.71	1.16–1.18	180	1795 1758	100b

[a] Dinitrogen in the complexes is able to be protonated and reduced.

It was interpreted as being essentially an oxidative addition to a W–N≡N fragment with nucleophilic attack of the coordinated dinitrogen on the carbon of the acetyl group.[94] Subsequently, this work was extended to protonation reactions in the same complex of the N_2 ligand and to alkylation, which generally proceeds via attack of an alkyl radical, generated by homolysis of an alkyl halide, upon coordinated dinitrogen. Further protonation and reduction of coordinated dinitrogen is difficult to achieve, but when dppe in tungsten and molybdenum complexes were replaced by monodentate ligands PMe_2Ph to form the complexes $M(N_2)_2(PMe_2Ph)_4$ the protonation was found to lead to high yields of ammonia.[95] Protonation and reduction of dinitrogen in the N_2 complexes was thoroughly studied in a number of laboratories.

Table I.9 presents typical examples of mononuclear complexes that can be protonated with accompanying dinitrogen reduction as well as those with inert dinitrogen. All the structural studies indicate that the N–N bond lengths in both classes of the complexes are only slightly increased as compared with free dinitrogen (about 1.12 Å in the complexes vs. 1.0968 Å in free N_2) and are not informative with respect to dinitrogen chemical activity.

The stretching frequency of the NN bond in the complexes, ν(NN), which presumably reflects the metal dπ donation to the N–N antibonding system, is much more indicative, low ν(NN) corresponding to more basic ligating dinitrogen. The data are in general qualitative agreement with the theoretical expectations (see Section I.2). The complexes are formed generally with electron-rich metal compounds. The reduction of dinitrogen in the complexes through their protonation proceeds only in those complexes that have strong basic properties. Thus the d^6 electronic configuration corresponds to dinitrogen reduction for V^{I-}, Mo^0, and W^0 complexes, sufficiently strong basic properties of coordinated dinitrogen for reacting with Lewis acids for Re^I complexes (however, without N_2 reduction), and neither addition of acids nor reduction for the Fe^{II}, Ru^{II}, or Os^{II} complexes. For platinum and palladium even the complexes of dinitrogen are unknown both for the M^{IV} state (d^6) and the M^{II} state (d^8). (The complexes are formed mainly with electron donors because dinitrogen in the complexes is a moderate acceptor and very weak donor.) For the same metal the ability to reduce coordinated dinitrogen rises with the decrease of the metal valency and the increase of ligands donor properties. For example, there is almost no reduction in the case of $[Fe^{II}H(N_2)(dmpe)_2^+]$ (dmpe = $MePCH_2CH_2PMe_2$) but dinitrogen can be at least partly reduced to ammonia under the HCl action in the complex $[Fe^0(N_2)(dmpe)]$.[93]

It is also easy to see that the ability for the coordinated dinitrogen to be reduced does not necessarily increase with the complex stability. Often it is the other way around. Let us consider, for example, two complexes, already mentioned $[Fe^{II}H(N_2)(dmpe)_2^+]$ and *trans*-$[V(N_2)(dmpe)_2]^-$. Stretching frequency ν(NN) in the IR spectra, which presumably reflects the metal dπ donation to the N–N antibonding system, is equal to 2094 cm^{-1} for the first complex and 1790 cm^{-1} for the second one. Accordingly, only in the second complex is dinitrogen sufficiently activated to be effectively reduced upon protonation.[96] At the same time the iron complex is relatively stable while the vanadium complex easily loses dinitrogen, and even N_2 can be pumped away at room temperature.

Metal–dinitrogen bond in a complex is formed essentially at the expense of the weakening of the NN bond in dinitrogen. We may consider the M–N_2 bond dissociation energy D(M–N_2) (M is metal with coligands) as the difference E(M–N) – Δ(N≡N), where E(M–N) is the M–N bond energy and Δ(N≡N) is the decrease in the NN bond energy in the complex as compared with triple NN bond energy in free dinitrogen. Ultimately, when D(M–N_2) = Δ(N≡N), the N≡N bond is cleaved and M≡N nitride is formed. It is clear that D(M–N_2) is small (the complex unstable) when both E(M–N) and Δ(N≡N) are small (dinitrogen nonactivated) or when both E(M–N) and Δ(N≡N) are large and therefore dinitrogen strongly activated. For a given E(M–N) the increase in Δ(N≡N) will correspond to the activation of dinitrogen and at the same time to the decrease of complex stability.

Mononuclear end-on dinitrogen complexes (e.g., with Fe^0 or Mo^0) are extremely unlikely to be the intermediates in biological dinitrogen reduction, particularly after the polynuclear structure of the iron–molybdenum cofactor, which presumably activates dinitrogen, became known. They are clearly not the intermediates in chemical reduction of dinitrogen in protic media described in the previous section, because higher oxidation states of the metal compounds are used in the latter systems. Nevertheless, the results obtained in investigations of dinitrogen reduction in the coordination sphere of mononuclear complexes and the comparison with other low-temperature dinitrogen reactions are very instructive for creation of a general picture of dinitrogen reactivity. Let us consider these reactions in some more detail.

I.5.2.1 Basicity of the Ligating Dinitrogen[91a]

Reduction of the coordinated dinitrogen under the action of protic molecules depends strongly on the basicity of ligating N_2, namely β-nitrogen atom in the molecule that bears more negative charge. The first demonstration of the basic character of ligating dinitrogen was provided by the already mentioned displacement of ligating water from $[Ru(NH_3)_5H_2O]^{2+}$ by dinitrogen complex $[Ru(NH_3)_5N_2]^{2+}$ to give a dinuclear μ-dinitrogen complex $[Ru(NH_3)_5(\mu\text{-}N_2)Ru(NH_3)_5]^{4+}$. However, Ru^{II} (d^6) is a soft acid: there is a back-donation of *d* electrons to the antibonding π*

TABLE I.10
The ν(N_2) and Relative σ-Donor Strengths of the Substances Against $AlMe_3$ (Et_2O = 1) in Benzene at 34 ± 2°C

Substance	ν(N_2) (cm^{-1})	K[a]
Tetrahydrofuran	—	70
trans-[ReCl(N_2)(PMe_2Ph)$_4$]	1923	20 · 6
trans-[Mo(N_2)$_2$(dppe)$_2$]	1979,	33
	2020w	(16 · 5)[b]
trans-[W(N_2)$_2$(dppe)$_2$]	1948,	15
	2015w	(7 · 5)
trans-[ReCl(N_2)(PMe_2Ph)$_2${P(OMe)$_3$}$_2$]	2000	5 · 5
trans-[ReCl(CO)(PMe_2Ph)$_4$]	—	3 · 3
Et_2O	—	1
mer-[$OsCl_2$(N_2)(PEt_2Ph)$_3$]	2063	0 · 3

[a] K: equilibrium constants of Reaction 30.
[b] Halved for comparative purposes to account for two basic sites on the molecule; dppe = $Ph_2PCH_2CH_2PPh_2$.

Adapted from Richards, R. L., in *New Trends in the Chemistry of Nitrogen Fixation,* Chatt, J., da Camara Pina, L. M., and Richards, R. L., Eds., Academic Press, London, 1980, 153.

orbital of dinitrogen. Another example is the formation of the complex $Cl(PMe_2Ph)_4Re^I(N_2)MoCl_4OMe$ with Mo^V (d^1). We are interested in protonation of the ligating dinitrogen, therefore in the interaction with a hard acid. $AlMe_3$ provides such an example. The σ-donor strength relative to diethyl ether of the terminal N atom in several N_2 complexes have been measured in benzene solution by determining the equilibrium constant K in equilibriums of the type:

$$M-N_2 + Me_3Al\cdot OEt_2 \overset{K}{=} M-N_2-AlMe_3 + Et_2O \quad (33)$$

These determinations, obtained by the NMR method, and similar data for analogous carbonyl complexes are shown in Table I.10.[91a] It can be seen that the basicity of ligating dinitrogen approaches that of tetrahydrofuran in some cases and the order of basicity approximately follows the order of decreasing ν(NN) values. The coligands of dinitrogen have a marked effect upon its basicity, e.g., if two PMe_2Ph ligands in *trans*-ReCl(N_2) (PMe_2Ph)$_4$ are replaced by less basic P(OMe)$_3$ ν(NN) is increased and K is lowered. Thus, the greater is the transfer of electron density from the metal into the π* orbitals of dinitrogen ligand, the more basic is the terminal N atom. Naturally, coligands also strongly influence the ability of dinitrogen to be reduced and the reduction rate.

It is of interest that in the equivalent complexes *trans*-[ReCl(XY) (PMe_2Ph)$_4$] (XY = CO or N_2) dinitrogen is more basic than carbon monoxide.

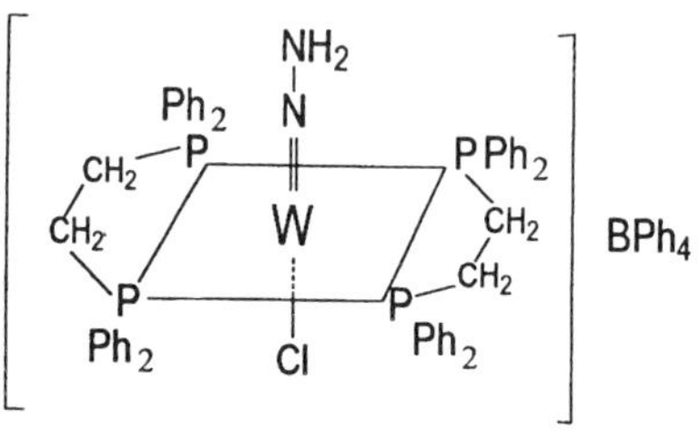

FIGURE I.21 The structure of *trans*- [$WCl(NNH_2)$ (dppe)] BPh_4. (Adapted from Richards, R. L., in *New Trends in the Chemistry of Nitrogen Fixation,* Chatt, J., da Camara Pina, L. M., and Richards, R. L., Eds., Academic Press, London, 1980, 153.)

I.5.2.2 Reduction of Coordinated Dinitrogen at Protonation[89-93]

When treated with an excess of halogen acid at 20°C the complexes *trans*[$M(N_2)_2$(dppe)$_2$] (M = Mo or W) undergo the reactions:

$$trans-\left[M(N_2)_2(\text{dppe})_2\right]+2HX\rightarrow\left[MX_2(N_2H_2)(\text{dppe})_2\right]+N_2 \qquad (34)$$

(M = W, X = Cl, Br, or I; M = Mo, X = Br or I)

The N_2H_2 group has the hydrazido(2–) form in which it acts as a four-electron donor, so that the metal retains the 18e configuration. The structure of *trans*[$WCl(NNH_2)$ (dppe)$_2$]BPh_4 (Figure I.21) shows that the M$\equiv$N–N unit is essentially linear and the short M–N distance indicates a bond order greater than unity, actually close to three.

All the evidence indicates that it is not a simple derivative of hydrazine, but rather a complex similar to dinuclear complexes MNNM (*vide infra*) with d^2-d^4 electronic configuration of each M. Thus, two H atoms may be considered to be partly isolobal[101a] with a metal, having both σ- and π-orbitals (Scheme 1.8). The analogy is not complete because the pair of hydrogen atoms has only one orbital isolobal to the π-orbital, whereas metal atom has two d_π orbitals. Nevertheless, this approach explains the existence of such complexes,[101b] with two hydrogen atoms replacing the metal. Flat configuration of the fragment $M\equiv N\equiv N<^{H}_{H}$ follows from this analysis.

SCHEME 1.8 Isolobality of *d* metal orbital and 2H σ orbitals.

Although there is no further reduction of the coordinated dinitrogen, the N–N bond is strongly activated as compared with the initial dinitrogen complex.

The hydrazido(2–) complex can be dehydrohalogenated with weak base to give diazenido complex

$$trans\text{-}\left[MX(NNH_2)(dppe)_2\right]BF_4 + NEt_3 = trans\text{-}\left[MX(N_2H)(dppe)_2\right] + \left[NEt_3H\right]BF_4 \quad (35)$$

Treatment of the latter complex with acids regenerates the hydrazido(2–) complex, therefore the diazenido complex may be regarded as the precursor to the NNH_2 complex in a stepwise protonation of ligating dinitrogen.

Attempts to reduce the NNH_2 ligand in the above complexes by different reducing agents failed, but moderate yields of ammonia (0.24 mol per atom) were obtained when $[MoBr_2(NNH_2)(dppe)_2]$ together with *trans*-$[M(N_2)_2(dppe)_2]$ was treated with aqueous HBr in *N*-methylpyrrolidone or propylene carbonate followed by removal of solvent and Kjeldahl distillation. Protonation and further reduction of NNH_2 ligand is probably the result of phosphine displacement, therefore monodentate phosphines were tried and high yields of ammonia (with little hydrazine in some cases) were observed on the treatment of the complexes *cis*-$[M(N_2)_2(PMe_2Ph)_4$ or *trans*-$[M(N_2)_2(PPh_2Me)_4]$ (M = Mo or W) with sulfuric acid in methanol at 20°C.[95] For the tungsten complex the reaction proceeds according to the equation:

$$cis\text{-}\left[W(N_2)_2(PR_3)_4\right] \xrightarrow{H_2SO_4/MeOH} N_2 + 2NH_3 + W(VI)\ \text{products} + 4\left[PR_3H\right]HSO_4 \quad (36)$$

For the molybdenum complex the yield of ammonia reaches only about 0.7 mol per metal atom, the remainder evolving as free dinitrogen, with molybdenum oxidizing to molybdenum(III).

A clear stoichiometry of protonation reaction has been observed[101c] in the case of the molybdenum triphosphine complex $[Mo(N_2)_2(triphos)(PPh_3)]$ (triphos = $PhP(CH_2CH_2PPh_2)_2$):

$$2\left[Mo(N_2)_2(triphos)(PPh_3)\right] + 8HBr \xrightarrow[THF]{} 2NH_4Br + 2\left[MoBr_2(triphos)\right] + 3N_2 + 2PPh_3 \quad (37)$$

On treatment with HBr in THF, the complex gives 0.7 mol of ammonia per mole of complex together with formation of 1.3 to 1.5 mol of dinitrogen per molybdenum atom.

Some indications of probable intermediates in these reactions were obtained when halogen acids HX (X = Cl, Br, or I) were used instead of sulfuric acid in methanol. Dihalido-hydrazido(2–) complexes have been isolated in these cases:

$$cis-\left[M(N_2)_2(PMe_2Ph)_4\right] + HX \xrightarrow{MeOH} \left[MX_2(NNH_2)(PMe_2Ph_3\right] + N_2 + \left[PHMe_2Ph\right]X \quad (38)$$

Substitution reactions carried out on the hydrazido(2–) complexes produce several complexes with NNH_2 ligand, with the N–N bond distance varying only a little with the nature of coligands. However, the further protonation reactions of these hydrazido(2–) compounds are strongly dependent on the ligand environment. Thus, on treatment with H_2SO_4/MeOH followed by base distillation, the compounds $[MX_2(NNH_2)(PMe_2Ph)_3]$ produce ammonia in essentially similar yields to those obtained from the parent dinitrogen complexes, confirming the suggestion that they are the intermediates in the reduction. Similar treatment of $[W(8\text{-hq})(NNH_2)(PMe_2Ph)_3]X$ (8-hq = 8-hydroxyquinoline) gives no ammonia but only hydrazine. Hydrazine is also produced upon treatment of the latter complexes with strong base alone, whereas for analogous molybdenum complexes only ammonia is formed in moderate yield. When treated with 40% KOH solution at 100°C, the tungsten hydrazido(2–) complexes $[WX_2(NNH_2)(PMe_2Ph)_3]$ (X = Cl, Br, or I) give ammonia and hydrazine in yields close to those obtained from the protonation reactions. The similar reaction of the parent dinitrogen complex *cis*-$[W(N_2)_2(PMe_2Ph)_4]$, however, results in the loss of all the coordinated dinitrogen as gas. The solvent used in protonation reactions of the parent dinitrogen complexes also affects the yields of ammonia and hydrazine and their ratio.

Protonation of the dinitrogen complexes beyond the NNH_2 stage has been reported to produce a hydrazido(–1) ($NHNH_2$) ligand:

$$\begin{aligned} &trans-\left[W(N_2)_2(PPh_2Me)_4\right] + 5HCl \xrightarrow{CH_2Cl_2} \left[WCl_3(NHNH_2)(PPh_2Me)_2\right] \\ &\quad + N_2 + 2\left[PMePh_2H\right]Cl \end{aligned} \tag{39}$$

Similar complexes have been prepared by protonation of *cis*-$[M(N_2)_2(PMe_2Ph)_4$ or of the corresponding hydrazido(2–) complexes $[WX_2(NNH_2)(PMe_2Ph)_3]$ (X = Cl or Br). They turned out to be hydrides containing a $WH(NNH_2)$ fragment. These complexes give rather high yields of hydrazine on further acid treatment.

Treatment of dinitrogen complexes with strong acid is not necessary for ammonia formation. *cis*-$[M(N_2)_2(PMe_2Ph)_4]$ gives almost as high a yield of ammonia (1.7 mol per W atom) on treatment with methanol alone, either at reflux (3 to 5 h) or under tungsten-filament irradiation (30 h at 20°C). With ethanol a lower yield is obtained (0.4 mol per W atom) whereas the molybdenum analogue gives very small yields with either alcohol.

Although no catalytic reduction of dinitrogen in mononuclear complexes has been observed so far, Pickett and Talarmin have developed a *cyclic* system based on the dinitrogen complex *trans*-$[W(N_2)_2(dppe)_2]$.[101d] Protonation and subsequent electrochemical reduction of the complexes yields ammonia with conservation of the $W(dppe)_2$ core. Under dinitrogen the initial complex is again recovered and the reaction cycle can be repeated on addition of an acid.

I.5.2.3 Kinetics and Mechanism of Protonation and Reduction[102]

The kinetics of protonation of *cis*-$[M(N_2)_2(PMe_2Ph)_4]$ (M = Mo or W) was investigated in methanol with an excess of acid which ultimately yields ammonia via complexes of the type $[MX_2(NNH_2)(PMe_2Ph)_3]$ isolated when X = Cl, Br, or I. Spectroscopic titration of a dilute solution of *cis*-$[M(N_2)_2(PMe_2Ph)_4]$ with HX (X =

Cl, Br, or HSO_4) in methanol shows that, for a given metal, a common product is formed with all three acids, and that 1 mol-eq. of acid is consumed per mole-equivalent of complex ($[M]/[H^+] = 1/1$). The stoichiometry corresponds to the equation:

$$cis\text{-}\left[M(N_2)_2(PMe_2Ph)_4\right] + H^+ + 2CH_3OH \rightarrow \left[M(NNH_2)(OCH_3)(PMe_2Ph)_3\right] + N_2 + \left[HPMe_2Ph\right]^+ \quad (40)$$

The kinetics of reaction 40 follows the first order with respect to the complex concentration and the second order with respect to the concentration of acid but is independent of the nature of the anion ($k_{Mo}^{app.} = 3.9(\pm 0.4) \cdot 10^5[H^+]^2 M^{-2}s^{-1}$ and $k_{W}^{app.} = 3.6(\pm 0.4) \cdot 10^8[H^+]^2 M^{-1}s^{-1}$, at 25°C; $k^{app.}$ = apparent rate constant).

These observations are consistent with the following mechanism.[102] Diprotonation of a coordinated dinitrogen in *cis*-$[M(N_2)_2(PMe_2Ph)_4]$ labilizes the *cis*-dinitrogen to yield the five-coordinate intermediate $[M(NNH_2)\ (PMe_2Ph)_4]^{2+}$. Rapid attack of methanol on the latter yields *cis*-$[M(NNH_2)\ (OCH_3)\ (PMe_2Ph)_4]^+$ and a mole-equivalent of protons. Subsequent dissociation of a phosphine results in the five-coordinate $[M(NNH_2)\ (OCH_3)\ (PMe_2Ph)_3]^+$, which is rapidly attacked by methanol to yield $[M(NNH_2)\ (OCH_3)_2(PMe_2Ph)_3]$ and a further mole-equivalent of protons. It is not clear whether loss of N_2 or PMe_2Ph is rate limiting; however, the latter seems more probable, and this is consistent with the kinetics if E is a steady-state intermediate.

Of interest is the greater reactivity of *cis*-$[W(N_2)_2\ (PMe_2Ph)_4]$ compared with its molybdenum analogue ($k_W/k_{Mo} = 9.2 \cdot 10^2$). This is a consequence of the greater basicity of dinitrogen coordinated to tungsten. The isotope effect observed in the reaction of *cis*-$[Mo(N_2)_2(PMe_2Ph)_4]$, $k_H/k_D = 0.3$, is consistent with a mechanism involving protolytic equilibria prior to the rate-limiting step. This mechanism shows that a protic solvent is advantageous for the protonation of coordinated dinitrogen in these complexes. The facile release of a proton upon coordination of a molecule of solvent is a process that is unique to protic solvents and has the result that protons are only consumed in the neutralization of liberated phosphine.

The reactions subsequent to the hydrazido(2–) stage depend upon the nature of the metal, the acid, and the solvent used. For molybdenum, further protonation of the hydrazido(2–) intermediate gives in general 1 mol of ammonia and 0.5 mol of dinitrogen per molybdenum atom, resulting in formation of a molybdenum(III) product (scheme in Figure I.22, pathway A).[92]

It is clear that the number of electrons of a single complex is insufficient to reduce N_2 to ammonia if molybdenum(III) is the product: some disproportionation with another particle of the complex is necessary. In contrast, when tungsten is the metal atom and H_2SO_4 is the acid, the third protonation probably occurs at the outer nitrogen atom of the hydrazido(2–) ligand, giving one mol of ammonia per tungsten atom together with tungsten nitride which is further hydrolyzed by acid to produce another mole of ammonia. Thus, all six electrons required to produce 2 mol of ammonia from dinitrogen come from tungsten(0), which is oxidized to tungsten(VI) (pathway B).

It would be, of course, presumptuous to claim that we could explain or even predict all the details of these results. Nevertheless, the results seem to agree with

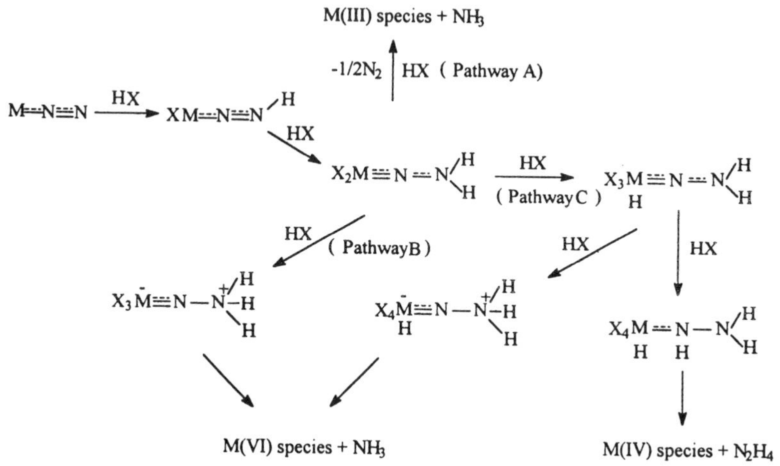

FIGURE I.22 Mechanism of dinitrogen reduction upon protonation in mononuclear complexes.

general arguments on dinitrogen reactivity in the coordination sphere of transition metals and they are a good supplement to the picture of N_2 reduction in protic media in polynuclear complexes given in the previous section.

Tungsten(0) in its complexes is apparently the strongest reductant and it possesses enough electrons to produce both hydrazine and ammonia. Coligands and the nature of the solvent largely determine the reaction selectivity. With stronger donors hydrazine is produced and we may compare this situation with dinitrogen reduction by polynuclear complexes in protic media at high pH, e.g., by the $V(OH)_2$–$Mg(OH)_2$ system. The hydrazido(2–) intermediate may be regarded as an analogue of a dinuclear MNNM complex with the same number of six *d* electrons per complex.

With weaker donors in the surrounding coligands, the complex prefers to use all six electrons and to produce ammonia (as in a number of polynuclear systems in protic media, e.g., in the homogeneous vanadium(II)–catechol system, as well as in the enzymatic N_2 reduction).

For molybdenum(0) complexes, additional support is needed to produce even ammonia, and when $MoNNH_2$ is formed, another molybdenum complex is used as a reductant and is oxidized (producing free N_2). The situation with molybdenum(III) complexes in protic media is somewhat similar, because a stronger reducing agent is needed to reduce dinitrogen. It must be remembered that polynuclear systems require generally milder reducing agents and it is easier to perform the reduction catalytically.

Among the most unexpected results is the reduction of dinitrogen in the complexes in methanol solution even without added acid, albeit at somewhat higher temperatures. The basic strength of dinitrogen in the coordination sphere is insufficient to be protonated by methanol. Probably hydrogen bonding occurs initially, which helps to diprotonate N_2, replacing phosphine by CH_3O^- at the same time.

I.5.3 PROTONATION AND REDUCTION OF DINITROGEN IN DINUCLEAR COMPLEXES[32,103,104]

Protonation of dinitrogen in mononuclear end-on complexes of the type M–N≡N requires a low-valent electron-rich M with strong reducing properties. As already mentioned, catalytic systems based on this mechanism are not very likely, particularly for protic media: an even stronger reducing agent must be available to return M to its initial reduced state and reactions with solvent protons to produce dihydrogen will be difficult to avoid.

Dinuclear complexes of the type M–N≡N–M are an evident alternative: both Ms will share their electrons and milder reducing agents could be expected to produce hydrazine or ammonia upon protonation. However, the first attempts to reduce bridging dinitrogen were disappointing. For example, in the dinuclear complex $[(NH_3)_5RuN_2Ru(NH_3)_5]^{4+}$ dinitrogen does not seem to be more activated than in the mononuclear $[(NH_3)_5RuN_2]^{2+}$ complex. NN bond vibration frequency in both complexes remain virtually the same and dinitrogen in the dinuclear complex is neither protonated nor reduced similarly to the parent mononuclear complex. Unsuccessful attempts were reported also for other dinuclear complexes of similar nature.

This is undoubtedly connected with the fact that the first complexes investigated were linear ones with a large number of *d* electrons on each M (usually at least six). According to the simple theoretical analysis presented above (Section I.2.3), they contain dinitrogen that is hardly more activated than in the corresponding mononuclear complexes because when the number of *d* electrons on each M is more than four the orbitals that are bonding with respect to the NN bond and antibonding with respect to MN bonds are occupied. Thus, the analysis carried out at the very early stages of the development of this field[6] showed that other dinuclear complexes with a smaller number of *d* electrons would be more promising as intermediates for dinitrogen reduction. However, the experimental results were lacking or nonconvincing for a long time.

I.5.3.1 Complexes $[\{(\eta^5\text{-}C_5H_5)_2TiR\}_2N_2]$

As mentioned above, these complexes were the first intermediate complexes observed in N_2 reduction in solution. They are formed upon the interaction of Cp_2TiCl_2 or Cp_2TiCl ($Cp = \eta^5\text{-}C_5H_5$) with RMgX in a dinitrogen atmosphere and can be observed at low temperatures. Thus, if *i*-PrMgCl reacts with Cp_2TiCl_2 at –60°C and then the temperature is lowered to –100°C, an intensely blue color appears in the presence of N_2 (even at very low pressures of dinitrogen).[52,53] The color disappears upon pumping of dinitrogen, showing that the complex is very unstable. Spectrophotometric investigation[105] of the equilibrium

$$2Cp_2Ti(i-Pr) + N_2 = [Cp_2Ti(i-Pr)]_2N_2 \tag{41}$$

shows that the formation enthalpy is only about –5 kcal/mol. This unstable complex can be reduced to another complex, $Cp_2Ti(MgCl)NNTiCp_2$ ($\nu_{[NN]} = 1100\ cm^{-1}$), when slowly reacting at –60°C with an excess of reducing agent (*i*-PrMgCl). The complex may be regarded as a hydrazine derivative because it produces hydrazine in the

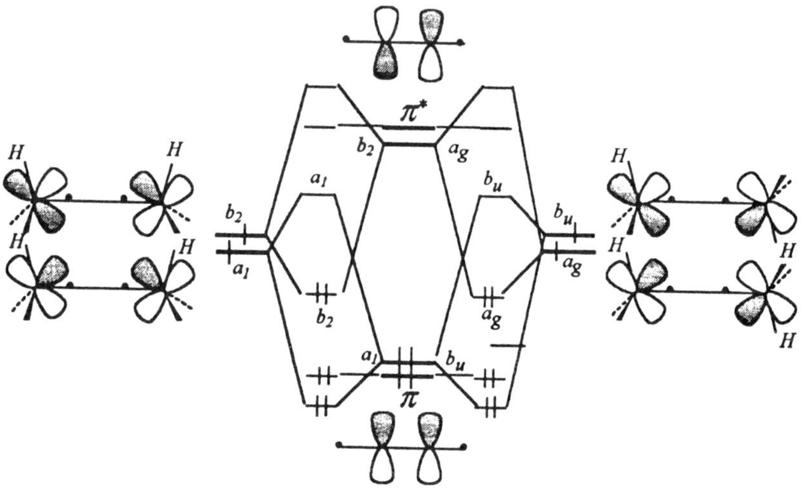

FIGURE I.23 Molecular orbital diagram for the model complex $(Cp_2TiH)_2N_2$. (Adapted from Hoffman, R. et al., *Koord. Khim.*, 3, 1260, 1977.)

reaction with an acid but also as an early example of multicoordinated polynuclear complex with strongly activated dinitrogen, which is confirmed by a very low NN bond stretching frequency. At higher temperatures further reduction in the presence of *i*-PrMgCl leads to the cleavage of the N–N bond and the formation of the nitride complex $Cp_2TiN(MgCl)_2$.[106]

Similar but much more stable dinitrogen complexes $(Cp_2TiAr)_2N_2$ with an aryl instead of alkyl group on titanium were prepared and investigated by Teuben.[103] The complexes are formed in the reaction of ArMgCl with Cp_2TiCl. Both alkyl and aryl complexes are diamagnetic, indicating electron coupling through bonding. A theoretical analysis[107] of the model complex $(Cp_2TiH)_2N_2$ was performed using the extended Hückel method. The distances Ti–N and N–N were assumed to be the same as in the complex $[(C_5Me_5)_2Ti]_2N_2$ (*vide infra*), i.e., 2.016 and 1.16 Å, respectively. It was found that the stable configuration of the complex corresponds to the flat unit C–Ti–NN–Ti–C with *trans* or *cis* configuration and small energy difference between the two. The MO diagram for both forms is presented in Figure I.23. Intense light adsorption responsible for the blue color of the complex corresponds to allowed transitions $a_1 \rightarrow b_2$ or $a_g \rightarrow b_u$ for the *cis* and for the *trans* forms, respectively. More recent elaborate analysis confirmed this conclusion.[108]

The X-ray study[103] of the single crystal of the complex $[\{Ti(\eta^5\text{-}C_5H_5)_2(p\text{-}MeC_6H_4)\}_2(N_2)]$ made later than the theoretical analysis[107] is in very good agreement both with the assumed data and obtained results. Flat *trans* configuration was indeed determined for the complex. The N–N distance (1.162 Å) shows that dinitrogen is more strongly activated than in mononuclear complexes with terminal dinitrogen, though each of the titanium atoms possesses only one *d* electron. However, the dinitrogen ligand still cannot be reduced upon protonation evidently because this would lead to thermodynamically unfavorable diazene N_2H_2. In order to reduce dinitrogen more electrons have to be added to the complex to form a hydrazine derivative.

Some less-well-characterized complexes with dinitrogen were reported in the systems Cp_2TiCl_2 + MeMgI and Cp_2TiCl_2 + K(naphthalene).[16,103] Divalent titanium appears to be an active species forming the complexes with N_2.

I.5.3.2 Complexes with Divalent Titanium and Zirconium

Bercaw et al. discovered and have studied dinitrogen complexes with titanocenes and zirconocenes with permethylated cyclopentadienyl rings. Ti $(\eta^5\text{-}C_5Me_5)_2$ forms a dinuclear complex with N_2 at 0°C with linear bridging dinitrogen as was shown by the X-ray structure determination.[109] The long N–N distance (1.160 Å) indicates activation of dinitrogen in the complex, but no protonation of the N_2 ligand to get hydrazine or ammonia was observed.

At low temperature (–80°C) the titanium dinitrogen complex takes up additional dinitrogen:

$$[\{Ti(\eta^5\text{-}C_5Me_5)_2\}_2N_2] + 2N_2 = [\{Ti(\eta^5\text{-}C_5Me_5)_2(N_2)\}_2N_2] \quad (42)$$

Presumably the complex contains two end-on N_2 molecules bound to titaniums and one bridging dinitrogen as follows from the similar zirconium complex with established structure (*vide infra*). In this case the titanium complex produces a nearly quantitative yield of hydrazine when decomposed by HCl.

Permethylzirconocene forms a dinitrogen-containing complex with bridging and terminal dinitrogen ligands with the structure seen in Figure I.24.[110] It reacts with HCl in toluene at –80°C to produce hydrazine (0.86 mol) according to the equation:

$$[\{Zr(\eta^5\text{-}C_5Me_5)_2(N_2)\}_2N_2] + 4HCl \rightarrow 2[\{Zr(\eta^5\text{-}C_5Me_5)_2Cl_2] + 2N_2 + N_2H_4 \quad (43)$$

A natural conclusion would be that it is bridging dinitrogen that is protonated and reduced, because according to the stretching frequencies and bond lengths the terminal dinitrogens are considerably less activated than the bridging one. However, a study of the reaction of HCl with the complex having terminal $^{15}N_2$ shows that half of the hydrazine is produced from terminal N_2 and the other half from the bridging dinitrogen.[109] The authors concluded that bridging N_2 was not protonated because, in particular, the related complex $[\{Zr(\eta^5\text{-}C_5Me_5)_2(CO)\}_2N_2]$ did not give hydrazine on reaction with HCl.[111] According to the mechanism proposed, a terminal N_2 is protonated, then the other terminal ligand is lost and the symmetric bis-diazenido complex $[\{Zr(\eta^5\text{-}C_5Me_5)_2(N_2H)_2]$ is formed. Further protonation of the latter gives diazene, which disproportionates to give dinitrogen and hydrazine.

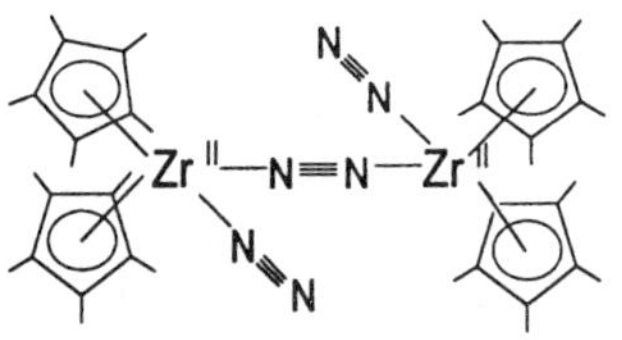

FIGURE I.24 The structure of the dinuclear complex $[\{Zr(\eta^5\text{-}C_5Me_5)_2(N_2)_2\}_2(\mu\text{-}N_2)]$. (Adapted from Manriquez, J. M. et al., *J. Am. Chem. Soc.*, 98, 8351, 1976.)

The mechanism with the intermediate N_2H_2 formation is by no means certain, however, taking into account its thermodynamic instability. There are definitely other alternatives. When HCl is added there is competition between the protonation of the ligating dinitrogen and the decomposition of the complexes with N_2 evolution due to their oxidation. Terminal dinitrogen ligands might increase the complex stability and that will help the protonation and reduction of dinitrogen, which actually could be a bridging one. The results with terminal $^{15}N_2$ may be explained by an exchange between the terminal and bridging dinitrogens and the absence of N_2 reduction in the carbonyl complex could be the result of the stronger electron-accepting properties of the CO ligands preventing the protonation of dinitrogen. In view of the general theoretical expectations and later findings (*vide infra*), the protonation of the bridging dinitrogen is perhaps more likely and Bercaw's results may be regarded in fact as the first example of the protonation of N_2 in well-defined bridging dinuclear complexes with d^2 electronic configuration of titanium and zirconium.

Nevertheless, the mechanism proposed by the authors and the absence of other definite results led to the conclusion[103] even as late as in 1980 that "in dinuclear N_2 complexes with a linear M–N–N–M unit (M = Ti, Zr), the ligated N_2 cannot be protonated directly despite the fact that the N–N distance is quite long and is intermediate between that found in free N_2 and MeN=NMe. On the other hand, terminal N_2 in Ti and Zr complexes may be just as reactive as in the other transition metal complexes..." This conclusion, which now has to be seriously revised, was definitely influenced by the spectacular results obtained by Chatt et al. for mononuclear complexes and described above.

I.5.3.3 Other Complexes with Linear MNNM Unit and d^2 and d^3 Electronic Configurations of M

More recently a new branch of dinitrogen complexes with N_2 capable of protonation and reduction appeared in the literature. The complexes are dinuclear and contain dinitrogen as a bridging ligand only. Initially they were considered unusual and even now they are sometimes described as a "special case".[93] However, their number is already quite considerable and continues to increase due to the works of Schrock, Henderson, Gambarotta, Floriani et al. (*vide infra*). Some of these complexes are presented in Table I.11. They include the complexes of vanadium(II), niobium(III), tantalum(III), molybdenum(IV), and tungsten(IV). The complexes often can be prepared directly from dinitrogen, in some cases via hydrazine or its derivatives. In all of them the metals have electronic configurations d^2 or d^3. The MNNM unit in the complexes is usually very close to linear. Examples of these complexes are presented in Figures I.25 and I.26.

According to the data on NN and MN bond lengths and NN vibration frequencies in dinuclear complexes, bridging dinitrogen is considerably more activated than in mononuclear complexes with a terminal dinitrogen ligand, even though the metals in the dinuclear complexes are in an essentially higher oxidation state. It is seen from Table I.11 that N_2 in the dinuclear complexes with d^2–d^4 electronic configurations is also much more activated than in dinuclear complexes with more reduced metals of d^6–d^8 electronic configurations. This is in agreement with the theoretical

TABLE I.11
Typical Dinuclear Complexes with Bridging Dinitrogen

Electronic configuration	Formula	Distances (Å) r(NN)	r(MN)	Ref.
d^1-d^3	$[\{Ti^{III}(C_5H_5)_2(p\text{-}MeC_6H_4)\}_2N_2]$	1.16	1.96	103
	$[\{Ti^{II}(C_5Me_5)_2\}N_2]$	1.17	2.0	109
	$[\{Zr^{II}(C_5Me_5)_2(N_2)\}_2N_2]$	1.18	2.08	110
	$(\mu\text{-}N_2)\{[o\text{-}Me_2NCH_2)C_6H_4]_2V^{II}(Py)\}_2\ (THF)_2$	1.23	1.83	112
	$[\{Nb^{III}(S_2CNEt_2)_3\}_2N_2]$	1.25	1.84	113
	$[\{Ta^{III}Cl_3(PBr_3)(THF)\}_2N_2]$	1.28	1.80	114
	$[\{MO^{IV}(C_5Me_5)Me_3\}_2N_2]$	1.24		115
	$[\{W^{IV}(C_5Me_5)Me_3\}_2N_2]$	1.33	1.76	116
d^6–d^{10}	$[\{Mo^0(C_6H_3Me_3)(dmpe)\}_2N_2]$	1.15	2.04	117
	$[\{Ru^{II}(NH_3)_5\}_2N_2]$	1.12	1.93	12
	$[\{Rh^IH(PPr^i_3)_2\}_2N_2]$	1.13	1.98	119
	$[\{Ni^0(PCy_3)_2\}_2N_2]$	1.12	1.78	120

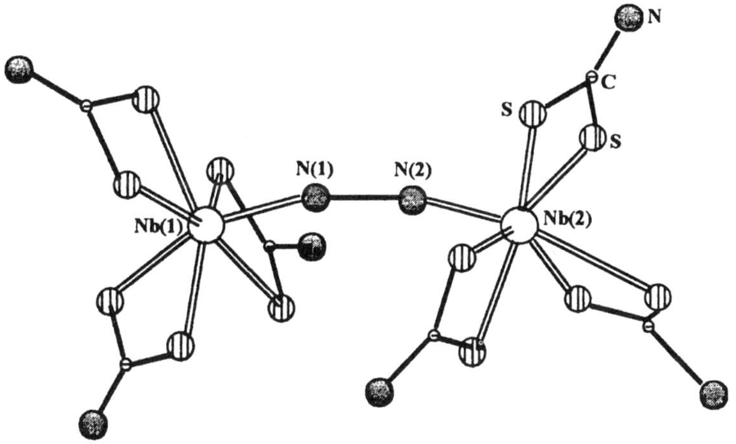

FIGURE I.25 The structure of the dinuclear complex $[Nb(S_2CNEt_2)_3]_2(\mu\text{-}N_2)$. (Adapted from Dilworth, J. R. et al., *J. Chem. Soc. Dalton Trans.*, 1077, 1990.)

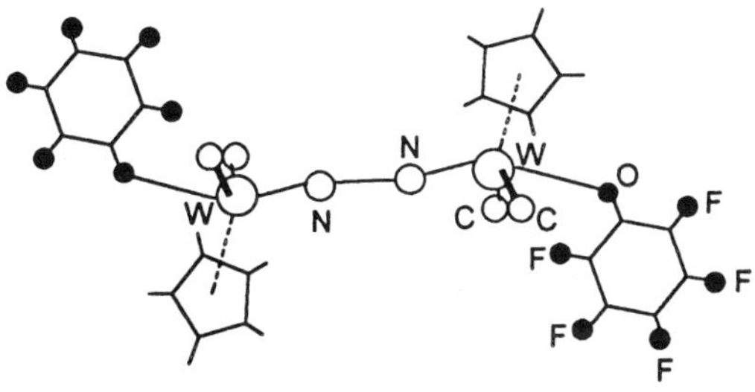

FIGURE I.26 The structure of the dinuclear complex $[WC_p{}^*Me_2\ (OC_6F_6)]_2(\mu\text{-}N_2)$. (Adapted from Churchill, M. R. and Li, Y.-J., *J. Organomet. Chem.*, 301, 49, 1986.)

TABLE I.12
Protonation[a] of Bridging Dinitrogen Complexes

Complex	N–N(Å)	Products (yields %)[b]	Ref.
$[Zr(C_5Me_5)_2(N_2)]_2(\mu\text{-}N_2)$	1.182 (5)	N_2(67), N_2H_4(33)	110, 111
$(\mu\text{-}N_2)\{[(o\text{-}Me_2NCH_2)C_6H_4]_2V(Py)\}_2(THF)_2$	1.228	NH_3 (33), V(III)	112, 121
$[Nb(S_2CNEt_2)_3]_2(\mu\text{-}N_2)$	1.25 (2)	N_2H_4(100)	113
$[Ta(S_2CNEt_2)_3)_2(\mu\text{-}N_2)$		N_2H_4(100)	113
$[Mo(C_5Me_5)Me_3]_2(\mu\text{-}N_2)$	1.236	NH_3 (16)	115
With Zn (Hg)		NH_3 (32–36)	
$[Mo(C_5Me_5)Me_3](\mu\text{-}N_2)$	1.235	NH_3 (16)	115
$[W(C_5Me_4Et)Me_3$			
With Zn(Hg)]		NH_3 (32–36)	
$[W(C_5Me_5)Me_3)_2\ (\mu\text{-}N_2)$		NH_3 (17)	115
With Na(Hg)		NH_3 (18)	

[a] HCl data. In the presence of Zn(Hg) or Na(Hg) lutidine-HCL or lutidine-$HOSO_2CF_3$ were used.
[b] Based on total N_2 content.

prediction that with a large number of electrons the orbitals that correspond to the bonding in NN and antibonding in MN will be occupied.

For some of the complexes the NN bond is so elongated that the structure M=N–N=M is more appropriate than the structure with NN triple bond M–N≡N–M. It is these complexes that illustrate the four-electron mechanism of dinitrogen reduction discussed earlier. Moreover, sometimes for the bridging dinitrogen it is more appropriate to consider the N_2^{4-} ligand donating its electrons to the metal atoms than the uncharged dinitrogen accepting electrons from the metals.[116]

The results obtained with the protonation of the bridging dinitrogen confirm the conclusion that the dinitrogen ligand is strongly activated (Table I.12).

Nb^{III} and Ta^{III} complexes form hydrazine quantitatively when protonated by HCL or HBr. Va^{II}, Mo^{IV}, and W^{IV} complexes produce mainly ammonia, the yields being less than quantitative: part of coordinated dinitrogen is evolved as free N_2. These differences in behavior can be reasonably explained in similar way as for N_2 reduction in protic media and in terminal dinitrogen complexes. Nb^{III} and Ta^{III} are sufficiently strong two-electron reductants to form hydrazine in four-electron reduction (cf. the reduction of N_2 in mononuclear W^0 complexes with the addition of strong donors, in hydroxide systems in strongly alkaline media, and in the catalytic systems in the presence of the sodium amalgam). The other complexes are unable to easily produce hydrazine and have to receive two more electrons to break the NN bond and to produce ammonia. These additional electrons are provided by other molecules of dinitrogen complexes present in solution and this results in the oxidation of the latter with free dinitrogen evolution. If the protonation is performed in the presence of another reducing agent (zinc or sodium amalgam) the yields of ammonia are considerably increased (see Table I.12), because additional electrons are provided by the reductants and therefore less dinitrogen is lost from the complexes (cf. the reduction in Mo^0 mononuclear complexes, reduction in homogeneous and in some heterogeneous protic media, and N_2 reduction in the enzymatic systems). Of course, the efficiency of this transfer of additional electrons is determined by the organization of the contact between the active center and the reducing agent. In the catalytic systems like nitrogenase or chemical model systems like those described above this contact is evidently very effective. For the complexes under consideration the yields do not reach 100% per the coordinated N_2 and no catalysis is observed.

The reduction of dinitrogen upon protonation in specially prepared complexes may be considered a model of the second stage of dinitrogen reduction in protic solutions in the stoichiometric and catalytic systems described above, the first stage being the coordination of N_2. In this connection it is of interest to compare metals (and their oxidation states) able to activate dinitrogen to reduction in protic media (1) with those able to form bridging dinuclear MNNM complexes with N_2 undergoing reduction upon protonation (2):

1. Ti^{II}, V^{II}, Cr^{II}, Nb^{III}, Mo^{III}, Ta^{III}
2. Ti^{II}, V^{II}, Zr^{II}, Nb^{III}, Mo^{IV}, Ta^{III}, W^{IV}

The resemblance of the groups is striking and hardly accidental. It supports the suggestion that metals of groups IV to VI with d^2 or d^3 electronic configurations capable of activating and reducing dinitrogen in protic media form linear intermediate MNNM complexes (usually bound to an additional electron donor, e.g., a metal cluster) in the process of the reduction, the final stage being the protonation of the ligating dinitrogen. The choice between the two possible products, hydrazine and ammonia, is determined mainly by the M reducing ability: hydrazine is formed in the case of sufficiently strong reducing agents, while less strong reductants require two electrons more to produce ammonia. Intermediate MNNM complexes are naturally much less stable than the model, specially prepared complexes, but they are no less active in N_2 reduction; rather, they contain more activated dinitrogen that can be reduced upon protonation even in alkaline media: apparently water is a sufficiently strong acid to protonate coordinated dinitrogen.

In connection with the intermediate complexes in dinitrogen-reducing systems it is interesting to consider, in particular, the dinitrogen complex with mesityl derivative of bivalent vanadium synthesized by Floriani et al.[121b] The complex was prepared by reducing of $[V(mes)_3(THF)]$ (mes = 2,4,6-C_6H_2) by sodium metal under dinitrogen. The structure of the complex is shown in Figure I.27. It is seen that the role of sodium is not only to reduce V^{III} to V^{II}, but also to coordinate two aryl groups, which results in formation of the cyclic complex with bridging dinitrogen.

The geometry of the complex is very close to that of the intermediate postulated in Section 1.4.2 for dinitrogen reduction in the system $V(OH)_2$–$Mg(OH)_2$ (see Scheme 1.4). There is a bonding in the complex between Na^+ and nitrogen atoms in bridging dinitrogen (Na–N distances are found to be 2.55 and 2.63 Å). This suggests a bonding similar to the bonding between dinitrogen and magnesium in the intermediate complex in the system involving vanadium–magnesium hydroxides. (This bonding increasing the number of contacts of dinitrogen with metal ions might strengthen the activation of N_2.) The structure of the complex may also throw some light on the role of the alkaline metals in the catalytic systems of ammonia synthesis

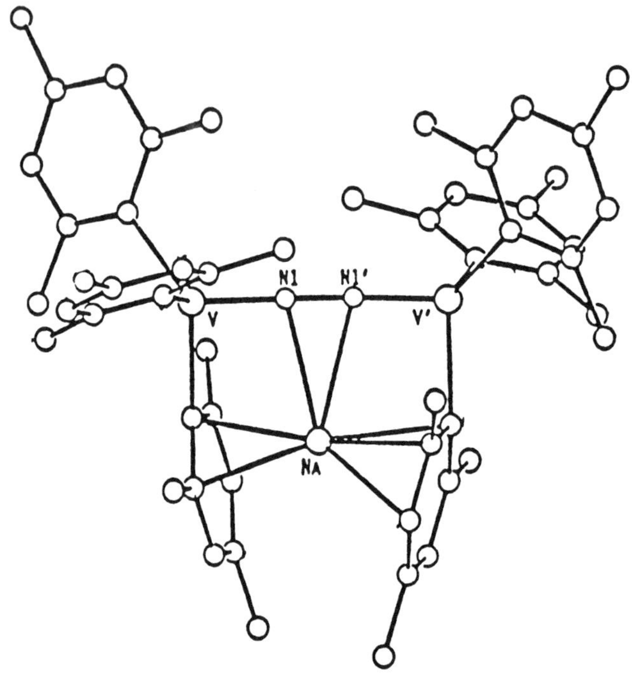

FIGURE I.27 The structure of the anionic part of the mesitylvanadium dinitrogen complex. (From Ferguson, R. et al., *Angew. Chem. Int. Ed.*, 32, 396, 1993. With permission.)

from dinitrogen and dihydrogen described in Section I.4.1. The complex produces hydrazine, ammonia, and dinitrogen upon protonation by HCl. It may be suggested that both protonation of dinitrogen and decomposition of the organometallic vanadium mesityl complex to form free dinitrogen proceed in parallel under the action of HCl and it is the combination of the respective rate constants that reflects the relative amounts of the produced molecules.

The mechanism of dinitrogen protonation and reduction in dinuclear complexes of niobium(III) and tantalum(III), $[\{M(S_2CNEt_2)_3\}_2(\mu\text{-}N_2)]$ (apparently more stable to acid) in CH_3CN was investigated by Henderson et al.[121c,d] The rate-limiting step was found to be in most cases the protonation of the bridging dinitrogen (by HCl or HBr) confirming the mechanism in which the protonation produces a hydrazine derivative from dinitrogen complex. In one of the cases (protonation of dinitrogen in dinuclear tantalum complex with HBr) the reaction was shown to rapidly afford an intermediate $[(Et_2NCl_2)_3Ta(\mu\text{-}NHNH_2)TaBr(S_2CNEt_2)_3]^{2+}$ that then produces $[TaBr(S_2CNEt_2)_3]+$ and $[Ta(NHNH_2)\ (S_2CNEt_2)_3]+$ in the rate-limiting step. The former complex rapidly picks up Br^- and the latter is further protonated to yield hydrazine and ultimately the second mole equivalent of the dibromo product.

I.5.4 DI- AND POLYNUCLEAR COMPLEXES OF OTHER TYPES

Many similarities between the dinitrogen reduction in protic media and N_2 reduction in the bridging dinitrogen in dinuclear complexes with the linear MNNM unit seem to confirm that the optimum mechanism of dinitrogen reduction in protic media with comparatively mild reducing agents involves such complexes as the intermediates. It was tempting to try to spread this mechanism to all N_2 reductions in protic media, the more so for biological nitrogen fixation, because two molybdenums (or two vanadiums) are present in the enzyme. Considering the reduction of dinitrogen, the author was inclined to support this idea for many years with a firm belief that two molybdenums take part in a polynuclear reaction center and a polyelectronic process takes place. Now it is clear that the mechanism of the enzymatic reduction does not involve the simultaneous participation of two molybdenums in a complex with dinitrogen, although at the same time the conclusion about the polynuclear complex is evidently confirmed and a polyelectronic mechanism of N_2 reduction is more likely than ever.

Moreover, a linear MNNM unit is hardly an intermediate taking into account the structure of the iron–molybdenum cofactor of nitrogenase: with iron atoms in the Fe^{II} or Fe^{III} state (d^6 or d^5 electronic configuration) such a complex is not a good candidate for an intermediate. Therefore other types of complexes must be considered as possible intermediates in the enzymatic process and in model systems.

I.5.4.1 Side-on Nickel Dinitrogen Complexes

The first example of the "side-on" binding mode of dinitrogen in a dinuclear complex was reported by Jonas et al. for a nickel complex $\{[(PhLi)_3Ni]_2(OEt_2)_2\}_2N_2$ and subsequently $[Ph\{Na(OEt_2)\}_2(Ph_2Ni)_2N_2NaLi_6(OEt_2)_4 \cdot OEt_2]_2$.[122] There are several contacts of the coordinated dinitrogen with lithium ions and a long N–N distance (1.35 Å) is reported, indicating strong activation of dinitrogen. Early attempts to protonate dinitrogen in the complex were disappointing, but later the conditions were found for the protonation and reduction of coordinated N_2. The products of the reaction are ammonia (33%) and free dinitrogen.[123] The authors suggest intermediate formation of diazene and its further disproportionation, which agrees with the stoichiometry and might be the case for this very strongly reducing system, but, as usual, formation of diazene is not the only alternative. Disproportionation may take place *before* liberating dinitrogen from the complex and formation of ammonia may proceed in the complex at the expense of other oxidations that produce dinitrogen. The transition metal nickel is not oxidized in the process of dinitrogen reduction and forms nickel metal when decomposed by HCl. The source of electrons is the phenyl ligands forming diphenyl in the process of the protonation. This behavior makes this complex rather unusual in dinitrogen-fixing systems. Let us recall that nickel compounds are among the least active in the Vol'pin and Shur systems.

I.5.4.2 Unusual Dinitrogen Coordination in Titanium Complexes

Another complex with very interesting coordination of dinitrogen was prepared by Pez et al.[124] The authors investigated dinitrogen reaction with a dinuclear titanium metallocene complex, $[\mu\text{-}(\eta^1\text{:}\eta^5\text{-}C_5H_4)]\ (\eta\text{-}C_5H_5)_3Ti_2$. Two molecules of the complex coordinate one dinitrogen, therefore four titanium atoms are situated around one N_2 molecule. In hydrocarbon solvents a deep-blue complex of composition $[(C_5H_4)]$ $(C_5H_5)_3Ti_2]_2N_2$ is formed. Reaction with N_2 (~10 atm) in 1,2-dimethoxyethane yields N_2 complex characterized by $\nu(N\text{–}N) = 1222\ cm^{-1}$, one of the lowest N–N vibrational frequencies yet observed for a coordinated dinitrogen ligand. Reaction with N_2 in tetrahydrofuran produces N_2 complex with ν(N–N) = 1296 cm^{-1}. Successive treatment of the complex obtained in glime with THF/glime and bis (2-methoxyethyl) ether (diglime) gives a crystalline dinitrogen complex with ν(NN) = 1282 cm^{-1}. Single-crystal X-ray crystallography of the complex revealed that dinitrogen is bound simultaneously to three titanium atoms. The N_2 is σ bonded to the one bivalent titanium atom in [μ-(η^1:η^5-C_5H_4)] (η-C_5H_5)$_3$$Ti_2$ and also to the two titanium atoms in (η^5:η^5-$C_{10}H_8$) (η-C_5H_5)$_2$$Ti_2$ in σ,π mode with one titanium bound end-on and the other side-on. As a result of multiple coordination, the N–N length (1.30 Å) is very much elongated as compared with free dinitrogen. Treatment of the solution of the latter complex with dihydrogen gas or with dry hydrogen chloride results in a loss of the coordinated dinitrogen as free N_2. However, aqueous hydrolysis of the complex in diglime yields mostly ammonia. As in similar cases, there is competition between the complex decomposition with N_2 evolution and the protonation (and reduction) of ligating dinitrogen. Therefore free N_2 formation could be the consequence of both weak dinitrogen activation and complex instability. But because at least in one case the complex yields ammonia upon hydrolysis, there is no doubt that dinitrogen is strongly activated. This result provides good evidence for the activation of dinitrogen by means of multiple coordination facilitating the subsequent polyelectronic reduction upon protonation. For unstable dinitrogen complexes this is possible only when several metal ions are united by common ligand(s) *before* coordinating N_2. Similar coordination may be suggested at least for some cases of dinitrogen reduction in solution in the presence of polynuclear transition metal complexes and it is very likely for the enzymatic dinitrogen reduction on iron–molybdenum cofactor.

The unusual side-on configuration of the dinuclear dinitrogen complexes with nickel could be explained by peculiarities of the complexes, particularly by participation of the lithium ions. More recent results indicate, however, that side-on dinuclear dinitrogen complexes are perhaps not very exotic. New examples of this type configuration appeared recently in the literature.

The reaction of *trans*-(TMEDA)$_2$$TiCl_2$ with 1 eq. of freshly sublimed (Me_3Si)$_2$NLi in toluene under N_2 at −40°C formed the complex {[(Me_3Si)$_2$N]TiCl (TMEDA)}$_2$(μ-N_2) (1).[125] The structure of the complex was determined by X-ray analysis. The TiNNTi unit is close to already familiar linear structure: ⟨Ti(I)–N(4)–N(4a) = 168.5(2)° and the N–N bond is strongly elongated with respect to N_2 (N–N bond distance is 1.289 Å). The Ti–N distance with coordinated dinitrogen is very short (1.762 Å).

The reaction of *trans*-(TMEDA)$_2$$TiCl_2$ with 2.5 eq. of (Me_3Si)$_2$NLi (instead of 1 eq.) in the presence of a small excess of TMEDA led, under the same reaction conditions to another stoichiometry (Scheme I.9).

The reaction is actually a disproportionation because from three molecules of bivalent titanium one trivalent titanium complex is obtained together with a dinuclear dinitrogen complex of mixed valency (Ti^{II}-Ti^{I}). The complex consists of two separate [Li(TMEDA)$_2$] and [{[(Me_3Si)$_2$N]$_2$Ti}$_2$(μ-η^2:η^2-N_2)$_2$] ionic fragments. Four of the six coordination sites of titanium are occupied by four N atoms of the two sideon coordinated coplanar dinitrogens. The N–N distance is very long (1.379 Å), significantly longer than in complex 1. Complex 1 is diamagnetic, while complex 2 is

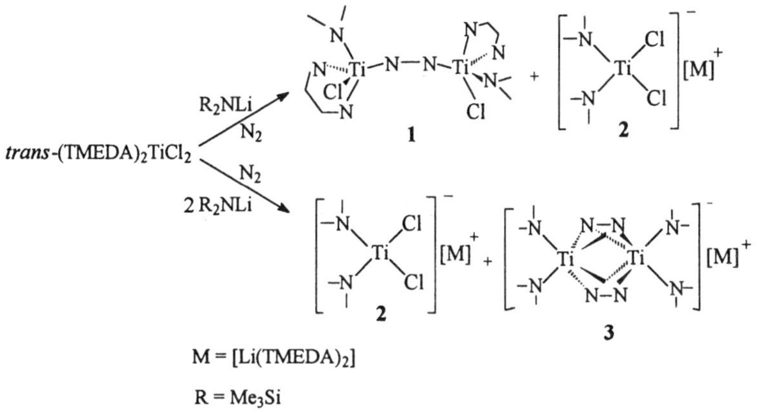

SCHEME 1.9 Two types of dinuclear titanium complexes with dinitrogen (**1** and **3**).

naturally paramagnetic (μ_{eff} = 1.37 μ_B) because it contains an uneven number of electrons. The authors suggest that the difference in dinitrogen bonding modes can be explained only by steric bulk in the two complexes. It seems more likely that in complex 2 titanium(I-II) in the anion is a stronger donor than titanim(II) in the neutral complex 1. Therefore σ-donor properties of N_2, which are more pronounced in the linear M–N≡N–M complexes, do not play an important role in side-on complex formation. It is clear that the energy of linear and side-on complexes are not very much different for such types of complexes and both types can therefore be intermediates in dinitrogen reduction.

I.5.4.3 Complexes with Flat $M<^{N}_{N}>M$ Unit

There have been very few examples of such complexes. The first complex of this kind was obtained in crystalline form after slow crystallization of $(C_5Me_5)_2Sm$ from toluene, presumably under dinitrogen.[126] The quantitative yield of the complex was obtained after a 4-week crystallization. The complex is $[(C_5Me_5)_2Sm]_2(\mu\text{-}\eta^2\text{:}\eta^2\text{-}N_2)$ and represents the first example of planar, side-on bonding of two metals and dinitrogen. The X-ray structural investigation revealed that while Sm–N and Sm–C distances are within a range typical for eight-coordinate Sm^{III} complexes, N–N bond distance is 1.088(12) Å, i.e., even shorter than in free dinitrogen (1.0975 Å). Perhaps this "shortening" is an artifact and is due to the X-ray measurement of a complex with dinitrogen rotation but it is clear that with two *d* electrons available (one electron per samarium atom), the value of the N–N bond distance shows dinitrogen to be only weakly activated. When the complex is dissolved in toluene an equilibrium is established:

$$2\,(C_5Me_5)_2Sm + N_2 \rightleftharpoons [(C_5Me_5)_2Sm]_2(\mu\text{-}\eta^2\text{:}\eta^2\text{-}N_2) \qquad (44)$$

The concentration of the complex increases at lower temperatures.

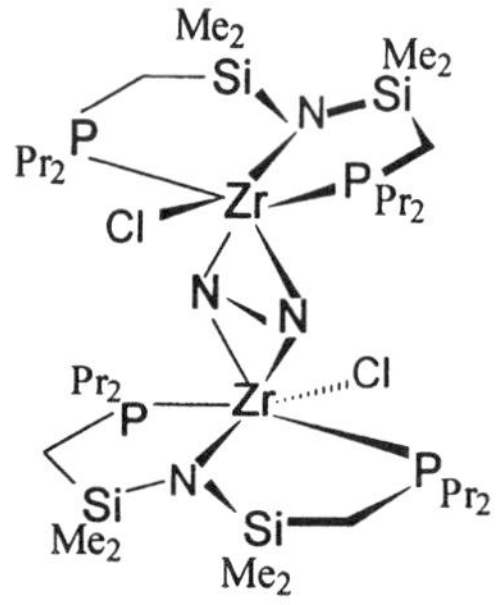

FIGURE I.28 The structure of the complex $\{[i\text{-}Pr_2PCH_2SiMe_2)_2ZrCl\}_2(\mu\text{-}\eta^2\text{-}N_2)$. (Adapted from Fryzuk, M. D. et al., *J. Am. Chem. Soc.*, 112, 8185, 1990.)

Another complex of this kind was obtained[127] with a Zr^{II} compound (d^2 electronic configuration) and has a formula $\{[(i\text{-}Pr_2PCH_2SiMe_2)_2ZrCl\}_2(\mu\text{-}\eta^2\text{:}\eta^2\text{-}N_2)]$. The structure of the complex is given in Figure I.28. It has flat configuration of the Zr<$^{N}_{N}$>Zr unit with extremely long N–N distance, about 1.55 Å, i.e., almost 0.1 Å longer than in hydrazine.

The complex produces 1 eq. of hydrazine when decomposed by acid and is considered by the authors[127] to be a compound of two Zr^{4+} with a bridging N_2^{4-} unit. Hence all four electrons available take part in N_2 activation.

More recently, a third example of a complex with side-on bridging dinitrogen was reported having a flat M<$^{N}_{N}$>M unit. This time it is of a quite different nature, the unit being part of the lithium dication. The complex was obtained upon reaction of Cp_2ZrCl_2 with the product of the interaction between $PPhH_2$ and *n*-BuLi.[128a] A crystallographic study revealed the complex to be a salt containing (as the authors[128] believe) zirconium(III) dianion, $[Cp_2Zr(\mu\text{-}PPh)]_2^{2-}$, and dication, $[\{(THF)_3Li\}_2\mu\text{-}N_2]^{2+}$. A crystallographically imposed center of symmetry relates the two pseudotetrahedral $Li(THF)_3$ fragments, which are bridged by a side-on N_2 moiety. The N–N distance in the cation (1.06[1] Å) is indistinguishable from that in free dinitrogen.

Considering the knowledge of low-temperature chemistry of dinitrogen presented in this chapter, coordination of dinitrogen with two lithium cations is extremely unexpected because dinitrogen is a very weak base and nucleophile and usually does not interact even with very strong acids and electrophiles. Moreover strong Coulombic repulsion would further weaken the bonding with two (!) lithium cations. It could be suggested therefore that the complex is actually a salt of *monoanion* $[Cp_2Zr(\mu\text{-}PPh)]_2^-$ (with mixed Zr^{III}-Zr^{IV} oxidation state of zirconium) and *monocation* $[\{(THF)_3Li\}_2\mu\text{-}N_2]^+$ and therefore dinitrogen is bound to Li_2^+. This would correspond to the reported extreme air sensitivity of the complex. Accordingly, in another work specially performed to check this idea the salt of composition $[\{(THF)_3Li\}_2(\mu\text{-}Cl)]^+[Cp_2Zr(\mu\text{-}PPh)]_2^-$ was obtained when the reaction was conducted in similar conditions but under argon instead of dinitrogen.[129] The X-ray structure of the anion, which is, no doubt, *monoanion* in this case because the cation has definitely unit charge, was found to be identical to the anionic part of the reported dinitrogen complex. Hence, if the existence of this complex is confirmed it is indeed the complex of N_2 with Li_2^+ and can be also regarded as the complex of N_2^- with two Li^+. Further investigation is needed to confirm the existence and clarify the nature of this unusual species.

Gambarotta reported the synthesis and isolation of another dinitrogen complex with two samarium(II) atoms bound to dinitrogen in the flat Sm_2N_2 unit, but in this case together with four lithium ions bringing with them additionally two electrons.[130] The reduction of the mixture obtained from the reaction of $SmCl_3(THF)_3$ with the tetralithium salt of octaethylporphyrogen (OEPG) $(OEPG)Li_4(THF)_4$ and subsequent reduction with 1 eq. of metallic lithium under argon in THF yields a deep-green, extremely air-sensitive crystalline material. Recrystallization of this from THF under dinitrogen produces orange-red crystals of $[(THF)_2Li(OEPG)Sm]_2(N_2Li_4)$. The structure of this complex is given in Figure I.29. It contains two $(THF)_2Li(OEPG)Sm$ units bridged by a planar N_2Li_4 moiety placed on a symmetry center. A dinitrogen molecule with an elongated N–N bond distance reaching 1.525 Å has been found inside the Sm_2Li_4 octahedron. The N_2Li_4 core is formed by one N_2 unit with two side-on lithium atoms and the two others coplanar and coaxial with the N–N vector. The two samarium atoms are symmetrically placed side-on with respect to the N_2 moiety above and below the N_2Li_4 plane. Thus, in these nonlinear di- and polynuclear complexes the necessity of four electrons for essential activation of dinitrogen is particularly clear: *with one and two electrons there is virtually no activation, whereas with four electrons dinitrogen becomes highly activated and forms N_2^{4-} ligands.* More examples are needed to make this generalization certain, but it is unlikely that these are merely incidental facts. The structure of the latter complex is evocative of the structure of iron–molybdenum cofactor of nitrogenase and the results agree with the suggestion that at least four central coordinately unsaturated iron atoms take part in the activation of N_2 in the enzymatic process.

Thus, a polynuclear complex with multicoordinated dinitrogen ligand that is likely to be the intermediate in catalytic N_2 reduction with mild reducing agents (including biological nitrogen fixation) can be formed via linear as well as side-on mono- and then dinuclear complexes with subsequent coordination of other metals. The order of the metal coordination may be different, but the final step should include formation of a maximum number of metal-to-nitrogen contacts and at least four electrons to achieve optimal activation for catalysis.

I.5.4.4 Iron–Dinitrogen Complexes[131]

The X-ray structure of nitrogenase strongly supports the hypothesis that the mechanism of dinitrogen activation on the enzyme involves iron atoms of the iron–molybdenum cofactor, with molybdenum (or vanadium) stabilizing the complex and/or perhaps optimizing its redox potential. Therefore we have to pay more attention to iron complexes coordinating and reducing dinitrogen.

The iron catalyst for conventional ammonia synthesis from dinitrogen involves a polynuclear iron cluster. The catalysis requires high temperature; however, dihydrogen, the weakest reducing agent among those reacting with dinitrogen, can be used. There is no doubt that dinitrogen is activated in a side-on fashion and the structure of a surface dinitrogen complex forming a bridge between two iron atoms

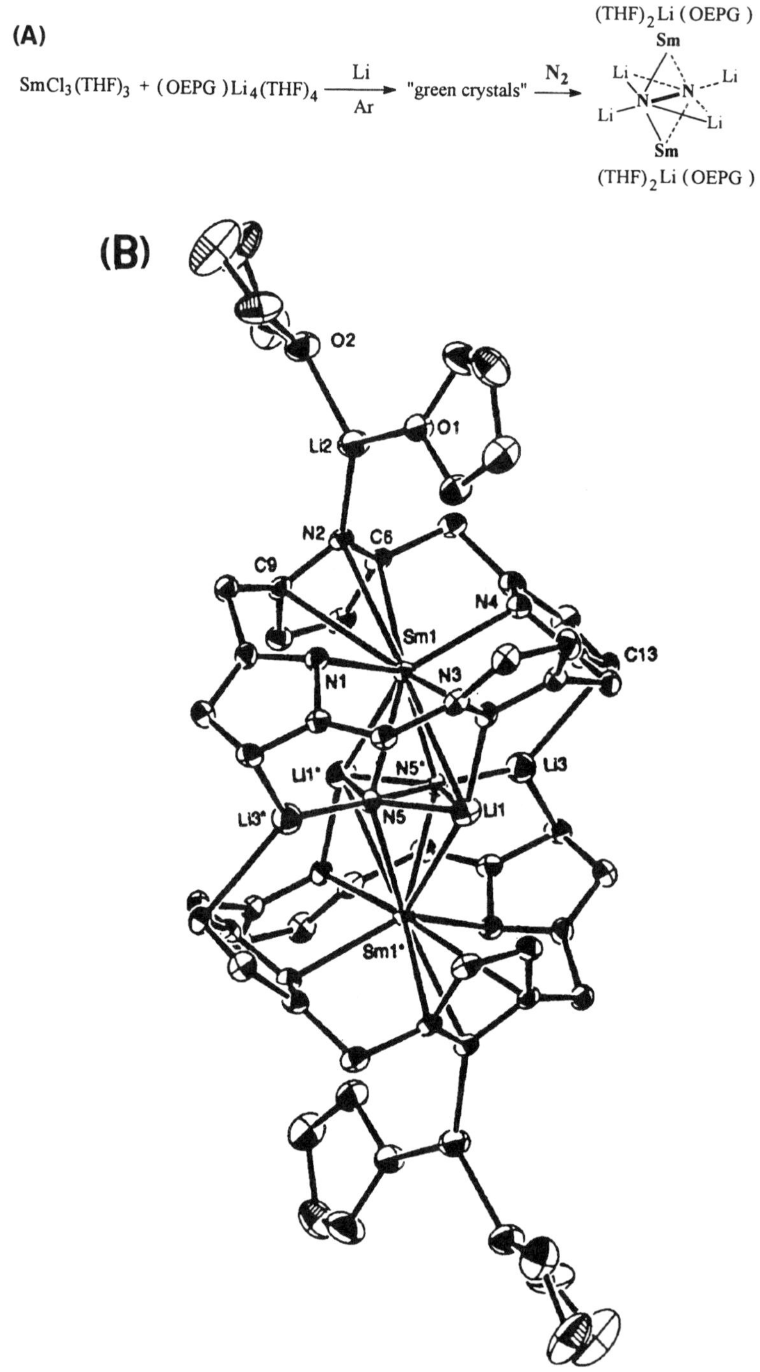

FIGURE I.29 (A) Formation and (B) structure of the $Sm_2Li_4N_2$ complex. (From Jubb, J. and Gambarotta, S., *J. Am. Chem. Soc.*, 116, 4477, 1994. With permission.)

has been proposed.[132] After all that we know now about nitrogen-fixing systems, paradoxically, the conventional catalyst for ammonia synthesis seems to be still the closest analogue of the nitrogenase active center, at least from the point of view of the cluster composition and the mechanism of dinitrogen activation and reduction.

Iron complexes were mentioned in the sections describing dinitrogen reduction in aprotic media. They reduce dinitrogen to nitrides with moderate yields by such reducing agents as RMgX and with almost 100% yields by lithium naphthalene. Dinuclear iron complexes of unknown structure were detected in dinitrogen-fixing systems in early works and, actually, this was the first indication that the reduction could be facilitated by acid (presumably by protonation). Dinitrogen reduction in protic media in the presence of iron compounds was also found (described above).

Iron(0) formed in LiAr reduction of $FeCl_3$ activate dinitrogen and reduce it upon protonation. The structure of iron complexes active towards N_2 has been elucidated.[133]

Dinitrogen iron complexes have been known for many years, since Aresta et al. published the first paper[134] on mononuclear dinitrogen iron(II) hydride phosphine complexes. These and many similar complexes were found to be inactive in N_2 protonation and reduction, but later the conditions were found to protonate a dinitrogen iron (0) phosphine complex with at least partial reduction to ammonia.[98]

From the viewpoint of analogy with the iron-molybdenum cofactor of nitrogenase, the complex may be considered that is produced in the reaction of tetraphenyliron(II) dihydride with tetrahydrofuran.[135] A dinuclear iron(II) complex is formed whose structure is presented in Figure I.30. The complex includes three μ^2-H and three μ^2-Li bridges and Fe–Fe distance in the complex (2.2 Å) is very close to that in the iron metal and in iron–molybdenum cofactor of nitrogenase.

Dinitrogen is coordinated by the complex with $\nu(N_2) = 1660\ cm^{-1}$ and is weakly bound: it is easily lost during the crystallization. At the same time dinitrogen can be reduced upon protonation, producing hydrazine and free N_2. It is of interest that the electrons to produce N_2H_4 are taken from iron (and not from LiPh ligand as in the case of the similar nickel complex), because the amount of hydrazine formed corresponds to the $Fe^{II} \rightarrow Fe^{III}$ transformation. Apparently, lithium ions take part in coordination with dinitrogen.

Many iron–sulfur complexes have been synthesized in the attempts to prepare models for the active center of nitrogenase since it became known that iron–sulfur clusters are involved in nitrogenase. Their activity towards dinitrogen is unknown, perhaps zero, but synthetic achievements are very significant. Some examples of the structures of the model complexes are shown in Figure I.31.[136-138]

Knowledge of the structure of the cofactor has encouraged syntheticists to prepare new complexes with structures closer to that of the cofactor. For example, a new class of clusters, that of double cubane clusters of the general formula $Mo_2Fe_6S_8(PR_3)_6(Cl_4\text{-cat})_2$ was recently synthesized (Cl_4-cat = tetrachlorocatecholate dianion).[139] They contain two Fe–S edge-linked $MoFe_3S_4$ cubanes with the Mo atoms at the periphery of the cluster (similarly to FeMoco). The cyclic voltammetry of the complex with R = Et (on a glassy electrode, in CH_2Cl_2) shows reductions at –659 and –1219 mV (vs. normal hydrogen electrode), whereas the starting cluster involving an $MoFe_3S_4Cl_3$ (Cl_n-cat) anion exhibits a reversible reduction at –709 mV, under identical conditions.[139]

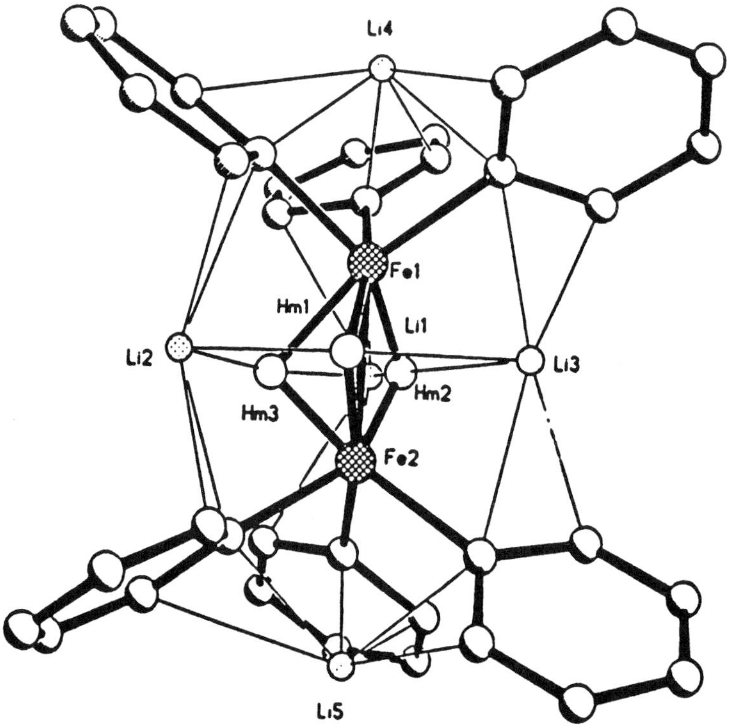

FIGURE I.30 The structure of the complex $Ph_3Fe[(\mu_2\text{-}H)_3FePh_3]Li_5(THF)_5$. (THF molecules are not shown.) (From Bazhenova, T. A. et al., *J. Organomet. Chem.*, 428, 107, 1992. With permission.)

Thus, an important property of large clusters is observed as compared with smaller clusters: when two smaller clusters are united their redox properties are changed. The redox potentials of the first electron reduction is less negative than for a one-electron reduction of the small cluster, whereas for the second electron reduction the potential is more negative; the fully reduced double cluster is a stronger one-electron reductant than the reduced single one. Apparently this property was used in construction of both P cluster and FeMoco in the natural system (see Section I.3.2).

Further development of this promising synthetic work is likely to produce close models of the natural enzyme, both structural and functional.

I.6 CONCLUSION

Low-temperature chemistry of molecular nitrogen developed entirely in the last 30 years includes now a number of systems acting in both aprotic and protic media as well as many dinitrogen complexes with transition metal compounds capable of reducing N_2 upon protonation. Each of these systems may be characterized by its own features, but some general characteristics may be formulated uniting all these systems, including nitrogenase complexes.

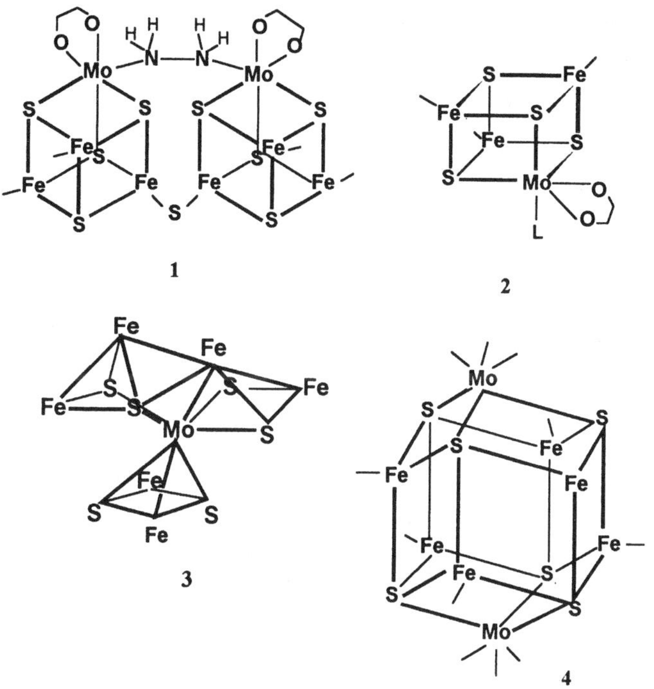

FIGURE I.31 Molecular structures of some FeMoS clusters. (Adapted from Eldredge, P. A. et al., *J. Am. Chem. Soc.*, 110, 5573, 1988 and Coucouvanis, D., *Acc. Chem. Res.*, 24, 1, 1991.)

These characteristics support and further develop the ideas that follow from considerations of dinitrogen properties and thermodynamics of its reactions with coordination compounds presented at the beginning of this chapter (Section I.2).

First of all, weak one- or two-electron reductants capable, however, of thermodynamically reducing dinitrogen require as a rule nitrogen-fixing systems of high nuclearity. Probably only such metals as lithium are capable of reacting with dinitrogen as a single atom with a full cleavage of one of the π-bonds:

$$Li + N{\equiv}N \rightarrow Li{-}N{=}N\cdot \qquad (45)$$

Even in this case Li^+ might be a catalyst in the reduction.

In aprotic media the reaction with a full or almost full cleavage of one of the π-bonds can be visualized in a dinuclear complex M–N=N–M requiring one electron at each M and strong reducing abilities of M. Diazene, N_2H_2, may be expected as an intermediate in reaction of this complex with an acid. In protic media with participation of comparatively weak reducing agents the reaction generally proceeds in polynuclear systems, e.g., hydroxides of vanadium(II), niobium(III), tantalum(III), chromium(II), titanium(III)–molybdenum(III), and polynuclear vanadium(II) complexes with catechols. The reaction rate is higher if the same reducing agent (e.g., sodium amalgam) functions with more polynuclear molybdenum(III) complexes. Nitrogenase cofactor, which seems to activate and reduce dinitrogen, includes eight transition metal atoms (eight iron atoms or seven iron atoms and one molybdenum or seven iron atoms and one vanadium) and is one of the most remarkable examples of these polynuclear systems. This phenomenon results from the peculiarities of dinitrogen molecule: with the same reducing agent a complex with higher nuclearity has the advantage as the catalyst for the reduction, ensuring four- or six-electron reduction of dinitrogen to hydrazine or ammonia.

In the reduction of dinitrogen upon protonation in mononuclear complexes of tungsten, molybdenum, or iron, four or six electrons are transferred from one metal ion. One has, therefore, to start with a zero-valent metal, and, evidently, the process is rather unlikely to be performed in a catalytic way in the presence of protons, because to return the metal to its initial zero-valent state a very strong reducing agent is required.

The reduction products in protic media are hydrazine or (and) ammonia; in virtually all cases diazene (N_2H_2) is apparently not an intermediate either in the free or the bound state: the process proceeds as a four-electron reduction with the simultaneous cleavage of both π-bonds. This mechanism, which was tentatively put forward in 1969, now has full experimental confirmation. However, this is only slowly becoming known; apparently many chemists are still unaware of the experimental confirmation.*

In di- and polynuclear complexes both π-bonds of dinitrogen participate in the interaction with the metal; in a number of cases dinitrogen in the complex is already close to the N_2^{4-} state and forms hydrazine upon protonation. The basicity of dinitrogen in its complexes with transition metal compounds is drastically increased and the protonation helps to transform dinitrogen into the hydrazido derivative, even in those cases when the initial state of the N_2 ligand is far from N_2^{4-}.

Hydrazine and ammonia are often produced in parallel reactions, though, of course, hydrazine may undergo further reduction. Formation of hydrazine from dinitrogen activated in dinuclear complexes requires the fulfillment of the three conditions: availability of at least four electrons, sufficiently strong reducing agent, and accessibility of protons. If at least one of these conditions is not met, the process will not proceed at all. If at least six electrons are available the reaction may proceed directly to ammonia, but presumably through a hydrazine derivative stabilized in the complex.

This property reveals itself both in dinitrogen-fixing systems in protic media and in the case of specially prepared dinuclear complexes capable of reducing dinitrogen upon protonation. Thus, upon protonation of dinitrogen in dinuclear niobium(III) and tantalum(III) complexes, N_2 produces hydrazine quantitatively. At the same time, in the cases of the N_2 complexes with weaker reductants (vanadium[II], molybdenum[IV], and tungsten[IV]) dinitrogen "awaits" two additional electrons to cleave the NN bond and subsequently form ammonia. Other molecules of a dinitrogen

* For example, the possibility of a four-electron mechanism in N_2 reduction was considered but *not accepted* in a recent review.[33a] At the same time, the likelihood of such a mechanism has already been mentioned carefully in an important modern biochemical paper:[32b] "One interesting possibility is that this barrier [two-electron reduction to diimide] could be side-stepped through reduction of bound dinitrogen by more than two electrons, possibly by electron donation from the P-cluster pair..."

complex may provide additional electrons (they are oxidized in this process, producing free dinitrogen) or specially added reductants can serve as electron donors, such as sodium or zinc amalgams or mercury cathode.

In enzymatic dinitrogen reduction these additional electrons are apparently provided by P clusters, which obtain them from the Fe protein. Uniting two Fe_4S_4 clusters in P clusters provides an opportunity to create a sufficiently strong assisting factor for the most difficult step in dinitrogen reduction in the coordination sphere of FeMo cofactor.

In chemical dinitrogen reduction in protic media the reaction proceeds mainly in the presence of the complexes of those transition metals that are able to form dinuclear dinitrogen complexes producing hydrazine or ammonia upon protonation. Electronic configurations d^2–d^4 are typical for metals in these dinuclear complexes. A larger number of *d* electrons hampers the possibility of N_2 activation and reduction in linear dinuclear complexes, in agreement with the theoretical predictions. This confirms the conclusion that similar complexes are the intermediates in the dinitrogen-reducing systems in protic media with intermediates too unstable to be isolated and investigated. Most active systems based on the complexes of V^{II} and Mo^{III} belong to this category, including the catalytic systems based on Mo^{III} complexes, which are so far the only nonbiological *catalytic* systems found to reduce dinitrogen at room temperature and atmospheric pressure. Presumably, dinuclear linear complexes M–N≡N–M with d^2–d^4 electronic configuration of M are the simplest systems sufficiently activating dinitrogen in protic media for reduction with the help of the action of protons. In aprotic media with strong donor properties of ligands at M the reaction of N_2 in such complexes can lead to the full cleavage of NN bond and formation of nitrides.

Although the number of known nonlinear di- and polynuclear complexes is still not sufficient for confident generalization, the necessity of four electrons is particularly clear: with two electrons available dinitrogen seems to be not activated at all, while four electrons transforms N_2 to the state that can be described as N_2^{4-} and produces N_2H_4 upon protonation. At the same time, four *d* electrons does not appear to be the upper limit for the complexes to activate and reduce dinitrogen. This important conclusion together with the knowledge of the structure of the iron–molybdenum cofactor of nitrogenase makes it likely that four (or less likely six) iron atoms are initially in direct contact with dinitrogen. The existing examples of model polynuclear dinitrogen complexes (hopefully more such complexes will be found in the near future) show that the number of direct contacts can be up to six and this corresponds to strong activation of dinitrogen.

The function of molybdenum and vanadium in the two corresponding nitrogenases, although important, apparently does not include direct interaction with dinitrogen (despite the fact that model chemical studies and chemical "intuition" indicated that both these elements could be directly involved in N_2 activation). This conclusion (which was difficult for the author of this book who for a long time was inclined to think otherwise) was confirmed by the discovery of nitrogenase with iron as the only transition metal, and consideration of the detailed structure of the cofactor makes it natural to believe that iron is likely to activate dinitrogen in all three nitrogenases. Although there are some differences in their behavior the similarities are more pronounced and molybdenum and vanadium seem to be on the periphery of the active center, rather than participate directly in dinitrogen activation and reduction. Although the problem of the site of N_2 activation has not been finally solved, the cofactor structure seems to be more suitable for the activation of N_2 on the central coordinatively unsaturated iron atoms than on the fully saturated molybdenum atom or in the dinuclear complex FeN_2Mo. The role of molybdenum (or vanadium) consists probably in the stabilization of the cluster, which is revealed in the shorter Fe–Fe distances at the side close to molybdenum (or vanadium). The differences are not very essential and not very difficult to explain, at least qualitatively. For example, the formation of some hydrazine together with ammonia in the case of vanadium nitrogenase, whereas only ammonia is the final product in dinitrogen reduction by molybdenum nitrogenase (and hydrazine derivative is formed only as an intermediate) may be explained by the stronger reducing properties of vanadium nitrogenase. This may also explain (though without certainty) partial reduction of acetylene to ethane in the case of vanadium and iron nitrogenases, ethylene presumably requiring a stronger reducing agent than acetylene. Shorter Fe–Fe distances and the stabilization of the cluster may be another reason for this very high selectivity of molybdenum nitrogenase towards acetylene. Apparently, the more compact structure of the FeMo cofactor as compared with FeV and FeFe cofactors prevents the coordination of ethylene on iron atoms and C_2H_4 leaves the coordination sphere without being further reduced, while in looser FeV and FeFe cofactors the lifetime of coordinated ethylene is increased and it is reduced partially to ethane.

In any case, in the model systems we have the examples of all different selectivities: formation of solely ammonia (from N_2) and ethylene (from acetylene), formation of hydrazine (together with ammonia) and ethane (together with ethylene), and also formation of only hydrazine (virtually without ammonia).

If the conclusion that the activation and reduction of dinitrogen proceeds on four (or six) central iron atoms is correct then the intermediate complex has possibly a bridging N_2 bound to two irons at the diagonal of the quadrangle in FeMoco; two other iron atoms may then stabilize the complex forming Fe–N σ-bonds. A further step would be the formation of a hydrazine derivative. The subsequent protonation and NN bond cleavage, possibly with participation of two more central iron atoms (in the sequence which is still unknown), would lead to the formation of ammonia.

We know now several examples of perpendicular coordination of dinitrogen in dinuclear complexes of the type $M<^{N}_{N}>M$, with strongly activated N_2 molecule. The dinuclear hydride complex of iron(II) described in Section I.4.5. turned out to be able to coordinate dinitrogen, perpendicular coordination of N_2 being very likely. It should be noted that the Fe–Fe distance in the complex coincides with that in FeMo cofactor. Protonation of the complex leads to hydrazine with the participation of four iron atoms, which increase their oxidation state from Fe^{II} to Fe^{III}.

Finally, the suggested mechanism of dinitrogen reduction on the cofactor makes us recall the classic Haber ammonia synthesis. Presumably, the activation of N_2 starts with formation of the bridging dinitrogen with perpendicular π coordination relative to the surface iron atoms. Although being a high-temperature process it requires the weakest possible reductant, dihydrogen, which evidently forms a surface hydride. The difference in activation energy is understandable taking into account more rigid and less optimal arrangement of surface iron atoms as compared with the complex structure of the cofactor.

Certainly any analogy must not be spread too far. Sulfides that are ligands in the cofactor are known to be a powerful poison of the heterogeneous catalyst. However, this merely demonstrates the affinity of the sulfides to iron. In the nitrogenase, sulfides do not interact with the active center but are a part of the catalytic molecular system.

Further generalizations may be too speculative and it is better to wait for new experimental results that will further clarify the mechanism of dinitrogen activation and reduction in the active center of nitrogenase.

New synthetic works may be suggested following the clarification of the structure of nitrogenase and new synthetic analogues of the cofactor will be prepared. We may expect that some of them will be catalysts for dinitrogen reduction including perhaps new catalysts of catalytic ammonia synthesis from dinitrogen and dihydrogen.

REFERENCES

1. Breslow, R., *Chem. Soc. Rev.,* 1, 553, 1972.
2. Henrice-Olivé, G. and Olivé, S., in *A Treatise on Dinitrogen Fixation,* Hardy, R. W. F., Bottomley, F., and Burns, R. C., Eds., Wiley, New York, 1979, 3.
3. Huber, K. P. and Herzberg, G., *Molecular Spectra and Molecular Structure. IV. Constants of Diatomic Molecules,* Dordrecht, New York, 1970.
4. Lias, S. G., Bartness, J. E., Liebman, J. F., Holmes, J. L., Levik, R. D., and Mallard, W. G., *J. Phys. Chem. Ref. Data,* 17, Suppl. 1, 1988.
5. Kondratiev, V. N., Ed., *Bond Dissociation Energies. Ionization Potentials and Electron Affinity,* Nauka, Moscow, 1974 (in Russian).
6. Shilov, A. E., in *A Treatise on Dinitrogen Fixation,* Hardy, R. W. F., Bottomley, F., and Burns, R. C., Eds., Wiley, New York, 1979, 31.
7. Rucic, B. and Berkourtz, J., *J. Phys. Chem.,* 95, 4378, 1991.
8. Mc Kee, M. L., Squillacote, M. E., and Stanbury, D. M., *J. Chem. Phys.,* 96, 3266, 1992.
9. Pople, J. A. and Curtiss, L. A., *J. Chem. Phys.,* 95, 4385, 1991.
10. Likhtenshtein, G. I. and Shilov, A. E., *Zh. Fiz. Khim.,* 44, 849, 1970; Borod'ko, Yu. G. and Shilov, A. E., *Usp. Khim.,* 38, 761, 1969.
11. Pelikan, P. and Boča, R., *Coord. Chem. Rev.,* 55, 55, 1984.
12. Treitel, I. M., Float, M. T., Marsh, R. E., and Gray, H. B., *J. Amer. Chem. Soc.,* 91, 6512, 1969.
13. Goldberg, K. I., Hoffmann, D. M., and Hoffmann, R., *Inorg. Chem.,* 21, 3863, 1982.

14a. Bothe, H., de Bruijn, F. J., and Newton, W. E., Eds., *Nitrogen Fixation: A Hundred Years After. Proc. 7th Int. Congr. N≡Nitrogen Fixation,* Gustav Fisher, New York, 1988.

14b. Young, J. P. W., in *Biological Nitrogen Fixation,* Stasey, G., Burris, R. H., and Evans, H. J., Eds., Chaman & Hall, New York, 1992, 21.

15a. Bortels, H., *Arch. Microbiol.,* 1, 333, 1930.

15b. Bortels, H., *Zentralbl. Bakteriol. Parasitenkd. Abt. II,* 95, 193, 1935.

15c. Winogradsky, S., *C.R. Acad. Sci. (Paris),* 190, 661, 1930.

16. Burris, R. H., in *A Treatise on Dinitrogen Fixation,* Hardy, R. W. F., Bottomley, F., and Burns, R. C., Eds., Wiley, New York, 1979, 383, 569.
17. Burris, R. H. and Miller, C. E., *Science,* 93, 114, 1941.

18. Carnahan, J. E., Mortenson, L. E., Mower, H. F., and Castle, J. E., *Biochim. Biophys. Acta,* 44, 520, 1960.
19. Bulen, W. A., Burns, R. C., and Le Compte, J. R., *Biochem. Biophys. Res. Commun.,* 17, 532, 1969.
20. Shah, V. K. and Brill, W. J., *Proc. Natl. Acad. Sci. USA,* 74, 3249, 1977.
21. Kim, J. and Rees, D. C., *Nature,* 360, 553, 1992.
22. Young, J. P. W., in *Biological Nitrogen Fixation,* Stasey, G., Burris, R. H., and Evans, H. J., Eds., Chapman & Hall, New York, 1992, 43.
23. Newton, W. E., in *New Horizons in Nitrogen Fixation,* Palacios, R., Mora, J., and Newton, W. E., Eds., Kluwer, Dordrecht, The Netherlands, 1993.
24. Evans, D. J., Henderson, R. A., and Smith, B. E., in *Bioinorganic Catalysis,* Reedijk, J., Ed., Marcel Dekker, New York, 1993, 89.
25. Burgess, B. K., *Chem. Rev.,* 90, 1377, 1990.
26. Newton, W. E., in *Biological Nitrogen Fixation,* Stasey, G., Burris, R. H., and Evans, H. J., Eds., Chapman & Hall, New York, 1992, 877.
27. Smith, B. E. and Eady, R. R., *Eur. J. Biochem.,* 205, 1, 1992.
28. Georgadis, M. M., Komiya, H., Chakrabarti, P., Woo, D., Kornuc, J. J., and Rees, D. C., *Science,* 257, 1653, 1992.
29. Bolin, J. T., Ronco, A. E., Mortenson, L. E., Morgan, T. V., Williamson, M., and Xuong, N.-H., in *Nitrogen Fixation: Achievements and Objectives,* Gresshoff, P. M., Roth, L. E., Stacey, G., and Newton, W. E., Eds., Chapman & Hall, New York, 1990, 117.
30. Hooever, T. R., Robertson, A. D., Cerny, R. L., Hayes, R. N., Imperial, J., Shah, V. K., and Ludden, P. W., *Nature,* 329, 855, 1987.
31. Kim, J. and Rees, D. C., *Science,* 257, 1677, 1992.
32a. Chan, M. K., Kim, J., and Rees, D. C., *Science,* 260, 792, 1993.
32b. Kim, J. and Rees, D. C., *Biochemistry,* 33, 389, 1994.
32c. Schultz, F. A., Gheller, S. F., Burgess, B. K., Lough, S., and Newton, W. E., *J. Am. Chem. Soc.,* 107, 5364, 1985.
33a. Stiefel, E. I. and George, G. N., in *Bioinorganic Chemistry,* Bertini, I., Gray, H. B., Lippard, S. J., and Valentine, J. S., Eds., University Science Books, Mill Valley, California, 1994, 365.
33b. Scott, D. J., May, H. D., Newton, W. E., Brigle, K. E., and Dean, D. R., *Nature,* 343, 188, 1990.
34a. Thorneley, R. N. F., Eady, R. R., and Lowe, D. J., *Nature,* 272, 557, 1978.
34b. Lowe, D. J. and Thorneley, R. N. F., *Biochem. J.,* 215, 393, 1983.
35. Lowe, D. J. and Thorneley, R. N. F., *Biochem. J.,* 224, 895, 1984.
36. Lowe, D. J., Fisher, K., Pau, R. N., and Thorneley, R. N. F., in *New Horizons in Nitrogen Fixation,* Palacios, R., Mora, J., and Newton, W. E., Eds., Kluwer, Dordrecht, 1993, 95.
37. Moser, C. C., *Nature,* 355, 786, 1992.
38a. Druzhinin, S. Yu., Syrtsova, L. A., Uzenskaja, A. M., and Likhtenshtein, G. I., *Biochem. J.,* 290, 637, 1993.
38b. Orme-Johnson, W. H., *Science,* 257, 1639, 1992.
38c. Deng, H. and Hoffmann, R., *Angew. Chem. Int. Ed. Engl.,* 32, 1062, 1993.
39. Vol'pin, M. E. and Shur, V. B., *Dokl. Akad. Nauk SSSR,* 156, 1102, 1964.
40. Vol'pin, M. E., Shur, V. B., and Ilatovskaya, M. A., *Izv. Akad. Nauk SSSR Ser. Khim.,* 1728, 1964.
41. Vol'pin, M. E. and Shur, V. B., in *New Trends in the Chemistry of Nitrogen Fixation,* Chatt, J., da Câmara Pina, L. M., and Richards, R. L., Eds., Academic Press, London, 1980, 67.
42. Vol'pin, M. E. and Shur, V. B., *J. Organomet. Chem.,* 200, 319, 1980.
43. Shur, V. B., Yunusov, S. M., Rummel, S., Wahren, M., and Volpin, M. E., *Dokl. Akad. Nauk SSR,* 299, 1409, 1987; Shur, V. B., Yunusov, S. M., Ilatovskaya, M. A., Rummel, S., and Vol'pin, M. E., *Dokl. (Rus.) Akad. Nauk,* 336, 631, 1994.
44. Bochkarev, M. N., Trifonov, A. A., Razuvaev, G. A., Ilatovskaya, M. A., Shur, V. B., and Vol'pin, M. E., *Dokl. Akad. Nauk SSSR,* 295, 1381, 1987.
45. Bell, L. G. and Brintzinger, H. H., *J. Am. Chem. Soc.,* 92, 4464, 1970.
46. Ozaki, A. and Aika, K., in *A Treatise on Dinitrogen Fixation,* Hardy, R. W. F., Bottomley, F., and Burns, R. C., Eds., Wiley, New York, 1979, 169.
47. Shur, V. B., Iunusov, S. M., and Vol'pin, M. E., *Dokl. Akad. Nauk SSSR,* 277, 884, 1984.

48. Yamamoto, A., Ookawa, M., and Ikeda, S., *Chem. Commun.*, 841, 1969; Yamamoto, A., Go, S., Ookawa, M., Takahashi, M., Ikeda, S., and Keii, T., *Bull. Chem. Soc. Jpn.*, 45, 3110, 1972.
49. Laplaza, C. E. and Cummins, C. C., *Science*, 268, 861, 1995.
50. Jubb, J., Scoles, L., Jenkins, H., and Gambarotta, S., *Eur. J.*, 2, 767, 1996.
51. van Tamelen, E. E. and Rudler, H., *J. Am. Chem. Soc.*, 92, 5253, 1970.
52. Shilov, A. E., Shilova, A. K., and Kvashina, E. F., *Kinet. Katal.*, 10, 1402, 1969.
53. Shilov, A. E. and Shilova, A. K., *Zh. Fiz. Khim.*, 44, 288, 1970.
54. van Tamelen, E. E., Fechter, R. B., and Scheller, S. W., *J. Am. Chem. Soc.*, 91, 7196, 1969.
55. Lorenz, B., Möbius, G., Rummel, S., and Wahren, M., *Z. Chem.*, 15, 241, 1975.
56. Shilov, A. E., in *Energy Resources through Photochemistry and Catalysis*, Grätzel, M., Ed., Academic Press, New York, 1983, 535.
57. Shilov, A. E., *J. Mol. Catal.*, 41, 221, 1987.
58. Shilov, A. E., in *Perspectives in Catalysis*, Thomas, J. A. and Zamaraev, K. I., Eds., Blackwell, 1992, 169.
59a. Shilov, A. E., *Pure Appl. Chem.*, 64, 1409, 1992.
59b. Luneva, N. P. and Shilov, A. E., *Kinet. Katal.*, 27, 328, 1986.
60. Denisov, N. T., Shuvalov, V. F., Shuvalova, N. I., Shilova, A. K., and Shilov, A. E., *Kinet. Katal.*, 11, 813, 1970; Shilov, A. E., Denisov, N. T., Efimov, O. N., Shuvalov, V. F., Shuvalova, N. I., and Shilova, A. K., *Nature*, 231, 460, 1971.
61. Denisov, N. T., Shuvalova, N. I., and Shilov, A. E., *Kinet. Katal.*, 35, 758, 1994.
62. Nikonova, L. A., Ovcharenko, A. G., Efimov, O. N., Avilov, V. A., and Shilov, A. E., *Kinet. Katal.*, 8, 1602, 1972.
63. Isaeva, S. A., Nikonova, L. A., and Shilov, A. E., *Nouv. J. Chim.*, 5, 21, 1981.
64. Becker, J. Y. and Posin, B., *J. Electroanal. Chem.*, 250, 385, 1988.
65. Folkesson, B. and Larsson, R., *Acta Chem. Scand. Ser. A*, 33, 347, 1979.
66. Luneva, N. P., Moravsky, A. P., and Shilov, A. E., *Nouv. J. Chim.*, 6, 245, 1982.
67. Cooper, S. R., Koh, Yu. B., and Raymond, K. N., *J. Am. Chem. Soc.*, 104, 5092, 1982.
68. Luneva, N. P., Mironova, S. A., Shilov, A. E., Antipin, M. Yu., and Struchkov, Yu. T., *Angew. Chem., Int. Ed. Engl.*, 32, 1178, 1993.
69. Denisov, N. T. and Shuvalova, N. I., *React. Kinet. Catal. Lett.*, 4, 431, 1976.
70. Didenko, L. P., Gavrilina, O. K., Yablonskaya, E. E., Shilova, A. K., and Shilov, A. E., *Nouv. J. Chim.*, 7, 605, 1983.
71. Didenko, L. P., Gavrilov, A. B., Shilova, A. K., Strelets, V. V., Tsarev, V. N., Shilov, A. E., Makhaev, V. D., Banerjee, A. K., and Pospišil, L., *Nouv. J. Chim.*, 10, 584, 1986.
72. Shilova, A. K., Strelets, V. V., Didenko, L. P., and Shilov, A. E., *Kinet. Katal.*, 28, 1239, 1987.
73. Shilova, A. K., Efimov, O. N., Makhaev, V. D., and Shilov, A. E., *Kinet. Katal.*, 36, 249, 1995.
74. Antipin, M. Yu., Struchkov, Yu. T., Shilov, A. E., and Shilova, A. K., *Gazz. Chim. Ital.*, 123, 265, 1993.
75a. Denisov, N. T., Shuvalova, N. I., and Shilov, A. E., *Kinet. Katal.*, 32, 1250, 1991.
75b. Denisov, N. T., Shuvalova, N. I., Shilov, A. E., Passynsky, A. A., Kolobkov, B. I., and Eremenko, I. L., *Kinet. Katal.*, 34, 858, 1993.
76. Abalyaeva, V. V., Denisov, N. T., Khidekel, M. L., and Shilov, A. E., *Izv. Akad. Nauk SSSR Ser. Khim.*, 196, 1973; 2638, 1975.
77. Gorodyskii, A. V., Danilin, V. V., Efimov, O. N., Nechaeva, N. E., and Tsarev, V. N., *React. Kinet. Catal. Lett.*, 11, 337, 1979.
78. Nikolaeva, G. V., Kulakovskaya, S. I., Efimov, O. N., and Shilov, A. E., *Kinet. Katal.*, 26, 473, 1985.
79. Schrauzer, G. N. and Schlesinger, G., *J. Am. Chem. Soc.*, 92, 1808, 1970.
80. Schrauzer, G. N., *Angew. Chem., Int. Ed. Engl.*, 14, 514, 1975.
81. Schrauzer, G. N., in *New Trends in the Chemistry of Nitrogen Fixation*, Chatt, J., da Câmara Pina, L. M., and Richards, R. L., Eds., Academic Press, London, 1980, 103.
82. Allen, A. D. and Senoff, C. V., *Chem. Commun.*, 621, 1965.
83. Dasent, W. E., *Nonexistent Compounds*, Dekker, New York, 1965.
84. Shilov, A. E., Shilova, A. K., and Borod'ko, Yu. G., *Kinet. Katal.*, 7, 768, 1966.

85. Harrison, D. F. and Taube, H., *J. Am. Chem. Soc.*, 89, 5706, 1967.
86. Bottomley, F., in *A Treatise on Dinitrogen Fixation,* Hardy, R. W. F., Bottomley, F., and Burns, R. C., Eds., Wiley, New York, 1979, 109.
87. Pombeiro, A. J. L., in *New Trends in the Chemistry of Nitrogen Fixation,* Chatt, J., da Câmara Pina, L. M., and Richards, R. L., Eds., Academic Press, London, 1980, 153.
88. Hidai, M., Tominari, K., Uchida, Y., and Misono, A., *Chem. Commun.*, 814, 1969; Hidai, M., Tominari, K., Uchida, Y., and Misono, A., *Chem. Commun.*, 1392, 1969.
89. Chatt, J., *J. Organomet. Chem.*, 100, 17, 1975.
90. Chatt, J., Dilworth, J. R., and Richards, R. L., *Chem. Rev.*, 78, 589, 1978.
91a. Richards, R. L., in *New Trends in the Chemistry of Nitrogen Fixation,* Chatt, J., da Câmara Pina, L. M., and Richards, R. L., Eds., Academic Press, London, 1980, 153.
91b. Chatt, J., Crabtree, R. H., Jeffery, E. A., and Richards, R. L., *J. Chem. Soc. Dalton Trans.*, 1167, 1973.
92. Hidai, M., in *Molybdenum Enzymes,* Spiro, T. G., Ed., Wiley, New York, 1985, 285.
93. Leigh, G. J., *Acc. Chem. Res.*, 25, 177, 1992.
94. Chatt, J., Heath, G. A., and Leigh, G. J., *J. Chem. Soc. D,* 444, 1972.
95. Chatt, J., Pearman, A. J., and Richards, R. L., *J. Chem. Soc. Dalton Trans.*, 1853, 1977.
96a. Rehder, D., Woitha, C., Priebsch, W., and Gailus, H., *J. Chem. Soc. D,* 364, 1992.
96b. Uchida, T., Uchida, Y., Hidai, M., and Kodama, T., *Bull. Chem. Soc. Jpn.*, 44, 2883, 1971.
97. Davis, B. R. and Ibers, J. A., *Inorg. Chem.*, 10, 578, 1971.
98. Leigh, G. J. and Jimenez-Tenorio, *J. Am. Chem. Soc.*, 113, 5862, 1991.
99. Bottomley, F. and Nyburg, S. C., *Acta Crystallogr.*, 24B, 1289, 1968.
100a. Ferdusson, J. E., Love, J. L., and Robinson, W. T., *Inorg. Chem.*, 11, 1662, 1972.
100b. Hammer, R., Klein, H.-F., Friedrich, P., and Huttner, G., *Angew. Chem. Int. Ed. Engl.*, 16, 485, 1977.
101a. Hoffmann, R., *Angew. Chem. Int. Ed. Engl.*, 21, 711, 1982.
101b. Shestakov, A. F., personal communication.
101c. Baumann, J. A. and George, T. A., *J. Am. Chem. Soc.*, 102, 6153, 1980.
101d. Pickett, C. J. and Talarmin, J., *Nature,* 317, 652, 1985.
102. Henderson, R. A., *J. Organomet. Chem.*, 208, C51, 1981.
103. Teuben, J. H., in *New Trends in the Chemistry of Nitrogen Fixation,* Chatt, J., da Câmara Pina, L. M., and Richards, R. L., Eds., Academic Press, London, 1980, 233.
104. Henderson, R. A., *Transition Met. Chem.*, 15, 330, 1990.
105. Borod'ko, Yu. G., Kvashina, E. F., Panov, V. B., and Shilov, A. E., *Kinet. Katal.*, 14, 255, 1973.
106. Shilov, A. E., Shilova, A. K., Kvashina, E. F., and Vorontsova, T. A., *Chem. Commun.*, 1590, 1971.
107. Hoffmann, R., Thorn, D. L., and Shilov, A. E., *Koord. Khim.*, 3, 1260, 1977.
108. Borshch, V. N., Thesis, Chernogolovka, 1992.
109. Sanner, R. D., Duggan, D. M., McKenzie, T. C., Marsh, R. E., and Bercaw, J. E., *J. Am. Chem. Soc.*, 98, 8358, 1976.
110. Manriquez, J. M., Sanner, R. D., Marsh, R. E., and Bercaw, J. E., *J. Am. Chem. Soc.*, 98, 8351, 1976.
111. Manriquez, J. M., Sanner, R. D., Marsh, R. E., and Bercaw, J. E., *J. Am. Chem. Soc.*, 98, 3042, 1976.
112. Edema, J. H., Meetsma, A., and Gambarotta, S., *J. Am. Chem. Soc.*, 111, 6878, 1989.
113. Dilworth, J. R., Henderson, R. A., Hills, A., Hughes, D. L., Macdonald, C., Stephens, A. N., and Walton, D. R. M., *J. Chem. Soc. Dalton Trans.*, 1077, 1990.
114. Churchill, M. R. and Wasserman, *Inorg. Chem.*, 20, 2899, 1981.
115. Schrock, R. R., Kolodziej, R. M., Liu, A. H., Davis, W. M., and Vale, M. G., *J. Am. Chem. Soc.*, 112, 4338, 1990.
116. Churchill, M. R. and Li, Y.-J., *J. Organomet. Chem.*, 301, 49, 1986.
117. Green, M. L. H. and Silverthorn, W. E., *J. Chem. Soc. Dalton Trans.*, 2164, 1974; Forder, R. A. and Prout, K., *Acta Crystallogr.*, 30B, 3778, 1974.
118. Zeigler, M. L., Weidenhammer, K., Zeiner, H., Skell, P. S., and Herrmann, W. A., *Angew. Chem. Int. Ed.*, 15, 695, 1976.

119. Yoshida, T., Okana, T., Thorn, D. L., Tulip, T. H., Otsuka, S., and Ibers, J. A., *J. Organomet. Chem.*, 181, 183, 1979.
120. Jolly, P. W., Jonas, K., Krüger, C., and Tsay, Y.-H., *J. Organomet. Chem.*, 33, 109, 1971.
121a. Leigh, G. J., Prieto-Alcon, R., and Sanders, J. R., *J. Chem. Soc. D,* 921, 1991.
121b. Ferguson, R., Solari, E., Floriani, C., Chiesi-Villa, A., and Rizzoli, C., *Angew. Chem. Int. Ed.*, 32, 396, 1993.
121c. Henderson, R. A., Morgan, S. H., and Stephens, A. N., *J. Chem. Soc. Dalton Trans.*, 1101, 1990.
121d. Henderson, R. A. and Morgan, S. H., *J. Chem. Soc. Dalton Trans.*, 1107, 1990.
122. Jonas, K., *Angew. Chem.*, 85, 1050, 1973; Jonas, K., Brauer, D. J., Krüger, C., Roberts, P. J., and Tsay, Y.-H., *Angew. Chem.*, 85, 1051, 1973.
123. Jonas, K. and Krüger, C., *Angew. Chem. Int. Ed.*, 19, 520, 1980.
124. Pez, G. P., Apgar, P., and Crissey, R. K., *J. Am. Chem. Soc.*, 104, 482, 1982.
125. Duchateau, R., Gambarotta, S., and Bensimon, C., *J. Am. Chem. Soc.*, 114, 482, 1992.
126. Evans, W. J., Ulibarri, T. A., and Ziller, J. W., *J. Am. Chem. Soc.*, 110, 6877, 1988.
127. Fryzuk, M. D., Haddad, T. S., and Rettig, S. J., *J. Am. Chem. Soc.*, 112, 8185, 1990.
128. Ho, J., Drake, R. J., and Stephan, D. W., *J. Am. Chem. Soc.*, 115, 3792, 1993.
129. Bazhenova, T. A., Kulikov, A. V., Shestakov, A. F., Shilov, A. E., Antipin, M. Yu., Lyssenko, K. A., Struchkov, Yu. T., and Makhaev, V. D., *J. Am. Chem. Soc.*, 117, 12176, 1995.
130. Jubb, J. and Gambarotta, S., *J. Am. Chem. Soc.*, 116, 4477, 1994.
131. Shilov, A. E., *New J. Chem.*, 16, 1, 1992.
132. Bauschlicher, C. W., Pettersson, G. M., and Siegbahn, P. E. M., *J. Chem. Phys.*, 87, 2129, 1987.
133. Bazhenova, T. A., Lobkovskaya, R. M., Shibaeva, R. P., Shilov, A. E., Shilova, A. K., Gruselle, M., LeNy, G., and Tchoubar, B., *J. Organomet. Chem.*, 244, 265, 1983.
134. Aresta, M., Giannocaro, P., Rossi, M., and Sacco, A., *Inorg. Chim. Acta,* 5, 115, 1971.
135. Bazhenova, T. A., Kachapina, L. M., Shilov, A. E., Antipin, M. Yu., and Struchkov, Yu. T., *J. Organomet. Chem.*, 428, 107, 1992.
136. Eldredge, P. A., Bryan, R. F., Sinn, E., and Averill, B. A., *J. Am. Chem. Soc.*, 110, 5573, 1988.
137. Ciurli, S. and Holm, R. H., *Inorg. Chem.*, 28, 1685, 1989.
138. Coucouvanis, D., *Acc. Chem. Res.*, 24, 1, 1991.
139. Demadis, K. D., Campana, C. F., and Coucouvanis, D., *J. Am. Chem. Soc.*, 117, 7832, 1995.

Chapter II

OXIDATION OF ALKANES

I believe that in twenty-five years at the conference commemorating 50th anniversary of the ICCC methane will be the most popular ligand in coordination chemistry.

J. Chatt
(Statement at the 17th International Conference on Coordination Chemistry (ICCC) commemorating the 25th anniversary of the ICCC, Hamburg, 1976.)

II.1 INTRODUCTION

Selective oxidation of hydrocarbons from oil, coal, and natural gas has been an important problem for many years and is now particularly pressing and urgent. This is due to the necessity of economically using natural raw materials, needed more than ever, as well as to ecological problems, i.e., necessity of minimizing side reaction products. This invokes the demand for new selective catalytic processes.

Alkanes are known to be chemically very inert, which is reflected in their other names, saturated hydrocarbons or paraffins (from the Latin *parum affinis,* i.e., devoid of affinity). This corresponds to high values of C–H and C–C bond dissociation energies and high ionization potentials (Table II.1). Saturation precludes addition reactions and seems to prevent protonation by ordinary acids as well as formation of complexes with various species including metal ions.

It looks natural, therefore, that the existing catalytic processes usually require difficult conditions and are not sufficiently selective. For example, the selective hydroxylation of a terminal methyl group in alkanes for the synthesis of primary alcohols, which are in great demand, is a difficult task because C–H bond dissociation energy exceeds that of CH_2 secondary groups and even more so of tertiary CH groups. Consequently, in usual chain oxidations radicals attack mainly tertiary and secondary C–H bonds, terminal methyl groups being the least reactive. It is particularly difficult to functionalize the simplest and most inert alkane, methane, in the presence of compounds containing weaker C–H bonds (e.g., in the molecules of a solvent). Another difficulty is to bring the reaction to a stop at intermediate oxidation products, such as alcohols, aldehydes, or ketones, which are as a rule more reactive than initial alkanes.,

The processes that lead to the functionalization of alkanes are usually oxidations. In living nature the problem of selective oxidation of alkanes, in particular with the cleavage of the strongest primary C–H bond and methane oxidation, was solved successfully with the use of special enzymes, therefore the biomimetic approach is in principle very promising. The cheapest oxidant, dioxygen, is utilized in living

TABLE II.1
Some Characteristics of Alkanes and Other Molecules

RH molecules	R	D(R–H), kcal/mol	IP, eV	EA, eV	PA, eV	pK_a
H_2	H	104	15.4	–3.6	4.4	25
CH_4	CH_3	104	12.7		5.3	40
C_2H_6	C_2H_5	98	11.5		5.6	42
C_3H_8	*n*-C_3H_7	97	11.1		6.1	
	iso-C_3H_7	94				44
C_4H_{10}	*tert*-C_4H_9		10.6			
C_6H_{12}	C_6H_{11}	94	9.9			45
C_6H_6	C_6H_5	109	9.2	–1.10	7.5	37
C_2H_2	C_2H	120	11.4			25
C_2H_4	C_2H_3	106	10.5	–1.81	6.9	36.5
CH_3OH	CH_3O	92	10.9		7.8	16
CH_3CN	CH_2CN	79	12.1		8.1	24
$CH_3CH{=}CH_2$	$CH_2{-}CH{=}CH_2$	86	9.7		7.9	35.5
$C_6H_5CH_3$	$C_6H_5CH_2$	85	8.8	–1.3	7.3	35
H_2O	OH	118	12.6	1.8	7.14	15.7

nature in aerobic oxidations and naturally it is also desirable to use it in model chemical systems.

II.2 GENERAL CONSIDERATIONS

II.2.1 ACTIVATION OF DIOXYGEN[1-3]

Dioxygen is known to be a weak oxidant and does not react with alkanes directly at ordinary temperatures. This is due, in particular, to the fact that in the ground state it is a triplet particle; its reactions with singlet organic molecules producing other singlet molecules are spin forbidden. Formation of two free radicals in the reaction:

$$RH + O_2 \rightarrow R^{\bullet} + HO_2^{\bullet} \quad (1)$$

is spin allowed, but it is usually highly endothermic and therefore very slow at ordinary temperatures. However, at elevated temperatures this reaction can initiate a chain process. The following classic scheme presents a typical mechanism of liquid-phase hydrocarbon oxidation for the early stages when the effect of reaction products may be neglected.

$$\left.\begin{aligned} &R^{\bullet} + O_2 \rightarrow RO_2^{\bullet} \quad (2)\\ &RO_2^{\bullet} + RH \rightarrow RO_2H + R^{\bullet} \quad (3)\end{aligned}\right\} \text{Chain propagation}$$

$$\left.\begin{aligned} &ROOH \rightarrow RO\cdot + \cdot OH \quad \text{or} \quad (4)\\ &2RO_2H \rightarrow RO\cdot + RO_2^{\bullet} + H_2O \quad (5)\end{aligned}\right\} \text{Chain branching}$$

$$\left.\begin{array}{l} R\cdot + R\cdot \rightarrow RR \\ RO_2^{\bullet} + R\cdot \rightarrow ROOR \\ RO_2^{\bullet} + RO_2^{\bullet} \rightarrow ROH + R'COR'' + O_2 \end{array}\right\} \text{Chain termination} \quad \begin{array}{r} (6) \\ (7) \\ (8) \end{array}$$

Metal complexes can accelerate the branching in the chain reaction by decomposing the peroxide:

$$M^{n+} + ROOH \rightarrow M^{(n+1)+} + OH^- + RO\cdot \quad (9)$$

Direct activation of dioxygen is also possible, O_2 forming complexes of different types with metal compounds that then can undergo various reactions (Scheme II.1). The oxygen-containing species can react with different substrates, the most reactive among these species being able to attack alkanes including methane.

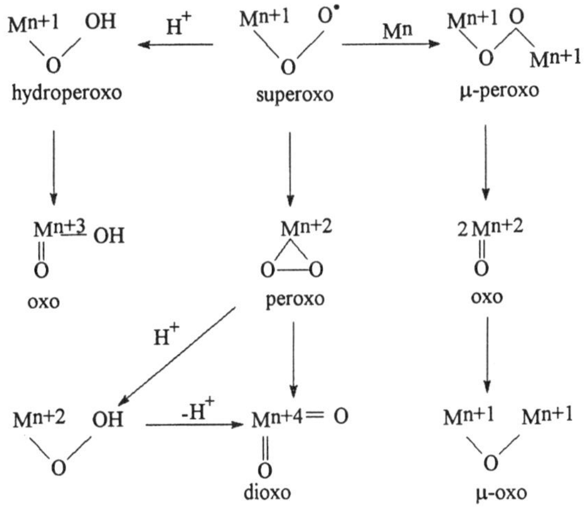

SCHEME II.1 Various dioxygen complexes and reactions with metal compounds.

This necessitates applying different catalysts and initiators. Transition metal compounds have been used traditionally for this purpose. Chain reactions are particularly widespread among alkane oxidations, active particles, such as atoms and free radicals, attacking alkane molecules. Coupled oxidation is also typical for alkanes, oxidation of comparatively chemically active material inducing the oxidation of inert hydrocarbons. These coupled reactions are often chain reactions as well and also involve formation of atoms or radicals easily reacting with alkanes. Thus, there is always a danger of nonselective radical attack on alkane molecules and the problem is to find the conditions for selective oxidation by certain active species. Another possibility for the participation of transition metal complexes is formation of high-valent metal compounds that can further directly react with alkanes. These reactions may have quite different selectivity and may not necessarily lead to formation of free radicals.

II.2.2 ACTIVE SPECIES REACTING WITH ALKANES AND THE MECHANISM

II.2.2.1 Nonmetal Species

II.2.2.1.1 Radical Reactions

Let us consider possible elementary reactions of hydrocarbon molecules. Among free radical reactions it is primarily the well-known radical abstraction of an H atom:

$$R\cdot + R'H \rightarrow RH + R'\cdot \tag{10}$$

In oxidation reactions oxygen-containing radicals react with alkane molecules:

$$RH + \cdot OH \rightarrow R\cdot + H_2O \tag{11}$$

$$RH + R'O\cdot \rightarrow R\cdot + R'OH \tag{12}$$

$$RH + R'OO\cdot \rightarrow R\cdot + R'OOH \tag{13}$$

To estimate the activation energy (E) as the function of the reaction heat (Q) one can use a simple equation (so-called Polanyi–Semenov rule):[4]

$$E = A - \alpha Q \tag{14}$$

Here A and α are constants, equal to approximately 11.5 and 0.25, respectively. Modern approaches give more complicated and more precise, although still approximate, formulae. For example, according to Denisov,[5] dependence of the activation energy of ΔH is expressed by the formula:

$$\Delta E_H = (\alpha'/1+\alpha')\cdot\Delta H_e + (\alpha'\Delta H_e/2br_e)^2 \tag{15}$$

Here α' is a coefficient that is constant for each pair of fixed bonds being cleaved and formed; br_e is a parameter also constant for a given series of reactions and not very different for different classes of reactions. It may be seen that the latter formula gives nonlinear dependence of activation energy on the reaction enthalpy whereas the simpler and less precise Polanyi–Semenov rule predicts a linear dependence: $\Delta E = \alpha\Delta\Delta H$.

Nevertheless, thermodynamics of these reactions remains an important basis of their kinetic characteristics. For example, strongly exothermic reactions 11 and 12 of OH and RO radicals proceed with low activation energies and are poorly selective as compared with less exothermic (or endothermic) reactions of type 13 with participation of radicals HO_2 and RO_2. Polar factors, while they should not be completely neglected, are not of great importance in reactions of neutral free radicals. If we have a series of substrates reacting via C–H homolytic cleavage we can expect that as a rule the stronger the C–H bond, the smaller the rate constant. Therefore, generally alkanes will react more slowly than other organic compounds formed as intermediates and there is little hope to obtain a high yield of an intermediate, particularly at high degrees of conversion.

II.2.2.1.2 Insertion of Singlet Carbenes and Oxygen Atom

In the cases of oxygen atoms, carbenes, and other similar particles, an insertion into the C–H bond is known, particularly for the singlet state:

$$CH_2(s) + CH_4 \rightarrow C_2H_6 \tag{16}$$

$$O(s) + CH_4 \rightarrow CH_3OH \tag{17}$$

whereas the same species in the triplet state often react with alkanes similarly to free radicals,[6a,b] because insertion is a spin-forbidden reaction for a triplet to form a singlet product.

Investigation of the singlet oxygen $O(^1D_2)$ reaction with alkanes in the gas phase has revealed[6c] that two distinct channels exist for these reactions. They were interpreted as an insertion into the C–H bond and the H atom abstraction. It is of interest that the insertion component dominates for small molecules (CH_4 and C_2H_6) with strong C–H bonds while the abstraction component is the dominant mechanism for production of OH from larger hydrocarbons (C_3H_8 and $C(CH_3)_4$) with strong steric hindrances and (or) weaker C–H bonds. The fraction of the abstraction component varies from ~0.05 for CH_4 to ~0.9 for $C(CH_3)_4$. It may also be significant that from the results obtained it follows that the abstraction is likely to result from a crossing from one potential energy surface to another one in the single entrance channel of the insertion path. Lifetimes for collision complexes vary from 0.35×10^{-12} s^{-1} for H_2 and 0.8×10^{-12} s^{-1} for CH_4 to 4.2×10^{-9} s^{-1} for $C(CH_3)_4$. The insertion mechanism for interaction of singlet oxygen with methane was confirmed in later experimental works and *ab initio* theoretical calculation.[6d] The minimum-energy path of the reaction was found to be an almost collinear approach of $O(^1D)$ to one of the hydrogen atoms of methane up to $r_{OH} = 1.66$ Å, and then the oxygen atom migrates off-axis. We shall see later that similar conclusions with respect to the mechanism (possibility of insertion as well as hydrogen atom abstraction) may be made for the nature of the reaction of activated oxygen ("oxene" bound to a metal) with alkanes in the enzymatic and model chemical systems.

II.2.2.1.3 Reactions with Acids and in Superacid Solutions

Alkanes are extremely weak bases and weak nucleophiles in reactions with usual acids and electrophiles under ordinary conditions. However, at elevated temperatures (about 100°C) industrially important reactions proceed in the presence of acid catalysts, such as isomerization, cracking, and alkylation of paraffins. Thus, protonation of alkanes leading to their activation is definitely possible. This is the consequence of their considerable proton affinity (see Table II.1), indicating that *five*

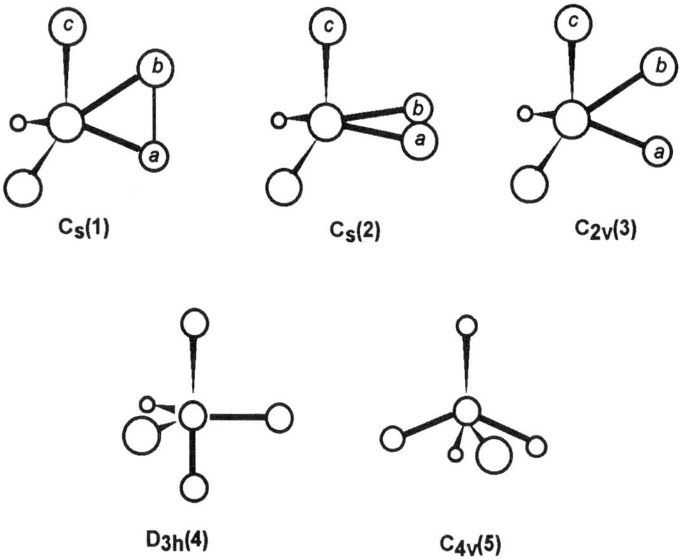

FIGURE II.1 Possible structures of CH_5^+. (Adapted from Schreiner, P. R. et al., *J. Chem. Phys.*, 99, 3716, 1993; Scuseria, G. E., *Nature*, 366, 512, 1993.)

coordinate carbon-protonated hydrocarbons, i.e., carbonium ions of the formula $C_nH_{2n+3}^+$, are sufficiently stable in the absence of stronger bases. The simplest ion, CH_5^+, was initially observed in the gas phase in the acid–base reaction, protonation of methane by CH_4^+ ion:[6e]

$$CH_4 + CH_4^+ \rightarrow CH_3^{\bullet} + CH_5^+ \tag{18}$$

As was discovered by Olah, very strong acids, so-called superacids, HF–SbF_5 or HSO_3F–SbF_5 ("Magic Acid") protonate alkanes in solution already at room and lower temperatures forming carbonium ions.[7a] As the result of the protonation, different reactions of alkanes may proceed: H–D exchange, isomerization, etc.; e.g.,

$$CH_4 + DF - SbF_5 \rightarrow \left[H_3C - < \begin{matrix} H \\ D \end{matrix} \right]^+ \rightarrow CH_3D + HF - SbF_5 \tag{19}$$

Theoretical investigations on the basis of the nonempirical calculations of the simplest carbonium ion, CH_5^+, produced upon protonation of methane, show[7b] that the ground state of this ion has C_s symmetry, $C_s(1)$, (see Figure II.1) which may be regarded as a complex of CH_3^+ cation with H_2 molecule (H–H distance is found to be 0.936 Å, which suggests formation of a bond).

The ground state is only insignificantly different (about 1 kcal/mol) with respect to its energy from the excited state of C_{4v} symmetry. Even closer are the transition states of $C_s(2)$ and $C_{2v}(3)$ symmetries (0.1 and 0.9 kcal/mol, respectively), which

TABLE II.2
Enthalpies of Dehydrogenation (ΔH_r) of Simple Carbonium Ions (RH_2^+) in the Gas Phase

Alkane (RH)	PA, eV	ΔH_r, kcal/mol
CH_4	5.3	36.2
C_2H_6	5.63	4.6
C_3H_8	6.12	9.2

correspond to hydrogen scrambling in CH_5^+ with extremely low activation barriers. Highest in energy is the D_{3h} isomer. Formation of dihydrogen from protonated hydrocarbons requires much less energy than for formation from the nonprotonated molecules (Table II.2).

For example, for CH_5^+ the enthalpy (ΔH_r) of the gas reaction

$$CH_5^+ \rightarrow CH_3^+ + H_2 \qquad (20)$$

is only 36.2 kcal/mol whereas for methane the decomposition to CH_2 and H_2 requires 109.2 kcal/mol. Dehydrogenation of protonated alkanes in polar solutions is probably even less endothermic than in the gas phase because carbonium ions, such as CH_3^+, $C_2H_5^+$, and $C_3H_7^+$, have to be more strongly solvated than the corresponding carbonium ions, CH_5^+, $C_2H_7^+$, and $C_3H_9^+$. In agreement with these considerations, H_2 is often evolved under the action of superacids on alkanes, particularly in the case of stable carbenium ions. In the case of apparently the most stable cycloalkyl cation yet obtained, the hydrido-bridged *in*-bicyclo-[4.4.4]-1-tetradecyl cation (see Figure II.2) dihydrogen is slowly evolved already under the action of glacial acetic acid on the parent hydrocarbon, and dihydrogen evolution is fast when a strong acid such as CF_3SO_3H is used. H_2 is evolved almost quantitatively under the action of CF_3SO_3H at 0°C, activation energy for this process being only 13.5 kcal/mol.[7c]

It should be mentioned that dihydrogen evolution from carbonium ion is actually an *oxidation* of an alkane molecule (it may be regarded as the abstraction of H^- by H^+), e.g., the oxidation state of CH_3^+ formed from protonated methane corresponds to that of methanol. Thus, protonation induces an oxidation of alkanes. Naturally, with the high rate of hydrogen scrambling in carbonium ions fast racemization of optically active hydrocarbons (e.g., isotopically labeled ethane, CH_3*CHDT) should be expected upon protonation.

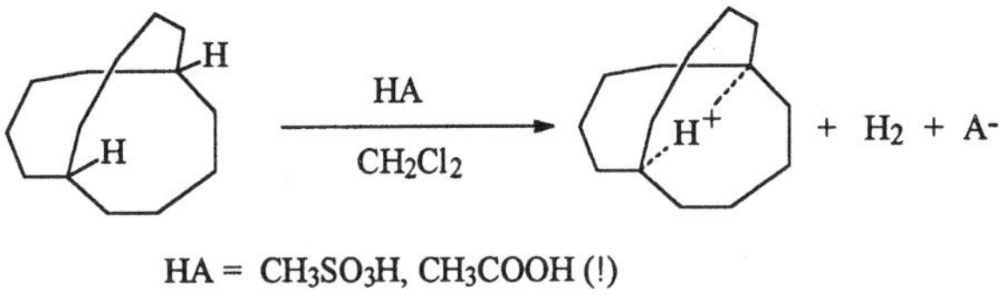

FIGURE II.2 Dihdyrogen evolution from bicyclo-[4.4.4.]-1-tetradecyl cation.

Other electrophiles are able to attack alkanes in superacid media via electrophilic addition to C–H bonds. It is particularly relevant that O atom insertion was observed by Olah et al.[7d] in the reaction of hydrogen peroxide in Magic Acid with methane above 0°C to produce methanol with very high (>95%) selectivity. ^+OH, which may be considered a protonated singlet oxygen atom, is the active species in the reaction. Methyl alcohol formed is simultaneously protonated to methyloxonium ion ($CH_3OH_2^+$) and this prevents further oxidation. Superacid-catalyzed oxygenation of methane with ozone affords predominantly formaldehyde.[7a]

II.2.2.1.4 Reactions with Other Electrophiles

Although acids and electrophiles of the strength of superacids cannot exist in biological systems it could be expected that other electrophiles forming complexes with alkanes will induce their various reactions possibly with low activation barriers.

Radical cations combine the properties of radicals and electrophiles. This leads, in particular, to strong influence of polar factors in reacting molecules and polarity of the medium in their reactions with C–H bonds. For example, aminium cation radicals, $R_3N^{\dot{+}}$, which can be produced in reactions of amino-N-oxides with iron(II) salts,[8]

$$R_3NO + 2H^+ + Fe^{II} \rightarrow R_3N^{\dot{+}} + Fe^{III} + H_2O \qquad (21)$$

react with C–H bonds in alkanes (yielding free radicals) but do not react with the molecules containing electron-accepting groups, such as CH_2OH, OR, or CO_2H. Alcohols produced at the hydroxylation with participation of aminium radicals are not further oxidized. If a molecule with a long hydrocarbon chain contains an electronegative group, the reaction will take place between the aminium radicals and the C–H bond most distant from this group. For instance, 1-octanol gets chlorinated (in the presence of N-chloramine) or hydroxylated (in the presence of Fe^{III} and water) mainly into position 7, with 92 and 70% selectivity, respectively. Cyclohexanol is chlorinated at the C-4 atom, cyclodecanone is attacked preferentially at positions C-6 and C-7, and cyclopentadecanone at the C-8 atom. These features are characteristic of biological oxidation. Proceeding from these results the authors of the work[8] consider the reactions between aminium radicals and C–H bond to be analogous to enzymatic oxidation.

Even more interesting from the viewpoint of comparison with biological oxidation is the reaction of cation radicals of aromatic amino-oxides[9a] capable to transfer O^+ to the acceptor molecule B:

$$\geqslant NO^+ + B \rightarrow \geqslant N + BO^+ \qquad (22)$$

In the case of hydrocarbons they can be oxidized. Thus, pyridine N-oxide oxidized at the anode to yield the cation radical ($C_5H_5NO^{\dot{+}}$) reacts with cyclohexane to ultimately produce cyclohexanol and cyclohexanone.

Cation radicals $C_5H_5NO^{\dot{+}}$ were produced in reaction of pyridine N-oxide with photoexcited chloranyl (tetrachloro-*p*-benzoquinone).[9b] Laser flash photolysis was used to follow the kinetics of excited-state decay. Spectral investigation of the cation radicals in the presence of toluene, cyclohexane, and methanol demonstrated formation of *molecular complexes,* whereas in the case of benzene no interaction was detected. Thus, the complexes are formed at the expense of C–H bonds rather than more nucleophilic aromatic rings. The weaker the C–H bond, the more stable the complex formed. It is of importance that the complexes were found to rapidly react with dioxygen, while the cation radicals without hydrocarbons are inert towards dioxygen. In the case of cyclohexane the products are cyclohexanone (the main product) and cyclohexanol together with pyridine. Hence, the following scheme may be suggested:

$$C_5H_5NO^{\dot{+}} + H_2C\langle \rightleftharpoons \left[C_5H_5NO\cdots H_2C\langle\right]^{\dot{+}} \tag{23}$$

$$\left[C_5H_5NO\cdots H_2C\langle\right]^{\dot{+}} + O_2 \rightarrow HO_2^{\bullet} + \left[C_5H_5NO-\overset{H}{C}\langle\right]^{+} \tag{24}$$

The subsequent reactions of alkoxyammonium salt produce pyridine and the oxidation products of cyclohexane. Formation of molecular complex with cyclohexane apparently presents an example of a *five-coordinate carbon intermediate* without, in this case, superacid media.

A similar mechanism was proposed for the reaction of the cation radical formed at the anode in electrochemical oxidation of phenazine di-N-oxide (OPO) with ethylbenzene.[9c] Alkoxyammonium salt was directly observed as an intermediate and the reaction investigated by voltammetry apparently proceeds according to the scheme shown in Figure II.3.

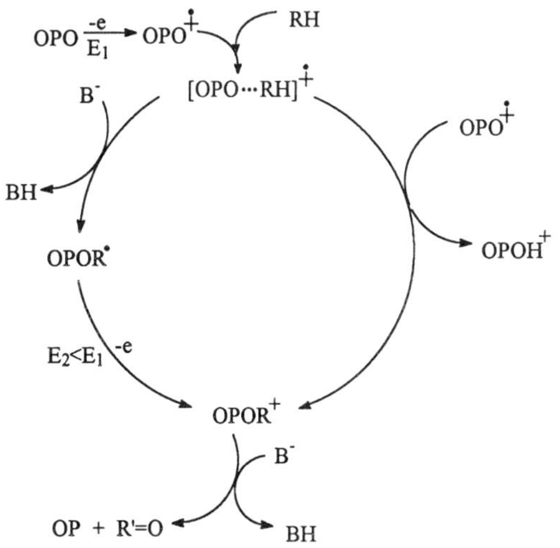

FIGURE II.3 Scheme of phenazine di-N-oxide (OPO) cation radical reaction with ethylbenzene.

Thus, in the total process there is an O^+ transfer from the cation radical, which is a three-electron oxidant, to the hydrocarbon molecule. Ketone is the product of a four-electron oxidation and is formed upon the additional one-electron oxidation of the radical intermediate at the anode followed by deprotonation of the alkoxyammonium salt produced. Another mechanism of ketone formation may be the hydrogen atom abstraction by dioxygen from the radical formed:

$$\geqslant N-O-\overset{|}{\underset{|}{C}}-H+O_2 \rightarrow HO_2^{\bullet}+\geqslant N+O{=}C\langle \qquad (25)$$

We shall see that an analogous mechanism may be suggested for the case of hydrocarbon oxidation with participation of metalloporphyrins (see Section II.3).

II.2.2.2 Metal-Containing Species

Biological oxidation of hydrocarbons involves metal-containing enzymes. Chemical models of these enzymes have to be also based naturally on metal complexes. These processes are treated in the following sections of this book. In this section only general problems of the reaction mechanism are briefly described.

The chemistry of alkane reactions with metal complexes is considerably richer than with nonmetallic particles, though alkanes usually remain more inert than most other organic materials. Dioxygen complexes and peroxo- and superoxo-complexes are sometimes active towards alkanes that contain not very strong C–H bonds, e.g., cyclohexane (see Sections II.4 and II.5), and they are often sufficiently active towards olefins. By increasing the metal electrophilic character one can strongly enforce the activity of oxygen atoms of peroxocomplexes to C–H bonds.

Various metal complexes can react with alkanes in different ways.[1-3] One may consider the following list (not comprehensive) of elementary reactions of metal complexes (M) with hydrocarbons (RH):[2,10a]

$$M^{n+} + RH \rightarrow M^{(n-1)+} + RH^{\dot{+}} \qquad (26)$$

$$M^{n+} + RH \xrightarrow{B} M^{(n-1)+} + R\cdot + BH^{\dot{+}} \qquad (27)$$

$$MO + RH \rightarrow [MOH + R\cdot] \rightarrow M + ROH \qquad (28)$$

$$MO + RH \rightarrow M-\overset{H}{O}R \rightarrow M + ROH \qquad (29)$$

$$MX + RH \rightarrow M-R + HX \qquad (30)$$

$$M{=}O + RH \rightarrow M\langle{}^{R}_{OH} \qquad (31)$$

TABLE II.3
Redox Potentials for the RH = RH† + e Processes in Aqueous Solution

RH	I(V)	$-\Delta G_s$ (kcal/mol)	E_0 (V)
CH_4	12.71	62	5.57
C_2H_6	11.5	54	4.68
C_3H_8	11.07	50	4.42
iso-C_4H_{10}	10.55	40	4.34
C_6H_6	9.25	56	2.45
$C_6H_5CH_3$	8.82	51	2.34

$$M + RH \rightarrow M\langle {R \atop H} \quad (32)$$

A simple one-electron transfer (reaction 26) has naturally the advantage of a long-distance process, provided it is thermodynamically allowed. This reaction is widespread and involves oxidations of many one-electron donors, including aromatic and alkylaromatic compounds. However, reaction 26 is mostly impossible for alkanes.

In Table II.3 redox potentials are presented calculated for water solutions. It may be seen that even for toluene the potential exceeds that of the oxidation of water (+1.229 V at pH 0) and for alkanes it exceeds the potential of formation of OH radical from water (+2.8 V). It is very difficult, if not impossible, to create such strong oxidating agents in aqueous solutions, and the potentials do not seem realistic for biological systems. Reaction 27 of one-electron oxidation of alkanes with *simultaneous* transfer of a proton to a base, e.g., water, is less difficult thermodynamically (see Table II.4). This is due to the fact that proton abstraction from the cation radical RH^+ is a thermodynamically favorable process at least for aqueous solutions, the difference between solvation heats of H^+ and RH^+ being considerably higher than the C–H bond dissociation energy in alkanes. Redox potentials of reaction 27 for hydrocarbons may be calculated from the comparison of the two reactions:

$$\tfrac{1}{2}H_2 = H^+ + e \qquad E_1 = 0 \quad (33)$$

$$RH = R\cdot + H^+ + e \quad E_2 \quad (34)$$

$$E_2 = \Delta H_{RH} - \Delta H_R \quad (35)$$

The values of E_0 obtained in this way are given in Table II.4. It may be seen that the potentials are much lower than for the formation of cation radicals $RH^{\dagger}$ and

TABLE II.4
Redox Potentials for Reaction
$RH \rightleftharpoons RH\cdot + H^+ + e$
in Aqueous Solution

RH	R	E_0 (V)
CH_4	CH_3	2.25
C_2H_6	C_2H_5	1.99
C_3H_8	*n*-C_3H_7	1.99
	iso-C_3H_7	1.84
iso-C_4H_{10}	*tert*-C_4H_9	1.78
C_6H_6	C_6H_5	2.48
$C_6H_5CH_3$	$C_6C_5CH_2$	1.44
C_6H_{12}	C_6H_{11}	1.85

they are already acceptable as strong oxidants in aqueous solutions because they do not exceed the potential of the one-electron oxidation of water. It may be mentioned that the formation of a free radical can be even endothermic in a real process and, nevertheless, ensure a sufficient reaction rate, particularly at elevated temperatures. It must be taken into account that when formed the radical may be further involved in a subsequent fast reaction, in particular provoking a chain reaction, and activation energies for chain initiation on the order of 10 to 20 kcal/mol (0.5 to 1 eV) may ensure a sufficiently high reaction rate. However, the potentials are still somewhat too high for biological systems oxidizing such inert molecules as methane and its homologues.

If a molecule contains a basic atom (e.g., O or N) electron transfer in reaction 27 can be expected to be coupled with *intramolecular* proton transfer from the C–H bond to this basic atom to form a more stable cation radical. Thus for the cation radical of methanol, the form $[CH_2OH_2]^{+\cdot}$ is found to be significantly more stable (0.5 to 0.6 eV) than $[CH_3OH]^{+\cdot}$.[11]

Reaction 28 is similar to the abstraction of the H atom by free radicals, and from the thermodynamic viewpoint is analogous to reaction 27, the distinction being that the proton from the RH molecule is added not to H_2O, but to $[MO]^-$, which is therefore functioning as a base. The next step could be the recombination of the radical produced with the OH group formed at the metal center. Alcohol will then be the final product. This mechanism, known as the oxygen rebound mechanism, is now widely accepted for biological C–H hydroxylations and their chemical models (*vide infra*), but, as we shall see, it apparently cannot explain all the facts observed and at least cannot be accepted as universal.

It is clear that if the compound MOH formed is a stronger acid than H_3O^+ then reaction 27 is more thermodynamically favorable than reaction 28. Therefore, reaction 27 may have definite advantages as compared with reaction 28, particularly if the approach of an alkane molecule to MO is accompanied by steric hindrances and there are base molecules, e.g., water, in the vicinity of RH.

Reaction 29 of insertion of the oxygen atom bound to M into the C–H bond is of particular interest because it is thermodynamically considerably more favorable

TABLE II.5
Redox Potentials for the Reaction
$RH + H_2O = ROH + 2e + 2H^+$

RH	ROH	E_0 (V)
CH_4	CH_3OH	0.59
C_2H_6	C_2H_5OH	0.48
C_3H_8	*n*-C_3H_7OH	0.44
	iso-C_3H_7OH	0.37
iso-C_4H_{10}	*tert*-C_4H_9OH	0.31
C_6H_6	C_6H_5OH	0.20
C_6H_{12}	$C_6H_{11}OH$	0.30

than reactions 27 and 28, and avoids free radicals that are dangerous for living organisms and are often responsible for side reactions in purely chemical systems. The oxygen atom insertion from metal complexes with dioxygen into the C–H bond ("oxenoid" mechanism) was proposed by Hamilton to explain the results of biological and some chemical oxidations as early as in the 1960s (see Section II.5.1) and is still considered to be an important mechanism for hydroxylation of C–H bonds (e.g., see the review by Drago[3]). The redox potentials for the two-electron oxidation of alkanes in water are given in the Table II.5. The values are seen to correspond to comparatively weak oxidants.

The mechanism of one-step oxygen atom insertion, although thermodynamically most favorable, is often rejected on several grounds, namely: (1) high H/D kinetic isotope effects (KIE) observed in many hydroxylations; (2) isomerization of a molecule in the process of the hydroxylation; and (3) positive results with the use of radical traps. However, we shall see that other reasons, e.g., proton tunneling (*vide infra*), is often responsible for high KIE values; therefore, a high KIE value does not necessarily indicate a radical abstraction mechanism. At least some of the isomerizations can be due to hydrogen scrambling in intermediate complexes formed in high-valent metal compounds reactions with alkanes via the oxygen atom bound to the metal. We have seen that such five-coordinate carbon intermediates are present in superacid solutions where alkanes can be protonated as well as in the reactions of alkanes with certain cation radicals. In high-valent metal complexes the oxygen atom is not as strongly bound as in superacids. Often they can be stable in water, which is rather strong base. However, the oxygen atom bound to a high-valent metal possesses electron pairs and can afford "back donation" of electrons to the empty antibonding orbitals of hydrocarbons, thus stabilizing the bond. The oxygen atom in the metal coordination sphere can be thus similar to the free singlet oxygen and at the same time to the metal center directly reacting with the C–H bond. Oxygen atomic orbitals in metaloxo derivatives will be mixed with those of metal, acquiring partially metallic character. In both cases (free singlet oxygen $O(^1D_2)$ and coordinatively unsaturated metal complexes) a five-coordinate carbon mechanism was established for reactions with alkanes. Therefore, a *five-coordinate carbon hydroxylation mechanism* of nonactivated C–H bonds by coordinated oxygen may be suggested, and, as we shall see, this is less contradictory than both oxygen rebound and one-step oxygen insertion mechanisms.[10b] The oxygen atom in the coordination sphere of a high-valent metal may be considered as an analogue of electron-rich (yet electrophilic) metal centers, which are certainly able to coordinate alkanes (see Section II.6). The complexes of alkanes with the oxygen center on the high-valent metal can, in principle, transfer radicals to the radical traps, thus imitating free radicals, but even if actually free radicals are produced in solution they can be formed as the results of *parallel* reactions proceeding alongside oxygen atom insertion into C–H bonds.

Although the results are still rather indirect and somewhat controversial we shall see that the five-coordinate carbon hydroxylation mechanism seems to be very likely and important particularly for small molecules with strong C–H bonds, such as methane, both in biological oxidation of alkanes and its chemical models. For branched hydrocarbons with large steric hindrances the necessity for an alkane molecule to approach the oxidant particle very closely may be an obstacle for this mechanism, therefore, it corresponds to a high activation energy and free radical formation may prevail. Hydrogen atom abstraction to form a free radical may be also expected for hydrocarbons with low C–H bond energies. However, lower C–H bond dissociation energy itself will also increase the rate of the oxygen atom insertion and therefore it is necessary to consider both mechanisms.

There is much evidence now that oxygen activated on a metal in purely chemical systems often reacts with methane and its analogues by a mechanism that does not include H atom abstraction (see Section II.5).

Formation of five-coordinate carbon intermediates opens the possibility of another important mechanism that may be considered as a particular case of the oxygen atom addition. As we have seen in CH_5^+, some bonding of two H atoms takes place and this increases the probability of H_2 abstraction, the more so as it becomes thermodynamically very favorable. It might be expected that a similar mechanism of H_2 abstraction will be possible for the intermediate containing an oxygen atom as one of the ligands at the carbon atom (or an organic molecule as a ligand at the oxygen atom). Oxygen insertion into a C–H bond is more thermodynamically favorable at low temperatures because its enthalpy is more negative than that of oxygen atom addition followed by H_2 elimination. For example, formation of methanol is 20 kcal more exothermic than formation of formaldehyde and dihydrogen. However, at high temperatures oxygen addition accompanied by dihydrogen elimination will become more favorable because of its larger entropy. In the case of the substances containing an electronegative atom X, such as nitrogen, oxygen, or halogen, a molecule of HX may be eliminated and this reaction may be thermodynamically favorable already at low temperatures:

$$M{=}O + CH_3X \rightarrow \left[MO\cdots CH_3X\right] \rightarrow M + O{=}CH_2 + HX \qquad (36)$$

We shall see that this mechanism of *oxygen addition accompanied by HX elimination* may be suggested for important biological as well as for purely chemical reactions. When X is an electronegative atom or a group, nucleophilic properties of the oxygen atom bound to metal may become important and, therefore, the highest oxidation state will not necessarily be the most active towards an organic substrate. Increase of nucleophilic properties may in such cases enhance the activity of the coordinated oxygen.

C–H activation by direct reaction with metal in the coordination sphere of the complex (reactions 30 to 32) involves the formation of the metal–carbon bond (in an intermediate or final product). Until recently the existence of such reactions was hypothetical, but now reactions of this kind are well documented for many examples (see Section II.6). Naturally these reactions must be very sensitive to the metal–carbon bond energy. It might be necessary to replace other ligands in the coordination sphere of M and this could require high activation energy. Therefore, reaction 31 is not very likely in the case of high-valent complexes ("hard acids") when donor molecules, such as water or alcohols ("hard bases") are present. In a number of known cases in living nature the oxidation involves the formation of high-valent complexes and therefore the process does not seem to occur via an intermediate metal–carbon bond. Reaction 32, the oxidative addition of alkane, requires the presence of at least two available electrons on M, which are used to form M–C and M–H bonds. Thus formally it is an oxidation with respect to the metal complex and reduction with respect to the alkane molecule. However, in subsequent reactions with an oxidant the formation of $M{<}^{R}_{H}$ may lead to the oxidation of alkane. Reaction 30 with the intermediate formation of metal–carbon bond, although so far without direct evidence, could be suggested for the anaerobic oxidation of methane and other alkanes as well as their chemical analogues (see Sections II.3 and II.6). There is increasing evidence that interaction of alkanes with metal complexes according to reactions 30 to 32 involves intermediate complexes with five-coordinate carbons similar in many aspects to the complexes formed via oxygen atoms bound to metals. In particular, they are able to isomerize definitely utilizing nonradical mechanisms (see Section II.6).

II.2.3 TESTS FOR IDENTIFICATION OF REACTION MECHANISMS[12a]

Distinct from dinitrogen whose reactions in solutions had been virtually unknown until very recently, alkane oxidation is a common reaction with a number of possible mechanisms. As will be seen later, for biological hydroxylation of the C–H bonds in alkanes with enzyme monooxygenases the mechanism includes the formation of the high-valent Fe^{V} or Fe^{IV} oxocomplexes that react further with C–H bonds. This reaction may be the formation of a free alkyl radical (by abstraction of the H atom or by the coupled electron and proton transfer) or a concerted process of insertion of the oxygen atom and formation of an alcohol. The closest chemical analogue is the reaction of high-valent metal complexes, such as those of Cr^{VI} or Mn^{VII}, which may also react with alkanes to form primarily free radicals or insert oxygen atoms into C–H bonds. The radicals, if formed, can react further without leaving the solvent cage.

At the same time other reactions of alkanes are possible in solution with active free radicals, e.g., ·OH or RO·, which are formed from hydrogen peroxide or alkyl hydroperoxides in their reactions with metal ions or complexes.

Alkanes, particularly those with weak C–H bonds, can enter various other reactions, e.g., with metal peroxocomplexes, cation radicals, etc. Metal–carbon bonds may be formed in some chemical reactions with alkanes, as was already mentioned, and new reactions may be discovered in future.

Despite the differences in mechanisms the products may be similar or even identical. For enzymatic oxidation the typical reaction is hydroxylation yielding alcohols. The latter may be the main products of the reactions with quite different mechanisms, e.g., of radical chain oxidations as well as of concerted oxygen atom insertion. In many cases when the oxidation is considered from the viewpoint of analogy with a biological process it is often difficult to choose between different possible mechanisms and to decide whether the reaction under investigation is really an analogue of a biological process. Therefore, a number of tests have been developed that help to determine the mechanism.

II.2.3.1 Correlation of Structure and Reactivity

The dependence of the active species' reactivity towards C–H bonds on the nature of the ligands surrounding the metal ion and also on the polar substituents in a substrate may be expressed in the form of the Hammett equation:[12b]

$$\log k_s/k_o = \rho\sigma \quad (37)$$

or in more general form, the Taft equation:[12c]

$$\log k_s/k_o = \rho^*\sigma^* + \delta E_s + \Delta n\varphi + n\psi \quad (38)$$

which may help to determine the rate-controlling step in the oxidation of alkanes. For example, if we have in solution a metal complex possibly active towards alkanes and at the same time suspect that an active species like OH radical may be formed and react with alkane, we can discriminate between these two possibilities by introducing polar substituents in the ligands of the metal complex and comparing the activities towards chosen substrates. If OH radical is the active species the ratio of activities towards different substrates will remain unaffected, whereas if the metal complex itself attacks the substrates directly the ratio of activities will be sensitive to the polar substituents. The sign (positive or negative of ρ or ρ^*) and its absolute value may give some information about the transition state.

II.2.3.2 Product Selectivity

It may be useful to compare the hydroxylation of alkanes with that of arenes and with the epoxidation of olefins. Hydroxylation of arenes is often accompanied by a so-called NIH shift (the shift of a D atom from the carbon that is attached to the carbon in neighboring *o* position). Aldehydes are often formed in olefin epoxydation in parallel with the main route producing epoxides, while the epoxides themselves do not isomerize or only slowly isomerize to aldehydes. Both the NIH shift and the aldehyde formation are indicative of the action of electrophiles to produce an intermediate complex followed by a concerted O atom transfer. The intermediate may undergo partial isomerization with an H atom transfer to the neighboring carbon atom (see Section II.3).

If hydroxylation of alkanes is observed in the systems where epoxydation of olefins and hydroxylation of arenes takes place, a concerted mechanism of O atom transfer via five-coordinated carbon intermediate in alkane hydroxylation for a given electrophile could be expected. The isomerization of isotopically marked alkanes with partial hydrogen scrambling could be visualized, which corresponds to an H atom shift to another hydrogen, as happens in CH_5^+ (*vide supra*). Therefore, in such cases the isomerization does not necessarily indicate the intermediate radical formation. Often the opposite is the case. When the distribution of the products is known a kinetic analysis may distinguish between the two mechanisms: abstraction of hydrogen followed by rotation of a radical formed and addition of oxygen followed by isomerization due to movements of H and D atoms.

Epoxidation is a widespread, important reaction of olefins with high-valent metal complexes, as are their enzymatic oxidation and their reactions with other electrophilic oxygen donors, such as metal peroxocomplexes, peracids, etc. Olefins are much stronger bases (and much stronger nucleophiles) than alkanes (see Table II.1). There are several possible mechanisms of epoxidation, but a concerted electrophilic mechanism is well established for many epoxidations.[13] Active free radicals, such as ·OH or RO·, do not form epoxides but rather abstract hydrogen in the allylic position to the double bond, and this may lead to alcohol formation in further reactions of the radical produced. Hydroxylation of a C–H bond in the allylic position proceeds in the case of high-valent metal complexes as well. Thus, the latter may perform both kinds of reactions. Therefore, if allylic hydroxylation is the only reaction observed then free radicals are likely to participate in the process. If, on the contrary, epoxidation is the only reaction (there is no hydroxylation) then it may be concluded that electrophiles (peroxocomplexes or high-valent metal compounds) are the active species in the reaction, but they are not sufficiently active to hydroxylate C–H bonds. Isomerization observed in the reaction product is in certain cases an indication of radical intermediate formation, provided the lifetime of the latter is sufficient for the isomerization. Using different substrates with known rate constant of isomerization, the lifetime of the radical formed may be estimated before its recombination with an OH group on a metal complex in a solvent cage (so-called "radical clock" method).[14] For example, methylcyclopropane produces cyclopropyl radical, which may undergo fast isomerization to the butenyl radical, finally forming butene-3-ol. Thus, if butenol is formed together with nonrearranged methylcyclopropanol, the ratio the products allows one to estimate the lifetime of the radical before its recombination with the oxygen atom bound to iron, *provided that all the product molecules are formed via the radicals.* If however, the radicals are formed *in parallel* with the concerted oxygen atom insertion or there is another (besides free radical) mechanism of isomerization then of course no such estimate can be made. Moreover, it was already mentioned that not all kinds of isomerization are indicative of the free radical intermediate formation. In particular, platinum complexes can react with cyclopropane rings to produce four-nuclear metallocycles and the latter form platinum butene complexes, apparently via nonradical mechanisms. Five-coordinate intermediates may be also expected to isomerize, therefore special analysis with several tests is needed to make a final decision.

II.2.3.3 Regioselectivity

Active free radicals are poorly selective when reacting with alkanes. In the oxidation of isopentane with the participation of OH radicals the ratio of the products formed corresponds to the following ratio of the rate constants of the primary, secondary, and tertiary bonds at 25°C:

$$1°:2°:3° = 1:5:12$$

(corrected for the number of C–H bonds in the molecule). In the reactions of electrophilic high-valent complexes the activity of the secondary and particularly the tertiary C–H bonds relative to that of the primary one is much higher, e.g., for MnO_3^+ the same ratio 1°:2°:3° = 1:60:2100.[15]

However, steric factors may become very essential for alkane reactions with metal complexes. Primary terminal CH_3 groups (ω) and secondary CH_2 groups next to the terminal ones (ω–1) are most open to the attack of the complex.

To estimate the importance of the steric factors it is convenient to use hexane as a substrate. C–H bonds of the C-2 and C-3 atoms of the hexane molecule have almost equal bond dissociation energy (94.7 and 95.5 kcal/mol, respectively) but are essentially different with respect to the steric factors. Taft steric constants (E_s for the C-2 and C-3 atoms are –1.87 and –2.70 and, therefore, the ratio of hexan-2-ol and hexan-3-ol can be used as a function of the steric factors in hexane oxidation.

The selectivity may change under the influence of the surrounding, in particular protein surrounding in enzymes, and may lead to the situation when only terminal (ω) methyl groups are hydroxylated (as in the case of the ω-hydroxylase) and methane becomes the most active hydrocarbon (as in the case of the methane monooxygenase).

Therefore, to draw conclusions from the selectivity, it is necessary to analyze the complex taking into account the strength and nucleophilicity of the C–H bond, electrophilicity and redox potential of the metal complex, steric factors of the hydrocarbon, and sensitivity of the active species to these factors, as well as the volume and the shape of the cavity formed by the surrounding (e.g., protein molecules in enzymes).

II.2.3.4 Stereospecificity

Stereospecificity is an important factor in determining the reaction mechanism. Usually, free radical reactions with alkanes are not stereospecific, because the lifetime of the radicals produced is sufficient to isomerize before a subsequent bimolecular reaction. For instance, the degree of retention of configuration at the hydroxylation of tertiary C–H bonds is *cis*- and *trans*-1,2-dimethylcyclohexane may be used as a test for the distinction between H atom abstraction by a free radical (e.g., OH) and by a metal complex. According to one estimate[16] the rate constant of the free radical transition from the pyramidal to the flat state is about 10^9 s^{-1}. Therefore only when dioxygen concentration is very high (corresponding to an O_2 pressure of about 100 atm) may the radical reaction with dioxygen be faster than its unimolecular *cis–trans* isomerization.

II.2.3.5 Kinetic Isotope Effect (KIE)[17,18]

If the C–H bond cleavage is the rate-controlling step, the substitution of hydrogen by a heavier isotope (deuterium or tritium) results in a change in the reaction rate, and the value of the KIE (k_H/k_D) may give some information concerning the reaction mechanism. According to the classical treatment[17] KIE is determined by the zero-energy difference in the initial and transition states. The maximum isotope effect is reached in this case if vibrations in C–H and C–D bonds in the initial state are fully transformed into translation movement in the transition state and the activation energy difference is totally determined by the difference in the zero-energies in the initial state. At room temperature this maximum KIE $(k_H/k_D)_{25°C}$ reaches 6 to 8. Higher values of the KIE (>9 at 25°C) may signify a contribution from proton tunneling, which is much less pronounced for the deutron due to the large mass ratio.[18] Temperature dependence of the KIE can be determined from the Arrhenius rate constant dependence

$$k_H/k_D = A_H/A_D \cdot \exp\left(E_D - E_H\right) \tag{39}$$

In the absence of noticeable contribution from tunneling A_H/A_D is in the range of 0.7 to 1.2. The tunneling contribution increases $E_D - E_H$ and decreases A_H/A_D. The evidence for tunneling contribution is if $A_H/A_D < 0.6$ provided, of course, that preexponential factors are determined with sufficient accuracy. The polarity of the medium may be essential for tunneling. If a proton is transferred in a rate-determining step it becomes "heavier" in more polar media (because of solvation) and the mass ratio between the proton and its isotopes, D^+ and T^+, gets smaller. A smaller KIE value should be expected therefore in polar than in nonpolar media.

Another reason for high isotope effects for *polysubstituted isotope-containing molecules* (besides tunneling) may be participation of more than one (hydrogen) atom in the transition state of a particular reaction. For example, if a five-coordinate carbon intermediate is involved in the rate-limiting step of methane reaction then CD_4 may be a much slower reagent than CH_4 because all four hydrogen atoms are involved in formation of the transition state.

II.2.3.6 Influence of the Size of the Cycle[19,20]

Some conclusions concerning the mechanism can be made considering the rates ratio of the reaction of five- and six-membered cycles as well as of six- and seven-membered cycles. For homolytic and heterolytic cleavage of the C–H bond (and in more the general case, of the C–X bond) the reaction rate is, as a rule, slower for the six-membered cycle than for the five- and seven-membered cycles.[19] This may be explained by the change of configurational tension for the carbon atom in the transition from sp^3 hybridization in the initial state to the intermediate between sp^3 and sp^2 in the transition state. High configurational tension in C_5H_{10} decreases upon the abstraction of an H atom. Contrary to this, in C_6H_{12} the transition $sp^3 \rightarrow sp^2$ is more difficult because it requires a change in the valent angle and an increase of the repulsion between the C–H bonds. Therefore the rate constant ratio for C_5H_{10}

TABLE II.6
Differences in Activation Energies ($E_6 - E_5$), Ratio of Preexponential Factors (A_5/A_6) and 5/6 Factors for Different Reagents

NN	Reagent	$E_6 - E_5$ (kcal/mol)	A_5/A_6	$(5/6)_{25°C}$
1	(·OH)gas	−0.14	1	0.7
2	MnO_4^-–H_2O	+1.4	0.21	2.2
3	CrO_3–H_2SO_4	+1.5	0.11	1.4
4	Hg(II)–H_2SO_4	(+2.4)	(0.01)	0.5
5	NO_2^+–H_2SO_4	−1.4	3.0	0.29
6	Pd(II)–H_2SO_4	−1.3	1.6	0.18
7	SO_3H^+–H_2SO_4	−1.3	3	0.33

and C_6H_{12} (so-called 5/6 function) corrected for the number of the H atoms in a cycle is more than 1 for the C–H bond cleavage as the rate-determining step and the difference $E_6 - E_5$ of the activation energies for cyclohexane and cyclopentane has a positive value. For very active particles, e.g., OH radical, 5/6 is close to 1 (really somewhat smaller than 1) and $E_6 - E_5$ is close to zero. Therefore the 5/6 function can be used to distinguish between different active species (e.g., between relatively slow reagents, such as high-valent complexes, and highly active OH radicals). It is of importance that for other mechanisms, for example, electrophilic substitution with the attack on carbon atom as the rate-determining step, the tension in the five-membered cycle increases more than in a similar process with the six-membered cycle. Consequently, 5/6 is smaller than 1 and the difference $E_6 - E_5$ has a negative value. Some examples of this mechanism (the systems 4 to 7) are given in Table II.6.[20]

II.2.4 CONCLUSIONS

There are two principally different ways of activation of alkane molecules to further reactions. One is the traditional radical formation from an alkane, generally by abstraction of the H atom to form a free radical. This way (which may be regarded as the activation of alkanes via three-coordinate carbon intermediates) usually leads to nonselective radical chain reactions.

Another, so far much less investigated, way involves formation of a five-coordinate carbon intermediate or at least a transition state in the interaction of the active particle with an alkane molecule. This mechanism was established for the reactions of alkanes with superacids as well as with carbenes and singlet oxygen atoms.

In the case of metal complexes, apart from free radical reactions that are still possible, alkanes can react directly with metal centers via oxidative addition or electrophilic substitution forming metal–carbon bonds. High-valent metal complexes with an oxygen atom bound to a metal can insert this atom into a C–H bond in a concerted fashion (presumably via a five-coordinate carbon intermediate).

These latter mechanisms (involving five-coordinate carbon intermediates) are expected to be much more selective than radical mechanisms. They seem safer than and are thus preferable to free radical processes in biological systems and are more promising for creation of new selective catalytic processes involving alkanes.

The higher the C–H bond energy and the less pronounced the steric hindrances, the more probable is this concerted mechanism, which requires minimum energy and leads to the selective production of alcohols (as compared with hydrogen atom abstraction to form a free radical).

Various tests have been developed that may help to distinguish between different possible mechanisms.

II.3 BIOLOGICAL OXIDATION OF ALKANES

II.3.1 INTRODUCTION

Various oxidation reactions play a great role in all living organisms. Enzymes participating in these reactions often involve metal ions, those of iron, copper, cobalt, nickel, manganese, molybdenum, vanadium, and more rarely some others. Substrates subjected to oxidation include molecules with activated C–H bonds (by neighboring polar groups, double bonds in allylic position, etc.) and also nonactivated C–H bonds. Among the latter are alkanes including methane and its homologues.

The study of biological mechanisms has a long history.[21] As early as the beginning of the 20th century Bach expanded his ideas of the mechanism of chemical oxidation to cover the field of biological oxidation, assuming the formation of intermediate peroxide compounds with the participation of enzyme oxygenase.[22]

$$A + O_2 \xrightarrow{\text{oxygenase}} A\left\langle \begin{matrix} O \\ | \\ O \end{matrix} \right. \tag{40}$$

Peroxides further oxidize substrates in peroxide-catalyzed reactions

$$A\left\langle \begin{matrix} O \\ | \\ O \end{matrix} \right. + X \xrightarrow{\text{peroxidase}} AO + XO \tag{41}$$

Later Warburg put this idea into more concrete form, assuming that iron ions involved in the heme molecule take part in dioxygen activation.[23]

$$\text{Enz}-(\text{Fe}) + O_2 \rightarrow \text{Enz}-(\text{Fe})-O_2 \tag{42}$$

$$\text{Enz}-(\text{Fe})-O_2 + 2X \rightarrow \text{Enz}-(\text{Fe}) + 2XO \tag{43}$$

There is much evidence, however, that the participation of dioxygen in many cases of enzymatic oxidation is not necessary, and O_2 can be replaced by other oxidants, such as methylene blue. This led Wieland to develop a scheme of biological oxidation[24] where dioxygen did not participate directly, but where substrate molecule HX_2 transferred electrons to a corresponding acceptor A:

$$XH_2 + A \underset{\text{reduction}}{\overset{\text{oxidation}}{\rightleftharpoons}} X + AH_2 \tag{44}$$

This scheme was confirmed for many cases and became widely accepted. The enzymes that catalyze these reactions are called dehydrogenases. In the cases of enzymatic processes where O_2 is itself an electron acceptor, the enzymes are called oxidases. Later, however, enzymes were discovered that catalyzed the reactions, not only with dioxygen as electron acceptor, but where oxygen atoms from O_2 entered into the molecule produced from the substrate. Such enzymes are called oxygenases. At present several groups of oxidases and oxygenases are known.[21]

The first group of oxidases involves enzymes that catalyze electron transfer to dioxygen to give a superoxide anion:

$$XH + O_2 \rightarrow X + O_2^- + H^+ \tag{45}$$

The second group catalyzes a two-electron transfer to produce hydrogen peroxide:

$$2XH + O_2 \rightarrow 2X + H_2O_2 \tag{46}$$

The third group includes oxidases that catalyze four-electron transfer to dioxygen to form two water molecules:

$$4XH + O_2 \rightarrow 4X + 2H_2O \tag{47}$$

Cytochrome oxidase, ascorbate oxidase, and ceruloplasmin are example of these oxidases. In this case, several metal ions are present, as a rule, in the enzyme molecule.

Oxygenases are divided into two groups, depending on whether one or two atoms of dioxygen enter the product molecules. They are called, respectively, dioxygenases and monooxygenases. Monooxygenases are undoubtedly the most important enzymes for the hydroxylation of alkanes. The reaction proceeds as a coupled process, an electron donor (NADH or NADPH) taking part:

$$RH + 2e + 2H^+ + O_2 \rightarrow ROH + H_2O \tag{48}$$

We shall first consider heme-containing monooxygenases and then nonheme monooxygenases. Among dioxygenases there are intramolecular dioxygenases that catalyze reactions where both oxygen atoms enter the molecules of the product, and intermolecular dioxygenases that catalyze the insertion of two oxygen atoms in two different substrates. The latter can insert one of the O atoms in a nonactivated C–H bond. In this case the other substrate is practically always α-ketoglutarate, which is oxidized to succinate with carbon dioxide evolution. In all cases where a C–H bond is hydroxylated one or two iron atoms are involved in the active center of an enzyme,

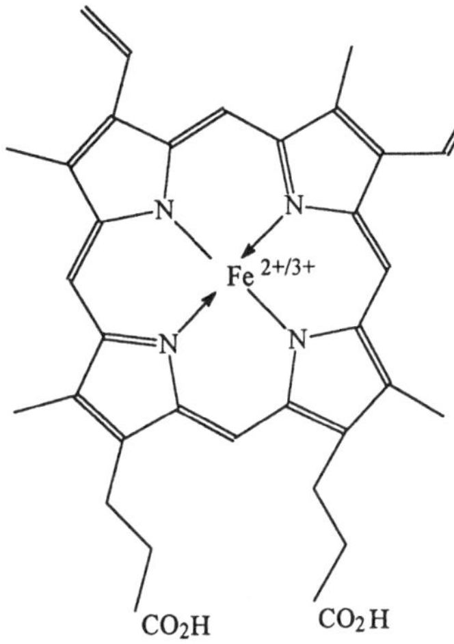

FIGURE II.4 Prostetic group of cytochrome P-450: heme (iron-protoporphyrin IX).

at least for those enzymes whose chemistry is sufficiently well documented.[25a] Some other enzymes are also considered for comparison.

II.3.2 HEME-CONTAINING ENZYMES

Heme proteins carry out a great number of various reactions including reversible dioxygen coordination (homoglobin and myoglobin), one-electron oxidation of organic substances (peroxidases), two-electron oxidations of organic substances, including alkanes with insertion of an oxygen atom (cytochrome P-450, secondary amine monooxygenase, and diamine oxidase), and halogenation (chloroperoxidase and thyroid peroxidase). In spite of their differences they have much in common and their reaction mechanism in alkane oxidation is better understood when similar enzymes are considered as well. Recently much work has been done in a number of laboratories and many principal features of mechanism has become clear.

Heme (Fe protoporphyrin IX, Figure II.4) is an iron-containing tetrapyrrol macrocycle; in proteins the iron atom is either five- or six-coordinated. The fifth position is occupied by a so-called proximal iron ligand. This ligand may be a sulfur ion from a cysteine molecule (cytochrome P-450 and chloroperoxidase), or a nitrogen atom of imidazol from a histidine molecule as in secondary amine monooxygenase, catalase, peroxidase, and oxygen carriers (hemoglobin or myoglobin), or an oxygen atom of phenoxide (from tyrosine). The sixth (distal) ligand may be exogenous or supplied by protein (dioxygen, water, peroxide, etc.).

The protein surrounding the heme is of great importance and determines the mechanism of oxidation to a considerable extent. In horseradish peroxidase organic substrates cannot reach the iron but interact with periphery of the heme, which results in one-electron transfer from substrates without interaction with an iron atom. In the case of cytochrome P-450, on the contrary, the approach of the substrate to heme periphery is precluded; however, a specific substrate-binding site near the heme atom

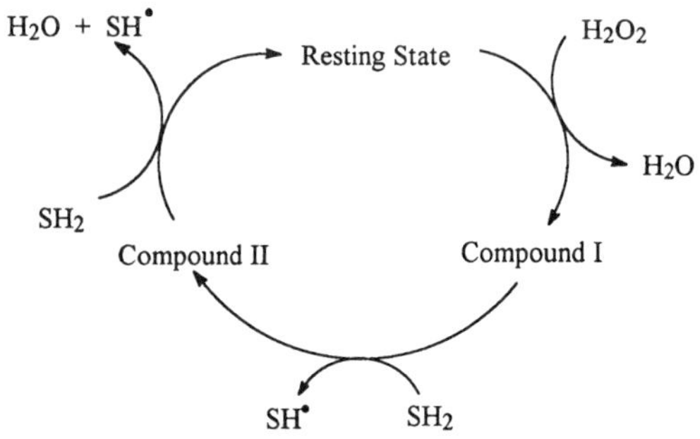

FIGURE II.5 Reaction cycle of peroxidases. (Adapted from Groh, S. E. and Nelson, M. J., in *Selective Hydrocarbon Activation*, VCH Publishers, New York, 1990.)

makes possible a direct interaction with the high-valent iron-oxo species leading to the hydroxylation.

II.3.2.1 Peroxidases[25a]

Peroxidases catalyze the oxidation of organic substances by peroxides, usually hydrogen peroxide. Horseradish peroxidase is particularly popular and has been thoroughly studied by many investigators. The mechanism was shown to involve three states of the enzyme (Figure II.5).

The substrate SH_2 forms free radical SH when reacting with compound I, which produces compound II. Compound II reacts with another SH_2 molecule, also producing SH. The resulting radicals SH react further to give final products.

Compound I is formed by two-electron oxidation of the Fe^{III} porphyrin and is formally Fe^{V}[porphyrin]. However, actually its structure corresponds to Fe^{IV}[porphyrin]† with one electron transferred from the porphyrin molecule. Spectroscopic results of ENDOR, EXAFS, and resonance Raman confirm this structure. Formation of compound I proceeds with the assistance of protonated imidazol of histidine from the distal side of the complex (Figure II.6).

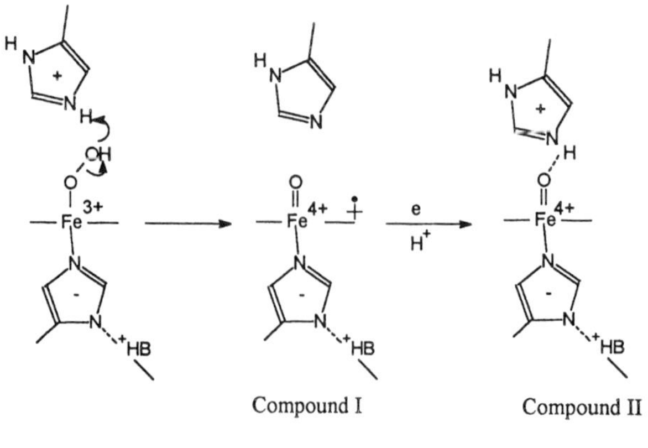

FIGURE II.6 Structures of peroxidase compounds I and II.

In peroxide-catalyzed hydroxylation oxygen is not transferred from compounds I or II. Instead, according to the results with $H_2{}^{18}O_2$, oxygen is originated from water. Presumably electron transfer from the substrate to the heme complex is accompanied by proton transfer of the substrate to water. The radical formed in this process then loses another electron and reacts with water to form hydroxylated product. Usually the substrates are one-electron donors with low redox potentials. Nevertheless, peroxide-catalyzed demethylation of dimethylaniline with one CH_3 group replaced by CD_3 proceeds with large deuterium isotope effect, $k_H/k_D = 8.6$ to 10.1, showing that the proton transfer is involved in a rate-determining step.[25b] One may think that the large H/D effect evidences for reaction 27, that is, simultaneous electron transfer from the substrate to the heme and intermolecular proton transfer to a base (presumably water) or intramolecular proton transfer to a basic nitrogen atom to form a cation radical $[C_6H_5NH(CH_3)(CH_2)]^{+\cdot}$ (see Section II.2.2).

II.3.2.2 Cytochrome P-450

II.3.2.2.1 General

The biological oxidation of various organic substances including alkanes has been studied particularly in detail for cytochrome P-450-containing monooxygenases (e.g., see References 21 and 25a,c,d).

These monooxygenases function in a variety of living organisms ranging from bacteria to mammals, including man.

Cytochrome P-450 is a hemoprotein, so named because of the abnormal position of the absorption maximum in the spectrum of the reduced complex with carbon monoxide. It has been found in tissues of various animals: in liver, lung, kidney, and cutis. Monooxygenases connected with cytochrome P-450 functioning in liver microsomes have been extensively studied. Their function involves the oxidation of various foreign compounds via oxygen atom insertion into products with subsequent removal of the product from an organism. The substrates include aliphatic and aromatic hydrocarbons, amines, alcohols, phenols, and thiophenols as well as steroids and fatty acids, which are harmful to living organisms. The reaction can proceed as hydroxylation, oxidative dealkylation, or deamination, formation of aminooxides and sulfoxides. Monooxygenases of aerobic bacteria can convert inert molecules by oxidation into more chemically reactive ones capable of undergoing further enzymatic reactions.

Thus, a number of monooxygenases are present simultaneously even in one organism and reveal different functions, possessing different substrate specificity.

Among P-450 monooxygenases there are those that are very selective. Thus, in the oxidation of alkanes and fatty acids catalyzed by P-450 in mammalian liver microsomes, the hydroxylation of a terminal (so-called ω) carbon atom occurs, i.e., the most stable C–H bond of a methyl group is attached. Besides the terminal C–H bond, the neighboring secondary C–H bond (ω-1) often also undergoes hydroxylation. This selectivity, which is very unusual from the chemical viewpoint, is definitely due to the steric effects because both terminal (ω) CH_3 and secondary (ω-1) CH_3 are most open in long-chain hydrocarbons and fatty acids. This is determined presumably by the protein surrounding (shape selectivity) because apart from P-450 monooxygenases with unusual selectivity there exist nonspecific monooxygenases, e.g., microsomes extracted from rat liver where the normal order of activity of the C–H bond hydroxylation is observed: 3° > 2° > 1°. Compared with model chemical systems (*vide infra*), somewhat lower selectivity is observed usually for the tertiary C–H bond and a somewhat higher selectivity (2 to 3 times) for the secondary C–H bond of the second (C-2) pentane carbon atom than of the third one (C-3). However, this difference in selectivity becomes apparent only with quantitative estimation, and in some modern and sophisticated models the artificial purely chemical systems can be even more sterically selective.

An important feature of cytochrome P-450, distinct from other heme-based enzymes, is the presence of a cysteine sulfur atom in the fifth (proximal) position of iron in heme. Thus cytochrome P-450 differs from molecular oxygen transfer agents (such as hemoglobin and myoglobin) as well as from peroxidases in which the proximal site of iron is occupied by a nitrogen atom of imidazol provided by histidine.

In resonance Raman spectra the O–O stretching frequency was found to be about 20 cm^{-1} lower than in hemoglobin.[26,27] The frequency found for ν(OO) in the $P\text{-}450_{cam}$ system (1140 cm^{-1})[28] is in agreement with the studies of thiolate heme model compounds. In particular, for the thyolato bound "picket fence" complex $Fe(T_{piv}PP)(C_6HF_4S^-)(O_2)$ (where $T_{piv}PP$ is *meso*-tetra(α,α,α,α-pivalamidophenyl)porphyrinato) the ν(OO) stretching vibration is strongly enhanced[29] and located at 1140 cm^{-1}. The position of ν(OO) is also similar in the nitrogen-based cobalt complexes (~1137 to 1145 cm^{-1}). This indicates that the quantity of electron density transferred to the $\pi_g^*(O_2)$ orbital due to the "extra" electron of cobalt(II) (as compared with iron(II)) is approximately the same as that transferred by the thyolate ligand in the iron complex.

All this may signify the weakening of the O–O bond in dioxygen in the presence of the fifth proximal ligand, and therefore provides an explanation for the importance of $-S^-$ ligand in O_2 activation by cytochrome P-450.

Cytochrome P-450-dependent monooxygenase contains, besides hydroxylase, a phospholipid that is essential for the catalytic activity of the enzyme because it is not active without the phospholipid, and the latter may be replaced by phosphatidyl choline whose addition to the other two components leads to the restoration of enzymatic activity. Thus, cytochrome P-450 is immersed in the phospholipid film. The efficiency of phosphatidylcholine depends on the nature of the derivatives of fatty acids involved in its structure. The highest activity is displayed by molecules with oleoyl residues and one or two lauroyl residues.

II.3.2.2.2 Mechanism of the Catalytic Cycle

The mechanism of P-450 catalytic action has been the subject of a large number of investigations by many authors, and is now certain in many aspects. It was studied in greatest detail for bacterial $P\text{-}450_{cam}$. The scheme of the most generally accepted mechanism is presented in Figure II.7.

In the resting state the enzyme involves six-coordinated Fe^{III} with low spin (S = ½). In the absence of a substrate the sixth ligand appears to be water or OH^-. The introduction of a substrate leads to the dissociation of the sixth ligand and the

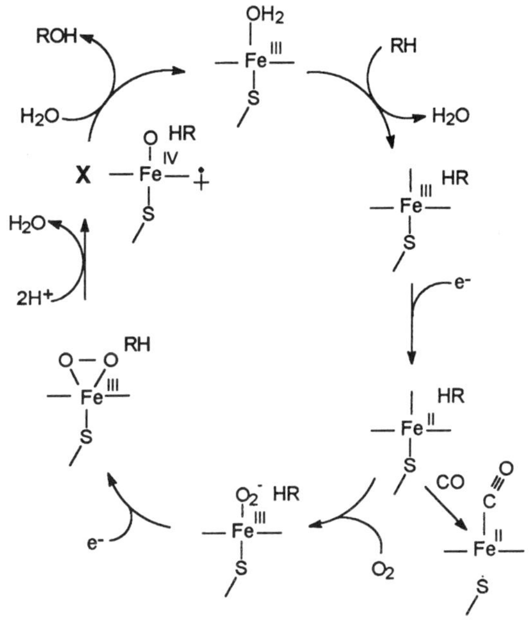

FIGURE II.7 Reaction cycle for cytochrome P-450.

formation of high-spin iron ($S = {}^5\!/_2$) and simultaneous shift of the iron redox potential to the less negative position. For P-450_{cam} the shift is from about –300 to –170 mV. It produces a triggering effect and leads to a reduction of PFe^{3+} to PFe^{2+} because the physiological reductant putidaredoxin (an iron–sulfur protein) has a redox potential of –196 mV and now becomes a stronger reductant than PFe^{2+}.

High-spin ($S = 2$) ferrous heme produced in the reduction coordinates the O_2 molecule, forming a metastable O_2 complex. Then the second electron is added to the complex from NADH and this event leads to the hydroxylation of a substrate and completion of the catalytic cycle.

Distinct from peroxidase, the active species formed in the catalytic cycle of P-450 have not been observed so far because of their high reactivity, but similarly to peroxidase one can expect the formation of products like compounds I and II in the reaction of $Fe^{III}O_2^{2-}$ complex with protons. The X-ray structure refinements were carried out for *Pseudomonas putida* CP450 both with[31a] and without[31b] bound substrate (camphor). Threonine residue was found to be close to the heme and may transform protons to the Fe^{III} peroxocomplex to form $Fe^{V}O$ (or, more likely, $P^{\dot{+}}Fe^{IV}O$) species as the product of the heterolytic O–O bond cleavage. Reduction of the complex by two electrons is a competitive process with a substrate hydroxylation and produces water and ferric enzyme. Therefore a complex similar to compound I of peroxidase (i.e., $P^{\dot{+}}Fe^{IV}O$) may be suggested. This suggestion is also supported by the possibility of hydroxylation without dioxygen and reducing agent by using iodosylbenzene and peroxides as oxygen transfer agents. It is not excluded, however, that for certain cases the active particle is $PFe^{IV}O$ (analogue of peroxocompound II) indicating that homolytic O–O cleavage may take place in PFeOOH in the catalytic process.

II.3.2.2.3 Hydroxylation Mechanism

Thus, the oxidation of a substrate, e.g., an alkane, is likely to be performed by the high-valent iron oxocomplex (ferryl complex). Formation of the alcohol molecule initially was often interpreted as a one-step concerted mechanism involving the insertion of the O atom into the C–H bond. There are some data indicating full retention of configuration in the process of hydroxylation. However, more recent data apparently contradict this suggestion. There is now much evidence of the existence of an intermediate that is usually interpreted as a tricoordinate carbon derivative, most likely a free radical. Isomerization is often observed in the reaction product, indicating that simple oxygen atom insertion cannot explain the experimental data. For example, Groves provided evidence that hydroxylation of *exo,exo,exo,exo*-2,3,5,6-tetradeuterionorbornane involved significant loss of the original stereochemistry. Part of deuterium (11%) turns out in the endo position (with an OH group in the exo position) as the result of the hydroxylation and part of hydrogen is found in exo position (5%) with an OH group in the endo position.[32a] Similarly, in the case of deuterium-substituted chiral phenyl ethanes, 23 to 40% of the hydroxylation events was shown[32b] to have resulted in alcohol with configuration opposite to that of the original hydrocarbon. The alcohols from singly substituted substrates were highly enriched in deuterium, indicating the operation of a large deuterium isotope effect on hydrogen removal. The k_H/k_D was found to be equal to 19 in the pro-S position and 4 in the R position of the substrate. At the same time, it was found that *pro-S*-hydrogen cleavage was favored 4:1 compared with pro-R cleavage. The results obtained seem somewhat controversial. For example, from the latter conclusion apparently it should follow that for both hydrogens in the CH_2 group mostly *S*-alcohol has to be formed. However, for both dihydrogen and for dideuterium phenylethane the ratio of *S*- and *R*-alcohols was found to be close to 1:1. Moreover, if pro-S cleavage *for hydrogen* is favored 4:1 and if at the same time the kinetic isotope effect is so different in the S and R positions then it is easy to see that the pro-S and pro-R positions must be equally active *for deuterium.* This difference in reactivities of the two isotopes is difficult to accept. Probably some of the results have to be reconsidered. Nevertheless, the main conclusions, i.e., large value of the isotope effect and the "crossover" events with isomerization of the molecules during hydroxylation, appear to be certain. In the case of d_2-substrate the relative yield of three minor products were reported to be increased. They are 2-phenylethanol, *o*-ethylphenol, and *p*-ethylphenol. The total relative yield of these products increased 18 times (from 0.2 to 3.6%) when both benzylic hydrogens were replaced with deuterium, confirming a large isotope effect. It should be mentioned that the formation of the phenols almost certainly proceeds as oxygen atom insertion (via an intermediate containing a FeO–C bond) because the carbon–hydrogen bond in arenes is too strong as compared with the benzylic one.

Another example of the evidence for a stepwise mechanism is the allylic hydroxylation of 3,3,6,6-tetradeuteriocyclohexene (I), methylcyclohexene (II), and β-pinene (III) with phenobarbital-induced liver microsomal cytochrome P-450

($P\text{-}450_{LM^2}$).[33] Epoxidation and allylic hydroxylation were the primary reactions of all the substrates. With I the major hydroxylation product (60 to 80%) was the result of hydroxylation of deuterated allylic site (70%) with significant allylic rearrangement (30%). Extensive rearrangement accompanied the hydroxylation of II and III. Similar results were obtained with iron porphyrin and chromium porphyrin model systems. The chromium porphyrin/iodosylvenzene model system (*vide infra*) showed complete scrambling of the allylic system of I and both model systems demonstrated rearrangement of II and III. Some authors[33] suggest a mechanism involving initial hydrogen atom abstraction from the allylic site and geminate, cage recombination of the incipient allylic free radical. This mechanism was already mentioned above and was termed "oxygen rebound mechanism" by Groves.[34] Now it is widely accepted for all the reactions of cytochrome P-450 with C–H bond-containing compounds. It may be thought that for weak C–H bonds free radical intermediates can indeed be formed in the hydroxylation and then can at least partially react in a fast recombination process with an OH substituent on iron. If the recombination is fast enough, faster than any isomerization of the radical, then stereospecific hydroxylation should be observed. In the opposite case of sufficient lifetime of the intermediate some isomerization is observed, indicating transient free radical formation. Hydroxylation of cyclic compounds was proposed to make an estimate of the lifetime of this intermediate. Formation of free radical from alkylcyclopropanes should be followed by a ring-opening process that leads to the formation of butenyl radicals. The relative amounts of hydroxylated cycloalkanes and butenyl products will depend on the rates of radical recombination and ring opening of the intermediates.[35] Ortis de Montellano and Stearns[36] have used this radical-clock method[37a] to probe the mechanism and kinetics of the hydroxylation process at the *in vivo* temperature of 37°C: methylcyclopropane, which would be hydroxylated via the cyclopropylmethyl radical, afforded only unrearranged alcohol, viz. cyclopropylmethanol. Cyclopropylmethyl radical opens with the rate constant $k_r = 1.2 \times 10^8\ M^{-1}s^{-1}$ at 37°C. Thus, lifetime for the radical before the recombination is definitely less than 10^{-8} s. Bicyclo[2.1.0]pentane produces both hydroxylated and ring-opened molecules in about 7:1 ratio. Special kinetic measurements[35] show that the rate constant for the rearrangement of the bicyclo[2.1.0]pentanyl radical is $k_r^o = 2.4 \times 10^9\ s^{-1}$, therefore, provided *free radicals are the only intermediates and the rearrangement mechanisms involves only free radicals,* the value for the oxygen rebound rate constant can be estimated as $k_{OH} \approx 7\ k_r^o \approx 2 \times 10^{10}\ s^{-1}$. Thus, the rebound step calculated in this way occurs more rapidly than many typical conformational and configurational changes. The reported cases of retention of configuration and stereochemistry and, in particular, the remarkable regioselectivity observed in the hydroxylation of bicyclo[2.1.0]pentane, which undergoes abstraction and rebound only at the endo face, seemed to be nicely explained by the oxygen rebound mechanism.

However, more recent results obtained with more sensitive radical probes using the same approach gave very different values for the oxygen rebound step, including also incredibly high values: $(2.3 \text{ to } 2.5) \times 10^{11}$, $(2 \text{ to } 7) \times 10^{12}$, and even $1.4 \times 10^{13}\ s^{-1}$![37b] The recombination of the radical and the oxygen atom on iron requires definitely much longer time than the time of one vibration; therefore these results evoke some doubts concerning the accepted scheme.

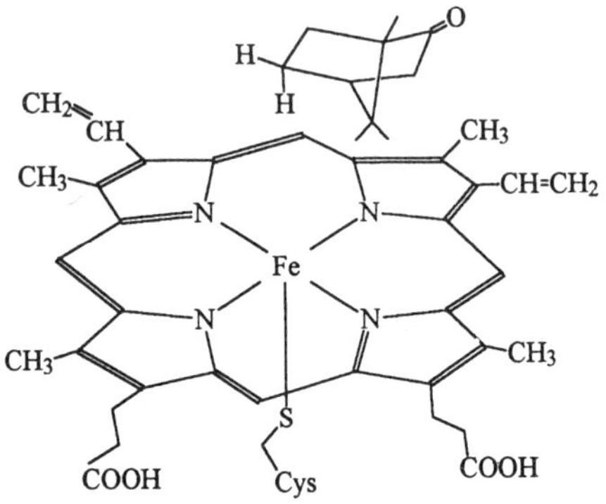

FIGURE II.8 Structure of the active site of the camphor-bound form of *Pseudomonas putida* CP-450. (Adapted from Armstrong, W.H., *Metal Clusters in Proteins,* ACS Symposium series, No. 372, 1988, chap. 1.)

We shall see in Section II.2.2 that similar results were obtained with nonheme methane monooxygenase; thus formation of free radicals does not seem to be the case in the biological oxidation of C–H bonds in hydrocarbons, at least in the most interesting cases of strong C–H bonds. (This is not to deny intermediate formation of free radicals in other biological oxidations; some examples of radical mechanisms are also given.)

Therefore, these data do not refute the possibility of the concerted oxygen atom insertion mechanism. Free radicals could be also produced *in parallel* with this concerted process, particularly in *in vitro* experiments and in this case the calculation of the rebound rate constant from the ratio of the isomerized and nonisomerized products is, of course, impossible.

How then to explain multiple examples of the isomerization definitely accompanying hydroxylation of C–H bonds?

It must be taken into account that the isomerization observed in the process of the reaction, while indicating that a *one-step* concerted mechanism does not work, does not contradict a *stepwise* mechanism with some intermediate, other than free radical, able to isomerize. Evidence obtained for the hydroxylation of camphor indicates that, on one hand, the simple oxygen rebound mechanism does not explain the observed isomerization, but, on the other hand, some intermediate must be formed. The results obtained with the hydroxylation of camphor catalyzed by P-450 are of particular interest because camphor is its own substrate.[38a] This enzymatic reaction yields selectively 5-*exo*-hydroxocamphor in accordance with the structure of the enzyme–substrate complex presented in Figure II.8.[38b]

When camphor contains deuterium only at the 5-exo or only at the 5-endo position the product in both cases contains some deuterium in both positions (Figure II.9), indicating some loss of stereochemistry at the carbon undergoing oxidation.

The results are explained by the authors[38] assuming intermediate free radical formation. If, however, iron-oxo species $Fe^{V}O$ both abstracts hydrogen and adds oxygen to the radical formed according to the oxygen rebound mechanism, only the 5-exo position would have been affected and hydrogen (or deuterium) at the endo

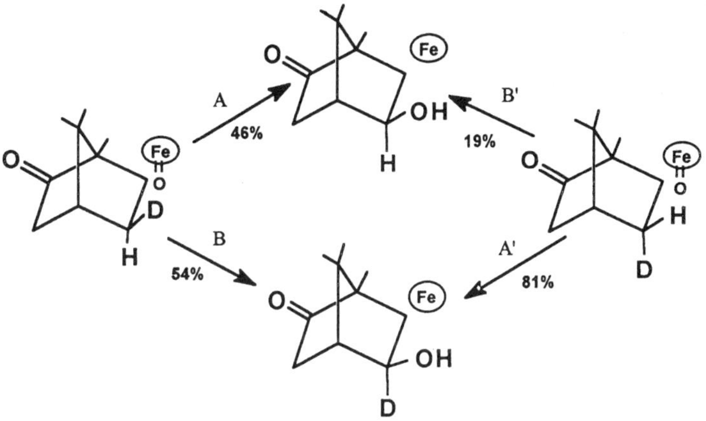

FIGURE II.9 Scheme for hydroxylation of d-camphor by P-450$_{cam}$. Paths B and B′ do not agree with oxygen rebound mechanism.

position would have remained untouched. Hence this result does not agree with the oxygen rebound mechanism.

Therefore, the authors[38] suggest a modified scheme assuming different identities for hydrogen-abstracting and oxygen-donating species. A radical center X· is postulated that can attack C–H in both exo or endo positions and then oxygen is added from the iron-oxo center from the exo position. The existence of the active radical center X· with high mobility to attack exo and endo positions of camphor would be very dangerous for the enzyme. Then it is clear from the distributions of H and D on exo and endo positions presented in Figure II.9 that X prefers to attack the C–H bond from the exo position, i.e., just from the side of the iron-oxo group. In principle the camphor free radical could be formed by an electron transfer mechanism coupled with proton abstraction from the C–H bond by a base, e.g., water molecule (see Section II.2.2). But in this case again the proton acceptor is likely to approach from the side opposite to that of the iron porphyrin complex and the results clearly contradict this suggestion. Besides, the electron transfer mechanism does not require the oxygen atom on the iron to be close to the substrate: direct attack of the oxygen atom on the C–H bond has to be assumed.

Another mechanism that seems to explain the obtained results, particularly taking into account the data on model systems for monooxygenases hydroxylation (*vide infra*), involves the hypothesis about an Fe–O–C intermediate with a five-coordinate carbon, the oxygen atom approach carbon only from the exo position.[10b] It is natural to expect that some isomerization will occur in the intermediate and this would explain partial loss of stereochemistry with preservation of only exo-selective hydroxylation in the product (see Section II.2.2).

The suggestion of this mechanism would bring hydroxylation of alkanes closer to the epoxidation of olefins and hydroxylation of aromatics. It was already mentioned that hydroxylation of aromatics is unlikely to involve free radicals. For hydroxylation of aromatic compounds, a so-called NIH shift is often observed of the H atom that is being hydroxylated to the neighboring carbon atom. The existence of this shift indicates the electrophilic attack of the aromatic ring by the high-valent iron-oxo complex to produce an Fe–O–C intermediate and is also present in non-enzymatic reactions of high-valent metal complexes with aromatic compounds.

Reaction of cytochrome P-450 and its chemical models with olefins produces epoxides, epoxidation proceeding with retention of the initial olefin configuration. This might be explained by a concerted mechanism, but the reaction clearly proceeds in more than one step. In some cases significant amounts of aldehydes (for terminal olefins) and ketones are formed together with epoxides, while little or no isomerization of specially taken epoxides is observed under the same conditions.[40a-c,41] This indicates that an intermediate is formed capable of inducing 1,2 migration of hydrogen in parallel with epoxide formation. The reaction is similar to the NIH shift mentioned above, in the case of hydroxylation of aromatic compounds.

Thus, the isomerization observed in the process of oxidation by cytochrome P-450 is not necessarily evidence for radical intermediates. For hydroxylation of aromatics and epoxidation of olefins the reaction almost certainly proceeds without free alkyl radicals and it is the Fe–O–C intermediate that undergoes the isomerization involving migration of hydrogen. Assuming similar (although of course less stable) intermediates for alkanes, we may suggest that at least some isomerizations could be explained without postulating free radicals.[10b] Depending on substrate molecules, the mechanism may be different, and even when a radical mechanism is confirmed for certain substrates it is premature to apply this mechanism for all the other cases. It is evident, for example, that for the substrates that do not suit the shape of the hydrophobic cavity of an enzyme the probability of the radical abstraction will be higher than for genuine substrates of the enzyme, particularly if the "unusual" substrate, as already mentioned, has a weak C–H bond and is sterically hindered. As we shall see later, the chemical mechanism in both enzymatic and chemical model systems can involve free radicals in solution, possibly with the radicals disappearing in the oxygen rebound step (actually, in the "cage" of the solvent), and in very important cases of strong C–H bonds the mechanism can proceed without free radicals, even short-lived ones.

The KIE for P-450 (k_H/k_D) vary from $k_H/k_D \approx 1$ to very high values. For example, the dealkylation of alkylamines and thioethers evince small isotope effects whereas larger values of isotope effects are observed for dealkylation of esters and ethers.

$$RCH_2OR' \rightarrow R\dot{C}HOR' \rightarrow RCH(OH)OR' \rightarrow RCHO + HOR' \qquad (49)$$

$$RCH_2NR'R'' \xrightarrow{-e} RCH_2\overset{+}{\dot{N}}R'R'' \xrightarrow{-H^+} R\dot{C}HNR'R'' \rightarrow RCH(OH)NR'R'' \qquad (50)$$

For thioethers and alkylamines the first step could be electron transfer yielding radical cation, and deprotonation of the latter is the fast subsequent reaction. For ethers and esters electron transfer is accompanied by proton transfer and may be similar to hydrogen atom abstraction.

Another possible mechanism is oxygen atom addition to carbon to form a five-coordinate carbon intermediate, followed by elimination of an amine or alcohol molecules that involves hydrogen atom transfer to nitrogen, oxygen, or sulfur. If the first step is rate determining no appreciable isotope effect will be observed. If the second step determines the rate (oxygen addition is reversible) the isotope effect might be considerable.

This mechanism could explain seemingly paradoxical results obtained for kinetic isotope effects for N-demethylation of N-methyl-N-trideuteriomethylaniline catalyzed by cytochromes P-450, on the one hand, and lactoperoxidase and horseradish peroxidase on the other.[25b] For peroxidase the ratio k_H/k_D was found to be large (8.6 to 10.1), whereas for two isosimes of P-450 low values of k_H/k_D were obtained (1.7 to 3.1). Similar results were obtained for oxidative N-demethylations catalyzed by secondary amine monooxygenase.[39] Meanwhile, for peroxidase, electron transfer seems to be the rate-limiting step and for cytochrome P-450, the cleavage of the C–H bond by an oxygen atom bound to iron is usually proposed to be the crucial step. Therefore, large KIEs should be expected for cytochrome P-450 and small KIEs seem to be likely for peroxidase. This apparent paradox may be explained by the mechanism in which the electron transfer is coupled with the transfer of proton (see Section II.2.2) and oxygen addition is the rate-limiting step in the cytochrome P-450 reaction with a rather small contribution from the cleavage of the C–H bond. Depending on the nature of the atom in the substrate molecule that forms a bond with a hydrogen atom, the KIE will be different. This hypothetical suggestion requires, naturally, careful consideration and experimental verification.

For hydrocarbons, particularly for alkanes, KIEs are often very large but also variable. Intermolecular KIE (H/D) is often absent, but it merely shows that in consecutive steps of dioxygen coordination, electron transfer and O–O bond splitting, the rate-determining step may be not the C–H cleavage. Intramolecular isotope H/D effect usually reaches high values (≥8). For example, for benzylic hydroxylation of 1,3-diphenylpropane-1,1-d_2 a value of $k_H/k_D = 11$ was obtained,[20] which is somewhat higher than the maximum classic isotope effect (see Section II.2.3) and probably indicates some proton tunneling. The evidence for proton tunneling is particularly clear for model systems and is considered later. High isotope effect (18 to 19) was already mentioned for monodeuteriophenylethane.

In summary, a high-valent intermediate similar to compound I of peroxidase ($P^{\dagger}Fe^{IV}O$) may be visualized as an intermediate in cytochrome P-450 action. Its reaction with a substrate may involve initially electron transfer (for strong electron donors) with subsequent deprotonation or H atom abstraction by $P^{\dagger}Fe^{IV}O$ to form free radicals (for substrates with weak C–H bonds), but in most important cases stepwise concerted O atom insertion (via a five-coordinate carbon intermediate) is likely, and in some cases O atom addition accompanied by elimination of another molecule (amine or alcohol) can be suggested.

II.3.2.3 Prostaglandin H Syntase[25a]

This enzyme is also called cyclooxygenase, involves a ferric heme that contains phenoxide from tyrosine as an axial ligand. The enzyme catalyzes dioxygenation of

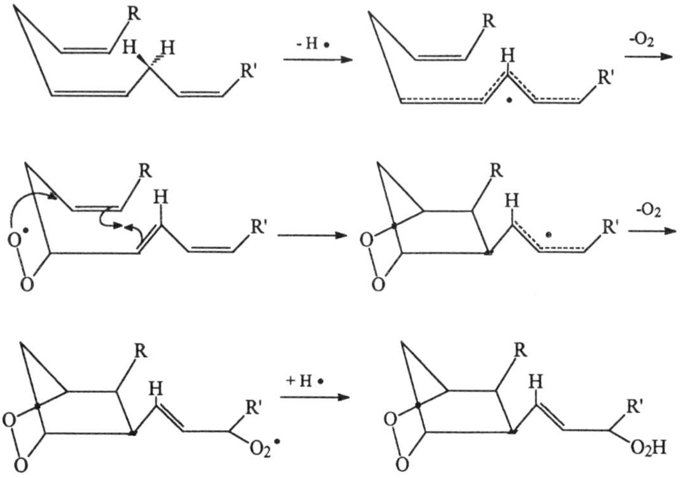

FIGURE II.10 Scheme for the mechanism of cyclooxygenase.

arachidonic acid to produce a cyclic fatty acid with a bridging peroxide and a hydroperoxide. Hydroperoxide is reduced to an alcohol in a peroxidase reaction.

The reaction of cyclooxygenase was studied by stopped-flow spectrophotometry and by EPR. Reaction of ferric enzyme with hydroperoxide (a substrate) results in the appearance of a visible spectrum similar to that of compound I of horseradish peroxidase. The intermediate with a lifetime of about 100 ms is transformed to a second intermediate with EPR spectrum similar to that of tyrosine radical in ribonucleotide reductase (*vide infra*). This is explained by the reaction of cyclooxygenase with hydroperoxides shown in Figure II.10.

The tyrosine radical initiates the reaction by abstracting an H atom from C-13 of arachidonic acid, forming a new allyl radical. The latter reacts with dioxygen to yield hydroperoxy radical, which evinces double cyclization, forming yet another allyl radical that reacts with O_2 and produces hydroperoxide. This mechanism shows how nature can use a peculiar free radical chain mechanism, similar in many aspects to the radical-chain mechanism of hydrocarbon oxidation in solution that keeps, however, the radicals inside the reaction center and therefore prevents damage to the surrounding proteins. At the same time the reaction mechanism is close to that catalyzed by ribonucleotide reductase (*vide infra*).

II.3.2.4 Chloroperoxidase[25a]

Chloroperoxidase catalyzes the chlorination of organic substrates by hydrogen peroxide and chloride. *In vivo* the CH_2 group of β-diketone is chlorinated, but olefins, acetylenes, and other substrates are also chlorinated. It can participate in reactions catalyzed by peroxidases and cytochrome P-450 too: N-dealkylation of alkylamines, epoxidation of olefins, etc. The mechanism is not very clear, particularly with regard to C–H bond cleavage. Similarly to cytochrome P-450, chloroperoxidase contains iron(III) porphyrin with an axial thyolate ligand, the evidence being obtained by such methods as EXAFS, MCD, and resonance Raman spectroscopies. Fe^{II} heme is not, however, involved in the catalytic reaction and from this point of view chloroperoxidase resembles other peroxidases. The reaction of Fe^{III} heme in peroxidase with H_2O_2 produces a compound I species similar to that proposed for horseradish peroxidase. $Fe^{IV}O$ porphyrin π-cation radical may react in various ways. With hydrogen peroxide it produces dioxygen and water as in a catalase reaction. It may also oxidize other organic substrates and epoxidize olefins similarly to cytochrome P-450. Reaction of compound I with chloride leads to an uncharacterized intermediate compound X capable of halogenating organic substrates. It may be thought that the function of the enzyme is to oxidize Cl^- to OCl^-. Indeed, OCl^- can be used to chlorinate substrates instead of H_2O_2 and Cl^-. However, OCl^- itself is not the active species because substrate-independent hypochlorite production is considerably slower than enzyme-catalyzed chlorination. Enzyme-bound hypochlorite, probably Fe^{III}–OCl may be suggested as an intermediate. Chloroperoxidase catalyzes the addition of HOCl across π-bonds. Therefore it may be suggested that the chlorination of β-diketones occurs at the enol tautomer.

II.3.2.5 Bleomicine[42-44]

Bleomicine is a natural antibiotic and is similar in its functions to heme-containing oxidases, oxidatively destroying DNA.

$$Fe^{II}Blm + O_2 \rightarrow O_2 - Fe^{II}Blm \xrightarrow{e} FeOOH \qquad (51)$$

$$FeOOH + H^+ \rightarrow Fe^{V}{=}O + H_2O \qquad (52)$$

When hydrogen peroxide is used as an oxidant there is evidence of Fe=O intermediate, whereas in reactions with O_2 hydroxyl radicals were reported to be formed. Hydrogen peroxide or iodosyl benzene epoxidize olefins in the presence of $Fe^{III}Blm$. Similarly to iron porphyrin complexes, cis-stilbene produces cis-epoxide.

The structure of bleomicine is known. There is an iron complex in the active center with pyridine and imidazol from aminoacids functioning as equatorial ligands. The fifth axial ligand is an aliphatic amine, an O ligand being able to replace the latter at different pH. A hydrophobic cavity is close to the active center that is formed by disaccharide. $Fe^{II}OBlm$ forms complexes with O_2 and CO. From the viewpoint of the chemical reactivity bleomicine is similar to cytochrome P-450. The catalytic cycle of bleomicine functioning is also similar to those of heme oxygenases and oxidases.

II.3.3 NONHEME MONOOXYGENASES

II.3.3.1 ω-Hydroxylase

Among nonheme monooxygenases, ω-hydroxylases are of particular interest because they selectively hydroxylate terminal methyl groups of *n*-alkyl chains of various compounds, including alkanes. They contain one iron atom in their active center. Similarly to other monooxygenases, ω-hydroxylase is able to epoxidize terminal olefins, produce sulfoxides from sulfides, and demethylate ethers. All these reactions are stereospecific.[45] Thus, hydroxylation of 1-^{2}H,^{3}H-*n*-octane by ω-hydroxylase in *Pseudomonas oleovorans* effects mostly achiral methyl groups (70 to 80%) and about 20% chiral methyl groups. (1R)-*n*-octane produces 38% (1R)-octanol and 62% (1S)-octanol, whereas (1S)-octane gives 82% (1R)-octanol and 18% (1S)-octanol, i.e., the reaction proceeds with 25 to 65% retention of configuration at the asymmetric carbon atom.[46] The substrate specificity of ω-hydroxylase was studied for the enzyme isolated from *Pseudomonas aeroginosa.*[47] The enzyme catalyzes the oxidation of *n*-alkanes from C_5 to C_{10} to linear acids with intermediate formation of *n*-alcohols. The activity decreases for branched alkanes. Methyl groups in 1,4-dimethylcyclohexanes are oxidized only for methyl substituents in the diequatorial position, i.e., only in compounds with flat conformation. The same effect is observed for oxidation of alkyl-substituted benzenes. The authors[47] make a conclusion that a hydrophobic cavity of the size 5.8 Å situated close to the reaction center is responsible for the shape selectivity observed. The presence of such a cavity is a general phenomenon for different monooxygenases.

Enzymatic epoxidation by the ω-hydroxylase of *P. oleovorans* of both *cis*- and *trans*-1-deutero-1-octene proceeds with about 70% inversion of olefinic configuration.[48] The system also produces aldehydes from olefins with deuterium migration from C-1 to C-2. The results provide support for a two-step mechanism involving enzyme-generated species with "cationic and/or radical character"[49] probably similar to those generated by active centers in cytochrome P-450 systems. The results obtained by the radical-clock method for the oxidation of *trans*-2-phenyl-1-methylcyclopropane agree with this conclusion.[47] 1-Phenyl-3-buten-1-ol was the only product, suggesting that the hydroxylation proceeds through a nonconcerted radical process. In this case the rate constant of the isomerization of the free radical formed is particularly high ($k = 2 \times 10^{11}$ s^{-1} at 30°C) and the rate constant for the rebound is estimated to be less than 4×10^9 s^{-1}. The KIE observed for the monodeuteromethyl derivative of the substrate was found to be 7.8 and this is the basis for the suggested mechanism. However, in view of the results obtained by the radical-clock method with cytochrome P-450 (*vide supra*) these results must be treated with a certain care. In particular, it must be taken into account that the isomerization may be in principle also the result of another stepwise mechanism with the addition of the oxygen atom bound to iron to one of the carbon atoms of the substrate followed by the cycle opening. More experimental evidence is needed to make a final decision.

II.3.3.2 Ketoglutarate-Dependent Dioxygenase

α-Ketoglutarate-dependent dioxygenase catalyzes various reactions in biological oxidation, in particular, the hydroxylation of chemically nonactivated C–H bonds.[50] The reaction with the substrate proceeds via the following general scheme:

$$\begin{aligned} &O_2 + X + HO_2C-CH_2-CH_2-CO-CO_2H \xrightarrow[(Fe^{2+})]{} XO + HO_2C-CH_2 \\ &-CH_2-CO_2H + CO_2 \end{aligned} \quad (53)$$

TABLE II.7
Examples of α-Ketoglutarate-Coupled Dioxygenase Hydroxylation

Enzyme	X	XO
Prolyl hdyroxylase	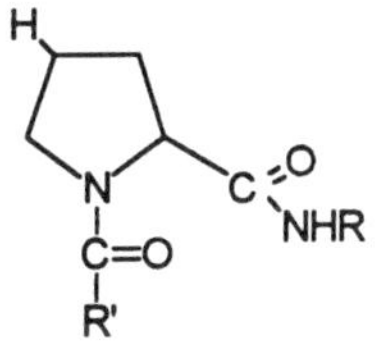	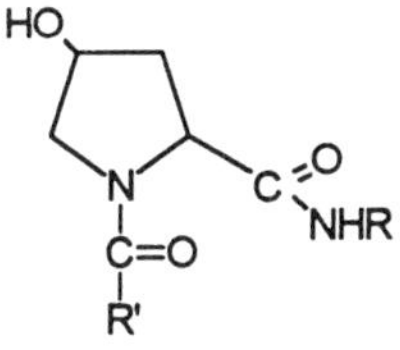
Lysyl hydroxylase	CH_2NH_2 | CH_2 | $(CH_2)_2$ | R – NH – CH – COR′	CH_2NH_2 | CHOH | $(CH_2)_2$ | R – NH – CH – COR′
γ-Butyrobetaine hydroxylase	$(CH_3)_3N^+$–CH_2–CH_2–$CH_2CH_2COO^-$ γ-Butyrobetaine	$(CH_3)_3N^+$–CH_2–CHOH–$CH_2CH_2COO^-$ Carnitine
Thymine-7-hydroxylase	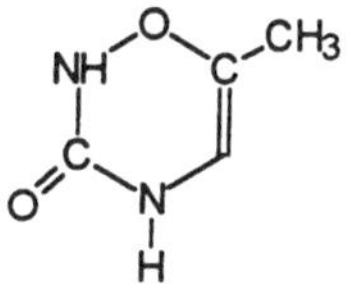 Thymine	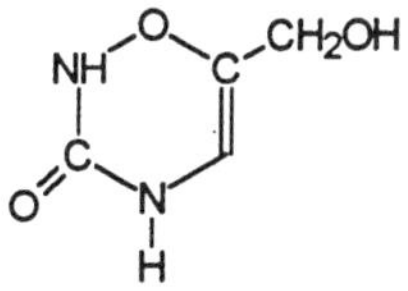 5-hydroxymethyluracil

The cases presented in Table II.7 are the examples of α-ketoglutarate-coupled hydroxylation of the aliphatic C–H bond.[50]

In all cases the reaction requires the presence of iron ions, which are highly specific and cannot be replaced by any other metal ions (such as Mg^{2+}, Mn^{2+}, Cu^{2+}, or Cd^{2+}) without complete loss of catalytic activity. Experiments with labeled dioxygen $^{18}O_2$ have shown that one oxygen atom of the O_2 molecule is inserted into the XO product, and the other into the succinate molecule formed from α-ketoglutarate. Ascorbic acid is required to keep the active center in the reduced state but does not participate in the reaction stoichiometry. All the reactions with the participation of ketoglutarate-dependent dioxygenase are highly regioselective and stereospecific. Presumably, the substrates are tightly oriented in the enzyme with a definite position of the C–H bond relative to the active center. The mechanism of dioxygen activation seems to involve a homolytic cleavage of the O–O bond due to the coaction of α-carbonyl group in ketoacid (Scheme II.2).

Formation of CO_2 makes this reaction thermodynamically favorable. The iron-oxo complex containing an Fe^{IV}=O fragment is apparently sufficiently chemically active to hydroxylate an adjacent C–H bond in a substrate. The intramolecular isotope effect for tritium-labeled γ-butyrobetaine was found to be $k_H/k_T = 15$ (which corresponds to $k_H/k_D = 6.5$) for the enzyme from liver.[51] Additional evidence for high-valent iron-oxo species follows from the observation that ^{18}O from $H_2{}^{18}O$ is included in the hydroxylation products.[52]

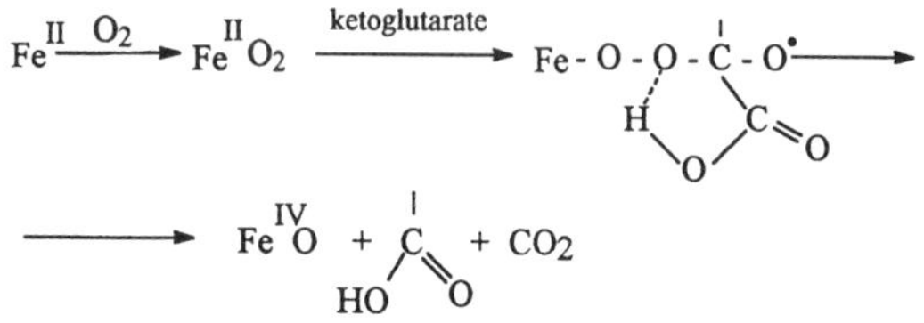

SCHEME II.2 Mechanism of dioxygen activation by ketoglutarate-dependent dioxygenase.

II.3.3.3 Hemerythrin and Rybonucleotide Reductase

Before discussing an important nonheme oxygenase oxidizing methane, methanemonooxygenase, it is desirable to consider briefly two related proteins, different in their functions, but apparently similar in the structure of their active centers. These are hemerithrin (Hr) and ribonucleotide reductase (RR). Both contain dinuclear iron μ-oxo complex in the active center. Hemerithrin[25a,53,54a] is a dioxygen-binding transport protein in the blood of sea worms, i.e., it performs the same functions as hemoglobin for mammals. High-resolution X-ray structures are known for both the unoxygenated di-iron(II) complex and its dioxygen adduct.[54b] Two iron atoms in the hemerithrin active center are bound by one μ-oxo bridge and two μ-carboxo bridges and contain altogether five imidazole residues of histidines. There is one coordination site to bind dioxygen. In the reduced state proton is bound to the oxygen bridge. Reversible addition of dioxygen proceeds with the participation of the proton from the OH bridge (Figure II.11).

Ribonucleotide reductase[25a,54a,55,56] is an important enzyme that catalyzes the reduction of ribonucleotide di- or triphosphate to 2′-deoxyribonucleotide di- or triphosphate, the first step in DNA synthesis. It is present in bacteria (e.g., *Escherichia coli*), certain viruses, and other living organisms, including mammals. The enzyme from *E. coli* contains two iron ions that form an oxo-bridged dinuclear iron cluster. Distinct from Hr, only two imidazoles are coordinated to iron atoms (one to each Fe). Other ligands are carboxylates and water molecules. The structure of the di-iron(III) site of the RR molecule from *E. coli* derived from its X-ray structure[54c] is shown in Figure II.12.

The cluster was characterized by resonance Raman, Mössbauer, and EXAFS spectroscopies and magnetic susceptibility measurements.

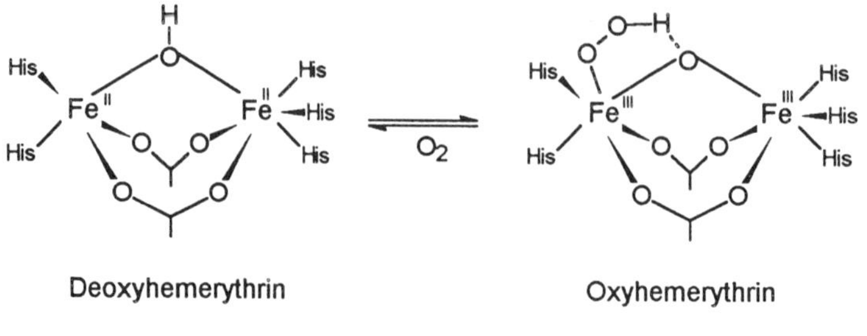

FIGURE II.11 Reversible addition of dioxygen to di-iron center of hemerithrin.

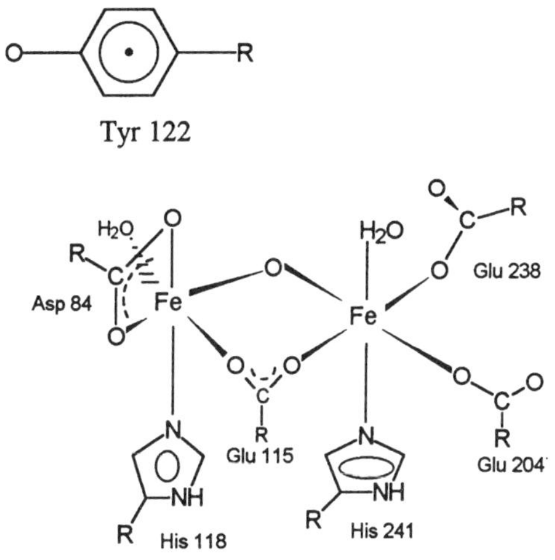

FIGURE II.12 Structure of the di-iron site of *E. coli* ribonucleotide reductase. (Adapted from Nordlund, P. et al., *Nature*, 345, 593, 1990.)

The mechanism of the RR functioning may be represented by the scheme shown in Figure II.13.[56a] When oxidized, RR forms a free radical from tyrosine that can be identified by its EPR spectrum. A peculiar chain reaction can be initiated: tyrosyl radical abstracts the 3′-hydrogen atom, which is followed by water formation from the OH group of C-2′ and the H of the SH, replacement of the OH by the H from the other SH, and return of the H from tyrosine.

The X-ray structure of the dimanganese(II)-substituted form indicates the presence of bis-(μ-carboxylato)Mn_2 nucleus without an oxygen bridge.[54d] The $Fe^{II}Fe^{II}$ form of the RR in the presence of dioxygen, as was already mentioned, produces tyrosyl radical and this could lead to the conclusion that RR functions similarly to heme-dependent peroxidases. Dioxygen could oxidize the iron(II) dinuclear center to iron(IV) dimer. The latter intermediate, which is analogous to compound I of peroxidase, may oxidize tyrosine by one-electron transfer to produce the tyrosyl radical. There is an observation in support of this mechanism, namely oxygen atom donors, such as iodosyl benzene and peracids, generate tyrosyl radicals with the diferri state of RR.[55] Spectroscopic studies have demonstrated the existence of two short-lived intermediates in the reaction of the Fe^{II}_2 form of RR with O_2.[56a] The first has a lifetime $\tau = 3$ s and is characterized by absorption at $\lambda = 570$ nm similarly to that of peroxy iron complexes. After a one-electron transfer from tyrosine or some other reductant, a second intermediate is produced ($\tau = 1$ s), which corresponds to the compound II of peroxidase. According to the Mössbauer spectrum both iron atoms in this intermediate are in the Fe^{III} state and therefore the unpaired electron is probably situated at a ligand (presumably at the bridging one: $Fe^{III}–X^{\cdot}–Fe^{III}$) because its EPR spectrum has a g-factor close to 2. Porphyrin ligands are known to be often present in the cation radical state, helping to transfer electrons from a substrate and stabilizing iron in the formally high-valent state (*vide supra*). Possibly the ligands may perform the same functions in a nonheme environment.

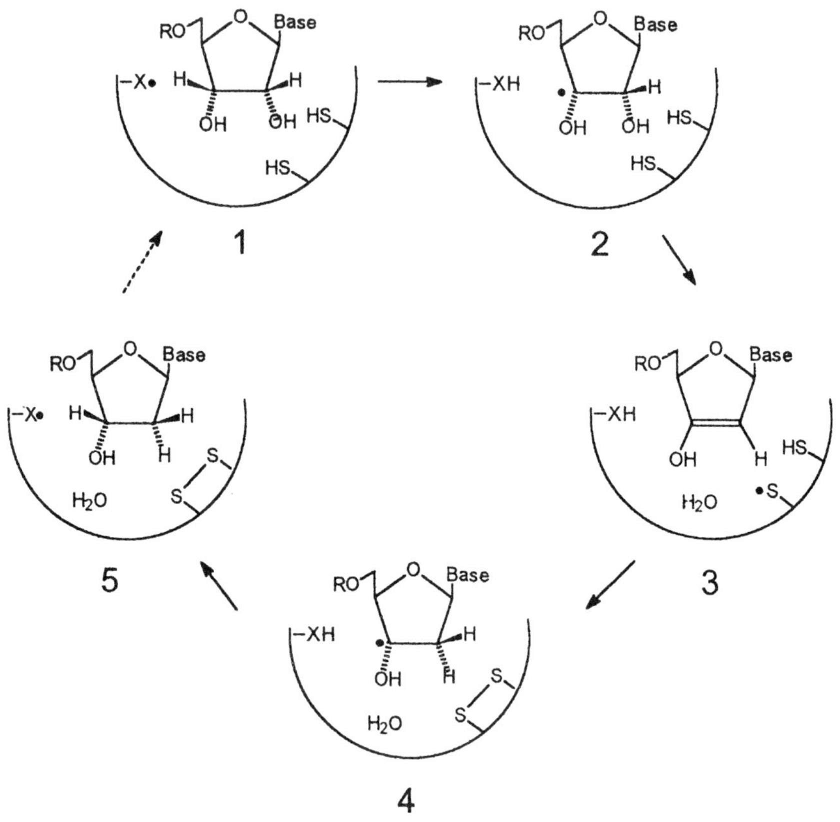

FIGURE II.13 Proposed mechanism for ribonucleotide reductase functioning.

RR is similar to methane monooxygenase in the sequence of aminoacids near the iron active center, demonstrating that these enzymes are closely related.[56b]

II.3.3.4 Methane Monooxygenase

II.3.3.4.1 General

Methane monooxygenase (MMO) catalyzes the oxidation of methane, the simplest and usually the most chemically inert alkane, to methanol in methanotrophic bacteria.[57a,b] MMO is present in various methane-oxidizing bacteria, numbering about a hundred. The first data on these bacteria appeared more than 80 years ago. Nevertheless, only a few of them have been thoroughly investigated and described. The first data on the chemistry of methane oxidation by cell-free preparations appeared in the literature fairly recently. However, the important results obtained in the last few years by several groups, particularly those of Dalton, Lippard, Lipscomb, and some others, have greatly improved our understanding of this important enzyme. The stoichiometry of methane oxidation follows the well-known monooxygenase reaction:

$$CH_4 + O_2 + 2e \xrightarrow{+2H^+} CH_3OH + H_2C \quad (54)$$

Hence the electron donor is essential (usually NADH, which can be replaced by NADPH or sodium ascorbate).

MMO can be isolated from different bacteria such as *Methylococcus capsulatus,*[58] *Methylosinus trichosporium,*[59] and *Methylobacterium.*[60] Depending on the growth conditions, the enzyme in all cases may be in soluble or membrane-bound form. The enzyme systems follow the order of reactivity (expressed in mmol/min/mg protein): *Ms. trichosporium* (1700) > *Mc. capsulatus* (150 to 36) > *Methylobacterium* (27). The substrate selectivity was studied thoroughly by Dalton et al.[57b] for soluble MMO from *Mc. capsulatus. n*-Alkanes are oxidized at terminal CH_3 and neighboring CH_2 groups (ω and ω–1 hydroxylation). For the branched hydrocarbons only the part remote from the branching is hydroxylated, showing the importance of steric factors. However, for sufficiently small molecules, e.g., *iso*-butane and *cis*-1,4-dimethylcyclohexane,[61] tertiary C–H bonds are hydroxylated preferentially, indicating that the steric influence is not induced by the active center but rather by the protein surrounding, its tertiary structure forming a hydrophobic cavity of a definite size. These data also indicate that the electrophilic center functions in MMO when reacting with C–H bonds. Accordingly, MMO hydroxylates aromatics and epoxidizes olefins.

II.3.3.4.2 Structure

MMO from *Methylococcus capsulatis* comprises three components designated A, B, and C.[62] The C component contains a flavin and an iron–sulfur cluster and transfers electrons from NADH to the site of hydroxylation. An important B component has a regulatory role and the A component, hydroxylase, contains the site at which substrates are oxidized.

Spectral data (EPR, EXAFS, and Mössbauer) indicate that the active site in MMO contains μ-oxo- or μ-hydroxo-bridged dinuclear iron cluster similar to that found in hemerythrin and ribonucleotide reductase.[54] EXAFS gives the distance between two irons as 3.42 Å and each Fe has from four to six O or N atoms at the distance of about 1.92 Å. No shorter distance is seen and therefore it may be thought that iron atoms are bridged not by an oxo, but by a hydroxo bridge.[63,64] In the oxidized form the cluster in the protein contains antiferromagnetically bound iron atoms [Fe^{III}–Fe^{III}]. Upon reduction by dithionite it can attain both the EPR active semireduced [Fe^{II}–Fe^{III}] and reduced [Fe^{II}–Fe^{II}] states.

The X-ray structure of dimeric hydroxylase protein of MMO from *Methylococcus capsulatus* (Bath) was recently determined by Lippard et al.[65] The hydroxylase consists of three polypeptides (α, β, and γ) of $\alpha_2\beta_2\gamma_2$ stoichiometry. Two dinuclear iron centers are positioned in α subunits (the distance between the centers is 45 Å). The dinuclear iron center is presented in Figure II.14.

Fe–Fe distance is 3.4 Å, i.e., consistent with the value determined by EXAFS spectrometry. The iron atoms are linked by two nonprotein ligands that were modeled as monodentate hydroxyl and bidentate acetate ions (the latter was present in the

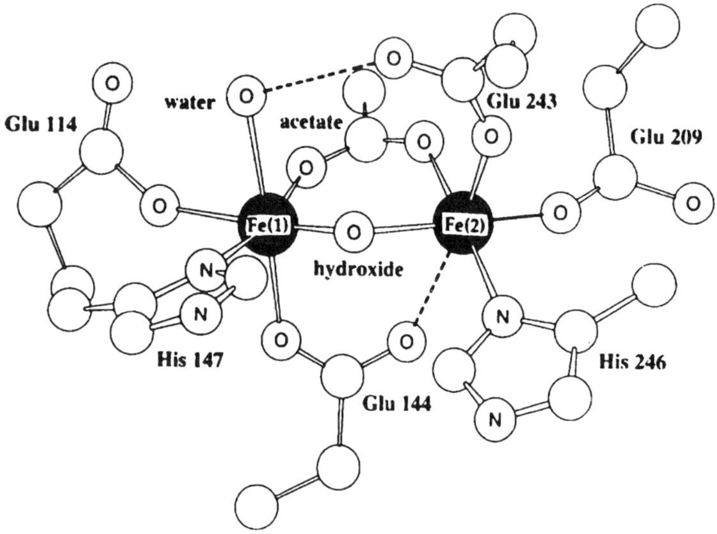

FIGURE II.14 Structure of the active center of MMO. (From Rosenzweig, A. C. et al., *Nature,* 366, 531, 1993. With permission.)

crystallization buffer). The space occupied by acetate probably shows where the oxidized substrate, e.g., CH_3O^-, resides before its protonation and release. Each iron atom is also coordinated to hystidine N atoms and Fe1 is ligated to a monodentate carboxylate of Glu 114 and a water molecule.

An expected hydrophobic cavity is revealed in the investigation, which is adjacent to the iron center and is filled by hydrophobic side chains formed by amino acid residues of protein helices. The cavity is apparently used by a substrate molecule approaching the iron center. There is also a cavity in ribonucleotide reductase protein near its dinuclear iron center, but it is smaller than the one in hydroxylase of MMO. This is consistent with the fact that RR catalyzes the oxidation of tyrosine to the tyrosyl radical and does not involve exogenous hydrocarbons. The residue corresponding tot tyrosine in RR is cysteine in MMO and therefore the participation of cysteine in the redox cycle is suggested.[65] The adjacent threonine (Thr 213) is the only protonated side chain other than cysteine in the hydrophobic cavity and can supply protons for catalytic reaction in the same way that has been proposed for threonine (Thr 352) in cytochrome P-450_{CAM}.

II.3.3.4.3 Functioning

Kinetic analysis of methane oxidation with soluble MMO[66] shows that a reduced [enzyme·CH_4] complex binds O_2, producing a triple complex that decomposes in the rate-limiting step to methanol and water.

The affinity of the substrate to enzyme that is determined by the ratio of maximum rate to Michaelis constant is about ten times higher for methane than for other alkanes. Although the center is hydrophobic this cannot explain the stronger affinity of methane to the enzyme as compared with other alkanes. Besides, the Michaelis constant for CH_4 increases with temperature whereas for a hydrophobic interaction it decreases with a temperature rise (this is observed for cytochrome P-450). Therefore a suggestion is made that CH_4 is chemically bound to a metal ion.[67] There has so far been no independent confirmation of this suggestion, but an intermediate complex with the oxygen atom bound to iron may be suggested as in the case of cytochrome P-450 (only, of course, in the oxidized state of the enzyme) and the chemical interaction of this small molecule could be more important than its hydrophobic interaction.

Until recently the following mechanism was considered as the most probable for MMO functioning.[66,68a] When an alkane is bound to the active center of the enzyme, fast two-electron reduction occurs of μ-hydroxo-Fe^{III}_2 to μ-hydroxo-Fe^{II}_2. One-electron reduction produces a μ-hydroxo-$Fe^{II}Fe^{III}$ intermediate that reacts slowly with O_2. The $Fe^{II}Fe^{II}$ complex undergoes fast reaction with O_2, producing the peroxo intermediate that could be, for example, similar to the hydroperoxo complex formed in the hemerythrin reaction with dioxygen. After subsequent heterolytic cleavage of the O–O bond a $Fe^{IV}Fe^{IV}$ active intermediate is formed, presumably containing an iron-oxo core capable of reacting with MMO substrates. Thus, the reaction scheme looks very similar to that of cytochrome P-450. The second Fe^{IV} in the active intermediate can activate the FeO center by electrophilic coaction.

Recent investigations of the intermediates in MMO reactions with dioxygen and hydrocarbons may bring some corrections to this scheme based mainly on analogy with cytochrome P-450.

The transient kinetics of the reduced $Fe^{II}Fe^{II}$ hydroxylase reaction with dioxygen was studied by Lipscomb et al.[68b] and also by Lippard et al.[68c] Three consecutive intermediates, termed compounds P, Q, and T, were identified during one catalytic cycle of the MMO for *Methylosinus trichosporium* OB3B.[68b] Colorless intermediate P is formed immediately after mixing of reduced hydroxylase with O_2 and then slowly transforms into the colored intermediate Q (λ_{max} = 330 and 430 nm, ε = 8000 M^{-1} cm^{-1}). Substrates were found to have little effect on the rate of compound Q formation, but they greatly accelerated its decay. The decay rate depended upon both the concentration and the type of substrate present. Therefore compound Q may be the activated form of the enzyme that leads directly to substrate hydroxylation or its immediate precursor. The compound Q was trapped by the rapid freeze-quench technique and characterized by Mössbauer spectroscopy. The results indicate that the iron atoms of compound Q are indeed in the Fe^{IV} oxidation level. The spectra recorded at 4.2 and 50 K show that Fe^{IV} sites of compound Q are diamagnetic and the Fe^{IV} spectrum originates from clusters containing two indistinguishable antiferromagnetically coupled Fe^{IV} sites. Analysis of high-field data suggests a lower limit for the exchange coupling constant of $J > 60$ cm^{-1}.

This work[68b] represents the first isolation and characterization of a high-valent iron-oxo species capable of hydrocarbon oxidation. Most interesting is the observation that the Mössbauer spectra of the two iron sites in the $Fe^{IV}Fe^{IV}$ cluster of compound Q are indistinguishable, suggesting that the activated oxygen is bound to the cluster *symmetrically*. Therefore, according to the authors,[68b] a single oxygen atom might bridge two cluster irons or two equivalent oxygen atoms may be present in a diferryl unit. As the result of the interaction of compound Q with the substrate, the third intermediate, termed compound T, is formed, which decomposes to liberate oxygenated product. The decomposition of T is believed to be the rate-controlling step of the process.

The nature of the first intermediate, designated P,[68b] was investigated by Lippard et al. by optical and resonance Raman and also by Mössbauer spectroscopies for hydroxylase protein of MMO from *Methylococcus capsulatus* (Bath).[68d] Rapid freeze-quench samples frozen 155 ms after mixing contained more than 95% of the first intermediate. The resonance Raman spectrum of the intermediate exhibits an isotope-sensitive (at the replacement of $^{16}O_2$ by $^{18}O_2$ and by $^{16}O{-}^{18}O$) feature at 905 cm^{-1}. This value as well as a visible adsorption (λ_{max} = 588 to 682 nm) fall within a range reported for metal *peroxo complexes.* The $^{16}O{-}^{18}O$ sample gives rise to a single Raman band, and this, according to the authors of Reference 68d, indicates that the peroxo intermediate is a *symmetrically bound species.* This conclusion is supported by the narrow quadrupolar doublet lines in the Mössbauer spectrum of the intermediate. This assignment suggests that the peroxo ligand coordinates in the μ-η^1:η^1 or the planar or nonplanar μ-η^2:η^2 binding mode.

These results may support a hypothesis put forward by Shteinman[68e] that the most probable structure of the MMO active center (distinct from ferryl in the case of cytochrome P-450) corresponds to bis-μ-oxo complex produced from the μ-η^2:η^2 intermediate, and the hydroxylation cycle may be represented as shown in Scheme II.3.

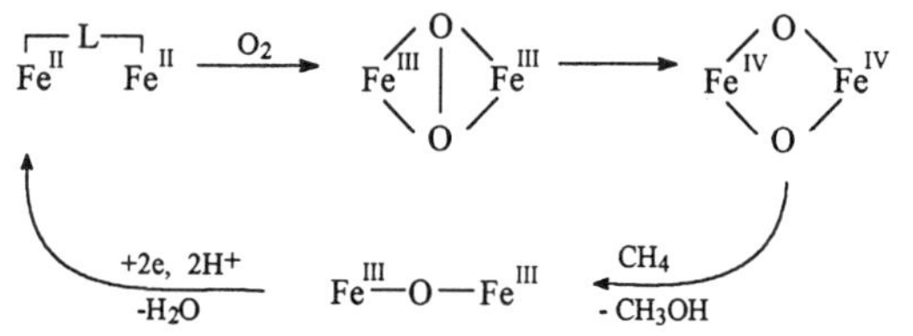

SCHEME II.3 Proposed catalytic cycle of the MMO.

II.3.3.4.4 Mechanism of C–H Bond Hydroxylation by MMO

Among various mechanisms of enzymatic hydroxylation the mechanism of direct oxygen atom insertion seems for methane *a priori* particularly preferable for several reasons.

First of all the C–H bond dissociation energy is among the strongest in methane as compared with other alkanes, and although the redox potential of the active oxidized MMO form is unknown, it is probably insufficient to break the C–H bond (see Section II.2). At any rate, if both the mechanisms (oxygen rebound mechanism starting with C–H bond cleavage and direct O atom insertion) were possible, methane would be among the first candidates for the second mechanism, the more so as it is sterically among the least hindered, and this is clearly more important for the insertion mechanism. Moreover, methyl radical is known to be chemically very active and if formed would be very dangerous for the enzyme even if the probability of the radical leaving the reaction "cage" is very small. As we have seen in Section II.2, the direct insertion of an atom into the C–H bond is definitely possible, including the insertion of both free and bound oxygen atoms, and we shall see more or less definite examples of this mechanism below in purely chemical systems.

However, paradoxically, until recently hydrogen atom abstraction to form a free methyl radical seemed more plausible as the first step of methane hydroxylation than the insertion, based on the experimental evidence obtained by several authors. Similarly to cytochrome P-450, the experimental evidence is usually based on three groups of results: (1) High kinetic isotope effect; for the insertion it is expected to be smaller. (2) Isomerization observed in the products in the process of the insertion; no isomerization is expected in the case of one-step insertion. (3) Results obtained with radical traps, indicating intermediate formation of carbon-centered radicals; the radicals are, of course, excluded for the insertion mechanism. Although the evidence seems convincing, we shall see that the results can also be explained, and even seem to be less controversial, particularly for methane, if (similarly to the results with P-450) we consider them from the point of view of a two-step mechanism with a five-coordinate carbon intermediate preceding the O atom insertion into the C–H bond. It does not mean that this mechanism should be accepted for any substrate other than methane. Hydrogen atom abstraction is a simple process and could be the case for weak C–H bonds provided it is thermodynamically allowed. However, for strong C–H bonds, particularly for those of methane, hydrogen atom abstraction is unlikely. Let us consider the experimental results obtained to disclose the reaction mechanism in more detail.

Mechanistic investigations of methane monooxygenase were started comparatively recently, after cytochrome P-450 had already been thoroughly studied. The oxygen rebound mechanism is widely accepted for P-450, and there is a natural tendency to apply the same mechanism to MMO. Let us remember that according to this mechanism the reaction with an alkane can produce primarily a radical intermediate, which then recombines with an OH ligand bound to iron to produce an alcohol. In many aspects MMO is similar to cytochrome P-450. The difference is mainly quantitative and may be explained by difference of lifetimes of intermediates in the cavity near the active center. For example, from the data presented below it may be suggested that lifetimes for the substrates mentioned are somewhat longer than for cytochrome P-450. However, there is also a qualitative difference. MMO epoxidizes propene in D_2O and D_6-propene in H_2O without exchange of substrate protons or deuterons with solvent,[70a] whereas in the case of cytochrome P-450 considerable exchange of the 1-*trans*-proton of propene with solvent protons is detected. Epoxidation by P-450 is believed to involve oxometallocycle and iron carbene intermediates. No such intermediates exist therefore for MMO. The MMO active center is probably less electrophilic than the P-450 center.

After intense investigation of the soluble methane monooxygenase *Methylococcus capsulatis* (Bath), Dalton et al. concluded that two pathways may take place, depending on the substrate.[68f] One of them is hydrogen atom abstraction from the substrate by a ferryl species to produce a radical and the other involves direct oxygen atom insertion. The first mechanism is proposed for methane hydroxylation while oxidation of pyridine and carbon monoxide proceeds as direct O atom addition to the substrate. Thus, oxygen atom transfer is considered possible, but just for methane this is not the case.

The KIE for C–H hydroxylation is definitely large and for methane ranges from 5[67] to 11.8,[61] perhaps depending on conditions. Recent investigation by Lipscomb and Nesheim of the intermediate Q (*vide supra*) reaction with methane and tetradeuteriomethane has resulted in an unusually large KIE, which amounts to 50 to 100.[70b] Very convincing evidence that the value is not due to any inhibitors that could be introduced with CD_4 was obtained with a 1:1 mixture of CH_4 and CD_4 producing nearly all CH_3OH. Compound Q-mediated oxidation of CH_3D, CH_2D_2, and CHD_3 shows intermediate second-order rate constants. Possibly, lower values obtained in previous works were due to the fact that other steps were involved in the reaction kinetics and C–H hydroxylation was not cleanly the rate-limiting step. It is interesting that other substrates demonstrate noticeable but not so large isotope effects. For example, the ratio of the second-order rate constants for compound Q reaction with C_2H_6 and C_2D_6 is about 4. According to the results,[69] $k_H/k_D = 3$ for the internal KIE of propane hydroxylation measured for the mixture of $CH_3CH_2CH_3$ and $CH_3CD_2CH_3$.

Involvement of tunneling is one of the possible explanations for the large isotope effect found for the hydroxylation of methane. However, if in all cases simple hydrogen atom abstraction is the rate-limiting step, the remarkable differences between the KIE values for methane and other substrates is difficult to explain.

Isomerization observed in the products in the process of the hydroxylation is important evidence for a stepwise mechanism. The hydroxylation of hydrocarbons in the allylic position to a double bond by MMO from *Methylosinus trichosporium* revealed close similarities to hydroxylations by cytochrome P-450.[70a] Allylic hydroxylation of 3,3,6,6-d_4-cyclohexene occurred with about 20% allylic rearrangement, while 33% was reported for cytochrome P-450. Similarly, hydroxylation of *exo,exo,exo,exo*-2,3,5,6-d_4-nobornane occurred with epimerization, although to a lesser extent than reported for cytochrome P-450.

In the oxidation of 1,1-dimethylcyclopropane catalyzed by MMO, the products are (1-methylcyclopropyl)methanol (81%), 3-methyl-3-buten-1-ol (6%), and 1-methylcyclobutanol (13%).[71] Formation of butenol can be rationalized on the basis of the (1-methylcyclopropyl)carbinyl radical, which undergoes partial rearrangement with ring opening to form the 3-methyl-3-buten-1-yl radical, the latter being oxygenated to give the final product. 1-Methylcyclobutanol is a ring-expansion product, whose formation is best explained on the basis that 1-methylcyclobutyl tertiary carbocation is an intermediate. Oxygenation, therefore, could proceed by abstraction of an H atom from the substrate to form a radical, followed by electron transfer from the radical to the oxidizing species to form a carbocation. The latter would be quenched by oxygen associated with the oxygenation cofactor to generate the main product or rearranged to give another cation with a four-membered ring. In experiments with $^{18}O_2$ and $H_2{}^{18}O$ approximately half of the oxygen atoms of methylcyclobutanol are originated from dioxygen and half from water, whereas in the main product (1-methylcyclopropyl)methanol all oxygen is originated from O_2. Thus, both possible intermediates, free radical and a cation, could be present in the reaction. In view of later findings, however, these conclusions should be dealt with care (*vide infra*). For example, it is not excluded that the products are formed in parallel reactions, each with its own mechanism.

TABLE II.8
Distribution of Stereoisomers (%) in Hydroxylation of S- and R-$[1\text{-}^2H,1\text{-}^3H]$ Ethanes

Substrate ethane	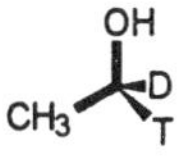	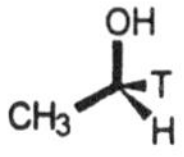	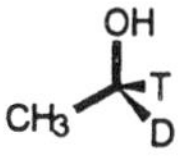	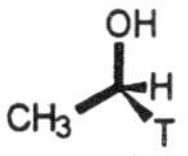
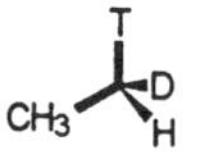 (S)	27	7	52	14
	(36% inversion)		(64% retention)	
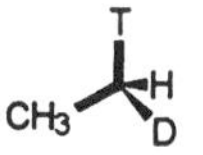 (R)	53	12	26	9
	(68% retention)		(32% inversion)	

Important evidence for an intermediate proposed to be a free radical was obtained by Lipscomb et al. in the oxidation of (R)- and (S)-substituted $[1\text{-}^2H,1\text{-}^3H]$ethane[72a] (Table II.8). Ethane is the closest analogue of methane and despite the observed difference in KIEs the hydroxylation mechanism could be essentially the same.

NMR analysis of the resulting ethanol samples gave an intramolecular primary KIE for hydrogen abstraction from the labeled methyl group of $k_H/k_D = 4.2 \pm 0.2$, i.e., the same as was found for rate constants ratio for the reactions of the intermediate Q with C_2H_6 and C_2D_6.

The results show that the hydroxylation of ethane proceeds with predominant retention of configuration (consistent with the mechanistic similarity to P-450-catalyzed reaction). Almost identical results were obtained for the reaction catalyzed by the MMO hydroxylase component alone in the presence of H_2O_2. The overall retention of configuration is accompanied by approximately 35% inversion (see Table II.8). The relatively high degree of inversion must be due to the formation of an intermediate that has a sufficiently long lifetime to undergo configurational inversion with appreciable frequency. This intermediate is suggested to be three-valent carbon derivative, i.e., ethyl radical, because another possible intermediate, ethyl cation, has an exceptionally high energy barrier to direct formation and is markedly unstable.

As was already mentioned, all these observations strongly support a stepwise mechanism of hydroxylation; however, they do not provide evidence for methyl radical as the only possible intermediate in the case of *methane*. Moreover, at least some of the results may be explained by other intermediates in the reaction. For example, for the latter case of ethane another alternative could be a five-coordinate carbon intermediate capable of rearrangement before the insertion of the oxygen atom. A similar rearrangement takes place in the case of selective hydroxylation of D-substituted camphor by cytochrome P-450 (see Section II.3.2) where no rotation of the molecule as a whole is allowed and the five-coordinate carbon hydroxylation mechanism provides a possible explanation for the epimerization process. Naturally, a similar mechanism is possible for ethane. Formation of allyl radicals is, of course, perfectly plausible in the case of weak C–H bonds in the allylic position (as well as nonradical insertion), but partial allylic isomerization could be also due to initial oxygen (bound to iron) addition to a double bond, which, besides further epoxidation, may lead to the shift of the double bond with proton abstraction from the carbon atom adjacent to the double bond. Thus direct allylic hydroxylation and the formation of the isomeric product could be the result of different parallel reactions.

The spin-trapping technique was used to investigate possible radical formation in soluble methane monooxygenase catalysis of the hydroxylation of methane and reactions of several other substrates.[72b] Water-soluble nitrones were used as spin traps. The radicals formed were indeed detected by the spin adducts formed and they were identified from the hyperfine splitting constants of their EPR spectra. In the case of methanol as the substrate isotopically labeled $^{13}CH_3OH$ was used apart from ordinary $^{12}CH_3OH$ and additional splitting due to ^{13}C was observed. The authors[72b] concluded that carbon-centered substrate radicals result from hydrogen atom abstraction by a ferryl species in a "cytochrome-P-450-like mechanism". On the other hand, no hydroxyl radical or another oxygen-based radical was detected in any of the spin-trapping experiments.

It is not clear, however, whether the whole reaction or only a small part of it involves intermediate free radical formation, because no quantitative measurements could be made so far. Moreover, according to the oxygen rebound mechanism, practically all the radicals initially formed presumably must recombine with the OH group formed on the iron atom and it is surprising that radicals can so easily leave the reaction center to react with the spin trap; again the danger of free methyl radicals appearing close to the protein C–H bonds should be remembered. It is interesting that alkenes, acetylenes, and aromatic substrates also yield spin adducts. But in all the latter cases the reaction definitely starts by addition of bound oxygen to the substrate molecules and no hydrogen atom abstraction takes place. So perhaps the spin traps can react not only with free radicals, but also with the intermediates formed in the process of initial oxygen atom addition. Then in the case of alkanes a similar reaction of the spin trap with the intermediate adduct of the oxygen atom bound to iron (i.e., with a five-coordinate carbon intermediate) can be suggested, which would transfer an alkyl radical to the spin-trap.

The very large KIE observed for the rate constants ratio of CH_4 vs. CD_4 and much smaller KIE for other substrates is also difficult to reconcile with the simple H atom abstraction as the initial step for all cases. It seems more likely that the mechanism of C–H hydroxylation is more complex. It might be, for example, that all three H atoms are involved in the formation of a five-coordinate intermediate with methane. The structure of intermediates with other alkanes should be definitely different. This could be a tentative explanation for both high rate constants and KIE values, particularly for methane.

Some experimental results definitely contradict the description of MMO only from the viewpoint of the oxygen rebound mechanism. Let us consider, for example, the results of the recent work by Lippard et al.[73a] The radical-clock method was used to find out whether radicals are the intermediates in the catalytic cycle of the MMO from *Methylococcus capsulatus* (Bath) in reactions conducted at 45°C. Four cyclopropyl derivatives were chosen for the investigation. Hydroxylation of *trans*-2-phenylmethylcyclopropane (**1**) gave two products in equal amounts, the alcohol *trans*-(2-phenylcyclopropyl)methanol and the phenol *trans*-2-(*p*-hydroxyphenyl)methylcyclopropane. The authors[73a] explain this by the discrimination by the enzyme of the two enantiomers of probe **1**, one of which leads to hydroxylation of the aromatic ring and the other to oxidation of the methyl group. In the other three cases only the products without the rearrangement were produced: compound **2** yields exclusively (2,2-diphenylcyclopropyl)methanol; **3** gives only *trans*-(2-methylcyclopropyl)methanol; and **4** produces mainly a mixture of *endo*- and *exo*-2-hydroxibicyclo[2.1.0]pentane. Apparently, the absence of the isomerization products cannot be explained by the fast rebound step, particularly for compounds **1** and **2** for which the rebound rate constants would reach the values of 4×10^{13} and 1×10^{13} s^{-1}, respectively, i.e., the rate constants corresponding to a single vibration; this would be too fast for any rebound. At the same time, for *Methylosinus trichosporium,* **1** gives 3 to 5% of the rearranged product, which corresponds to the rebound rate constant of 6 to 9×10^{12} s^{-1}. This is also too high a value and is better explained by both radical formation and nonradical insertion proceeding in parallel. (There is, of course, some probability that the rearrangements observed are also the result of an intermediate other than a free radical, e.g., a five-coordinate carbon adduct.) But the conclusion may be made that for MMO (as well as for cytochrome P-450 and ω-hydroxylases) at least different mechanisms are possible; therefore the results explained by the single-oxygen rebound mechanism have to be reconsidered: with two reactions in parallel no conclusion can be made about the rebound rate constant from the amount of the rearranged product. One of these mechanisms of hydroxylation is definitely nonradical oxygen atom insertion, presumably via a five-coordinate carbon intermediate. We have concluded earlier that if there is competition between different mechanisms for substrates other than methane the probability of nonradical mechanisms for methane have to be higher than for almost all other alkanes. It follows from these considerations that methane is almost certainly hydroxylated by MMO via a nonradical mechanism.

Isotope effects for derivatives of **1**, mono-, di-, and trideuterated at the methyl position, were also measured in the work.[73a] No intermolecular isotope effect was found by comparing the rates of conversion of the undeuterated and trideuterated probes, indicating that C–H bond cleavage is not involved in the rate-determining step of the enzymatic reaction. Intramolecular KIEs of $k_H/k_D = 5.15$ and 5.03 were obtained from studies of mono- and dideuterated probes. Taking into account that a nonradical mechanism is obviously functioning in the hydroxylation, these values show that a high isotope effect is not an indicator of the radical mechanism of the C–H bond breaking.

Other evidence against free radical oxygen rebound mechanism in MMO functioning in the case of methane is demonstration that dihydrogen is apparently much less active than methane as the MMO substrate.[73b] No obstacles could be expected for dihydrogen to reach the MMO active center. The H–H bond in dihydrogen and the C–H bond in methane have almost equal bond dissociation energies and both H_2 and CH_4 react with close rate constants with OH radicals and O atoms. Ferryl reactions leading to hydrogen atom abstraction from CH_4 and H_2 may be expected to require similar activation energies. For the reactions beginning with oxygen atom addition to C–H and H–H bonds the reactivity naturally can be certainly very different. Indeed, the reaction of bare FeO^+ with H_2 in the gas phase to form H_2O, which is strongly exothermic and spin-allowed, is a comparatively slow process compared with the reaction with CH_4.[74]

The general conclusion may be drawn that the MMO enzymatic system (similarly to cytochrome P-450) can react in different ways with C–H-containing compounds, depending perhaps on the nature of a substrate, the size and shape of the hydrophobic cavity containing the substrate, and the conditions. A large variety of reactions exists for dimeric iron centers interacting with dioxygen. Such a center may be merely a dioxygen carrier, as in the case of hemerythrine; it may form a high-valent di-iron center to induce electron transfer to form free radical, as in the case of ribonucleotide reductase; and it may hydroxylate a C–H bond, as in the case of MMO. For the latter reaction, radical formation is possible followed by its further oxidation, as well as nonradical insertion of the oxygen atom into the C–H bond. Nonradical oxygen atom addition or insertion is the preferable mechanism not only for such substrates as pyridine and carbon monoxide, but also for substrates with strong C–H bonds, particularly for methane, the natural substrate of the MMO (very likely via a five-coordinate carbon intermediate). The evidence for a five-coordinate carbon non-radical hydroxylation mechanism of the MMO for the most important case of methane is growing, although for other substrates, particularly those with weak C–H bonds, radical mechanism is by no means excluded. As we shall see, the same conclusion can be made for chemical model systems (*vide infra*).

II.3.4 ANAEROBIC ALKANE OXIDATION

II.3.4.1 The Process[74,75a,75b]

The oxidation of alkanes described above requires dioxygen, but there is evidence that even without O_2, oxidation of methane and other alkanes occurs in living nature. The microorganisms involved in this process, which are evidently more ancient than those using dioxygen, are sulfate-reducing bacteria. One example is the oxidation that proceeds in marine sediments in a thin subsurface anaerobic zone. Below this zone methane is produced anaerobically and diffuses upwards, reaching the zone of the anaerobic oxidation. Different from freshwater sediments, a major part of the upward methane flux appears to be consumed before it reaches the interface between water and the sediment. The scale of this process on the planet may be quite considerable, but because methane disappears before reaching the surface it is largely unnoticed. The zone of methane anaerobic oxidation coincides with a depth where the produce $[CH_4] \times [SO_4^{2-}]$ is maximum, therefore it is concluded that the zone location is kinetically controlled. The net reaction

$$CH_4 + SO_4^{2-} \rightarrow SH^- + HCO_3^- + H_2O \qquad (55)$$

is thermodynamically allowed (free energy is –6 kcal/mol), but clearly sulfate is much less powerful an oxidation agent than dioxygen. The enzymes involved in the anaerobic methane oxidation are apparently still unknown, and even the very existence of the process of methane anaerobic oxidation evoked some doubts. At present, however, very good evidence has been obtained that hydrocarbons in crude oil are used by sulfate-reducing bacteria growing under strictly anaerobic conditions.[75b] A pure culture of moderately thermophilic (55 to 65°C) bacteria (strain TD3) was isolated and further cultivated on *n*-decane. The strain TD3 was found to oxidize *n*-alkanes from C_6 to C_{16}, with best growth in the C_8 to C_{12} range. Sulfate is reduced to sulfide, and the authors[75b] suggest that sulfate-reducing bacteria utilizing aliphatic and aromatic hydrocarbons may be a significant source of sulfide in oil deposits and oil-producing plants.

The reaction mechanism for C–H activation and oxidation is a problem for future investigation. However, the low absolute value of the process free energy could lead to important consequences. Whereas for aerobic oxidation exothermic stages make the process strongly nonequilibrium, in anaerobic oxidation we may expect that the whole process will be close to equilibrium and each step, including perhaps that of methane reaction with an active center, will be reversible. Therefore anaerobic methane oxidation may be closely related to the opposite natural process of methane production by methanogens, which is much better understood.

Moreover, there is evidence that enzymatic anaerobic methane production is accompanied by some methane oxidation.[76] Methanogenic bacteria were found to form and oxidize methane anaerobically at the same time. This was found using $^{14}CH_4$ and investigating carbon-containing substances present. Compared with the quantity of methane formed, the amount of methane simultaneously oxidized varied between 0.3 and 0.001%, depending on the strain used. All the strains of methane producers tested reoxidized methane to carbon dioxide.

The results show that in all cases the increase of methane oxidized paralleled the increase of methane formation, suggesting that the same active centers might be involved in methane production and reoxidation. The major oxidation product was carbon dioxide, but in addition two of the strains, so-called acetate organism and *Methanosarcina barceri,* produced acetate (the former) and acetate together with methanol (the latter). Interestingly, these were the same organisms that, distinct from the others, could use these compounds as methane precursors. The oxidation is clearly not simply back-reaction because, for example, the acetate organism produces both acetate and carbon dioxide from methane but is not able to convert CO_2 to CH_4. Nevertheless, most interesting stages of methane activation and first steps of methane oxidation appear to be reverse reactions of the final steps of methane production.

Therefore, to find a possible mechanism of methane oxidation we may use the results obtained for methane formation by methanogenic bacteria, which are now sufficiently well understood.

II.3.4.2 Mechanism of the Final Stages of Methanogenesis[77]

The reduction of CO_2 to CH_4 in methanogenic archaea proceeds via carrier-bound C_1 units.[78] Formylmethanofuran, N^5-formyltetrahydromethanopterin (N^5-formyl-H_4MPT), N^5,N^{10}-methenyl-H_4MPT, N^5,N^{10}-methylen-H_4MPT, N^5-methyl-H_4-MPT, Co-methyl-5-hydroxybenzimidazolyl cobamide, methyl coenzyme M, and apparently Ni-methyl coenzyme F_{430} are the intermediates identified. Most methanogenic archaea can grow on H_2 and CO_2 as sole energy sources. The net reaction is

$$CO_2 + 4H_2 \rightarrow CH_4 + 2H_2O \qquad \Delta G = -31 \text{ kcal/mol} \tag{56}$$

Methane is ultimately produced in methanogenic bacteria by the reaction of the methyl thioether 2-(methylthio)ethanesulfonate (methyl coenzyme M; MeCoM) with the thiol N-7-mercaptoheptanoyl-O-phospho-L-threonine (HS-HTP). The mixed disulfide HTP-S-S-CoM is produced together with methane in the reaction shown in Scheme II.4.[78]

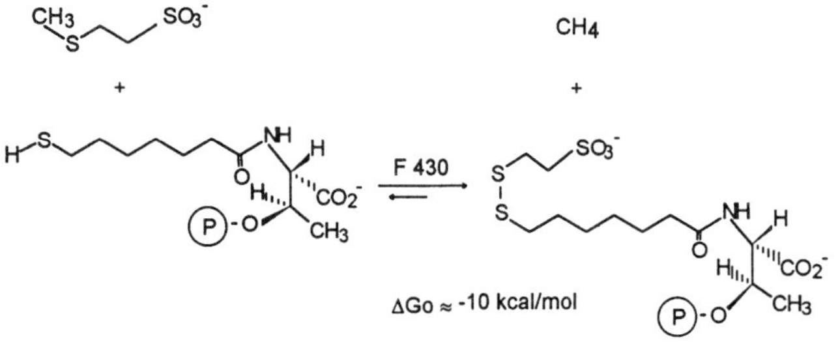

SCHEME II.4 Final step in methane production by methanogenic bacteria.

The enzyme methyl coenzyme M reductase catalyzes this transformation and contains as cofactor the nickel tetrahydroporphynoid "factor 430" (F_{430}) of unique structure (Figure II.15).[79]

The soluble enzyme (300 kDa) is composed of three different subunits in $\alpha_2\beta_2\gamma_2$ arrangement and contains 2 mol coenzyme F_{430} tightly but not covalently bound.

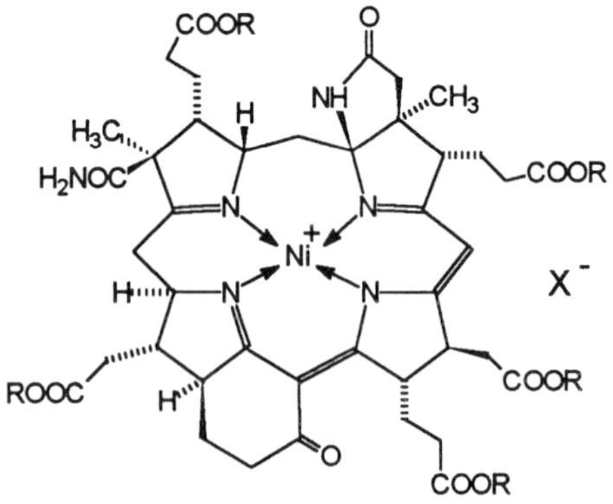

F 430

FIGURE II.15 Structure of the F_{430}. (Adapted from Lin, S.-K. and Jaun, B., *Helv. Chim. Acta,* 74, 1725, 1991.)

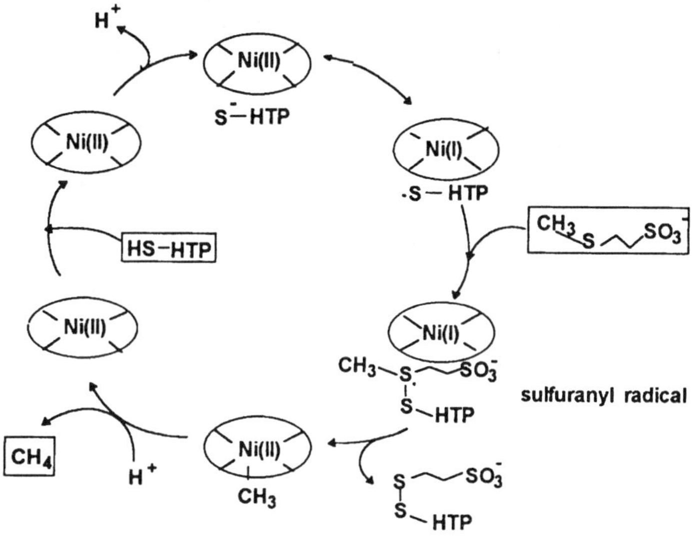

FIGURE II.16 Proposed cycle for methane production by methanogens. (Adapted from Berkessel, A., *Bioorganic Chem.*, 19, 101, 1991.)

EPR studies indicate that in the active enzyme, coenzyme F_{430} is in a reduced Ni(I) state. Ni-methyl coenzyme F_{430} has been proposed as an intermediate and the catalytic cycle shown in Figure II.16 is postulated.[80]

If a similar but reverse cycle is responsible for anaerobic methane oxidation the activation of methane may be visualized as interaction of a methane molecule with the Ni^{II} complex to form an $Ni^{II}–CH_3$ fragment:

$$Ni^{II} + CH_4 \rightarrow Ni^{II} - CH_3 + H^+ \qquad (57)$$

A mild oxidating agent, similar to disulfide produced in the methane formation, can participate in the subsequent reaction.

Model studies support principal features of the proposed scheme for methane formation in the enzymatic methanogenic process. For example, the Ni^{I}-state of isolated F_{430}[Ni(I)] reacts with methyl iodide or methyl sulfonium salts to afford methane:

$$2F_{430}[Ni(I)] + CH_3X + H^+ \rightarrow 2F_{430}[Ni(II)] + CH_4 + X^- \qquad (58)$$

A paramagnetic $CD_3–Ni^{II}$ derivative of the pentamethyl ester ($F_{430}M$) of coenzyme F_{430} was prepared by *in situ* methylation with $(CD_3)_2Mg$ and characterized by its ^{2}H-NMR spectrum. At –40°C the very broad D-signal of the axially coordinated CD_3 group is found at –400 ppm.[79]

Thermodynamics of the first step of methane oxidation is somewhat unfavorable ($\Delta G \approx +10$ kcal/mol), but the absolute value of ΔG is not very large and small changes in the porphinoid substituents may change it to acceptable values for the oxidation process. Thus Ni^{II} with d^8 electronic configuration in porphinoid surrounding may be responsible for methane activation in anaerobic conditions. The question arises of why this mechanism was replaced by a principally different one in living nature when dioxygen became abundant in the atmosphere, although a reducing agent (NADH) is required for monooxygenase type of dioxygen activation. The answer may lie in the possibility to use iron, which is more accessible than nickel, in the coenzyme. It is also probable that dioxygen would destroy the Ni(I) complex irreversibly and a special mechanism to prevent it would have to be developed. However, it is not always obvious why nature prefers certain ways of performing chemical reactions, and clearly more has to be learned about anaerobic methane oxidation and its chemical analogues to try to obtain a more definite answer.

II.4 CHEMICAL ALKANE OXIDATION WITH PARTICIPATION OF METALLOPORPHYRINS

We have seen in the preceding section that the enzymes monooxygenases have the ability to hydroxylate nonactivated C–H bonds in hydrocarbons, including C–H bonds in terminal methyl groups. The reaction is coupled with the oxidation of an electron donor (NADPH) and proceeds according to Eq. 48 (Section II.3.2):

$$RH + 2e + O_2 + 2H^+ \rightarrow ROH + H_2O$$

The active center of cytochrome P-450-based monooxygenases, which are particularly well known and thoroughly investigated, contains porphyrin iron complexes similar to those that function in such O_2 transfer agents as hemoglobin and myoglobin and in such redox enzymes as peroxydase and catalase. The stability of the porphyrin system is apparently one of the reasons for its widespread use in living nature. Furthermore, the porphyrin ring can play an important role of electron donor forming a radical cation in the oxidation, which makes it an active participant in the process.

It was natural to try to use metal porphyrin complexes, including those of iron, as catalysts for hydrocarbon oxidation. In the beginning these attempts did not produce particularly spectacular results. However, by using different approaches the investigators eventually came upon effective catalysts for epoxidation of olefins as well as hydroxylation of C–H bonds in alkanes. In some cases they are very close to the natural systems from the viewpoint of mechanism and reaction rate. Particularly important was the use of various synthetic porphyrins. New methods for efficient synthesis of these porphyrins were reported in a number of important papers by the groups of Groves, Mansuy, Traylor, Meunier, and some others, and at least three generations of porphyrins have been tried as catalysts for epoxidation of olefins and hydroxylation of alkanes. The activity of the porphyrins as catalysts was greatly enhanced with the introduction of halogen substituents in several sites of the porphyrin molecules. This field, already very much advanced, continues to develop rapidly and demonstrates the fruitfulness of the biomimetic approach in the search for purely chemical catalysts (see References 1 to 3, 13, and 81 to 83).

This section describes the recent advances in this field with emphasis on the reaction mechanism.

II.4.1 THE SYSTEMS OF ALKANE OXIDATION

II.4.1.1 Metal Porphyrin Complexes in Homogeneous Solutions

II.4.1.1.1 With Donors of Active Oxygen

An important step in the investigations of both biological and model systems was the use of active oxygen donors instead of comparatively nonactive dioxygen. These include iodosylbenzene and its perfluoro derivative, sodium hypochlorite,[84] hydrogen peroxide and alkyl hydroperoxides,[85] potassium persulfate, amino N-oxides, and nitrous oxide.

Ullrich et al. were the first to use iodosylbenzene as an active oxygen donor in the biological system.[86] Groves et al. used it for the first time in a simple system, FeTPPCl–PhIO.[87a] The system can oxidize a number of substrates, including alkanes. For example, cyclohexane is oxidized to cyclohexanol with a yield of 8% with respect to PhIO. Recently, many similar systems oxidizing alkanes have been described in literature. They are based on metalloporphyrins and various donors of active oxygen in organic solvents or in water–organic solvent mixtures in the presence of the catalysts of interphase transfer. The most successful systems are given in Table II.9 (see Reference 83b and references therein). Alkanes are oxidized in these systems under ambient conditions to produce corresponding alcohols and ketones (in some cases alkyl halogenides) with rates comparable to those observed for cytochrome P-450 (6 to 70 turnovers per minute[25a]), the yields being more than 60% per oxidant and about 90% per substrate. The rates observed are presented in Table II.10.[83b]

The products yields, selectivity and reaction rates depend on the nature of the metalloporphyrin, the axial ligand, the oxygen donor, and the substrate. Ruthenium porphyrin complexes turned out to be particularly active. For example, the system including 2,6-dichloropyridine N-oxide as the oxidant and $RuTMP(O)_2$ as the catalyst (TMP = tetramesitylporphyrinato) is considered to be the most efficient catalytic system for hydroxylation of alkanes known so far[87b] (see Table II.9). Recently Groves and Roman found that nitrous oxide can be used as the oxygen atom donor for the porphyrin complex of ruthenium. The reaction of O transfer is facilitated by the formation of dinitrogen complex with another ruthenium porphyrin particle.[87c]

II.4.1.1.2 With Dioxygen and a Reducing Agent

To construct a full model of biological oxidation it is necessary to find systems using dioxygen and a reductant. The systems have to involve the reductive activation of dioxygen, which, reacting with a reduced metalloporphyrin, transforms it to a state that is able to oxidize a chosen substrate. The first system based on this monooxygenase reaction principle was that proposed by Tabushi and Yazaki with dihydrogen on colloidal platinum as a reducing agent.[88] Other reducing agents, such as sodium ascorbate, electrons of cathode, dihydropyridine, and zinc (powder or

TABLE II.9
Oxidation of Alkanes in Systems with Active Oxygen Donors

Catalyst	Oxidant	Substrate	Products (ratio)	Yield (%)	Ref.
MnTFPPCl	NaOCl	Cyclohexane	Cyclohexanol Cyclohexanone Cyclohexyl Chloride	24[a]	8
MnTDCPPCl + Im	H_2O_2	Cyclohexane	Cyclohexanol Cyclohexanone (3:1)	40[a]	9
$MnCl_{12}TMPCl$	$KHSO_5$	Adamantane	Adamantanol-1 Adamantanol-2 Adamantanone (12:5:1)	65[a]	10
MnTDCPPCl+ 4-*t*-BuPy	$KHSO_5$	Cyclohexane	Cyclohexanol Cyclohexanone	90[a]	11
FeTPPCl	PhIO	Cyclohexane	Cyclohexanol Cyclohexanone (10:1)	26[b]	12
FeTPPCl	PhIO	Adamantane	Adamantanol-1 Adamantanol-2 Adamantanone (39:5:1)	45[b]	13
$FeTF_5PPCl$	PhIO	Cyclohexane	Cyclohexanol	67[b]	3e
FeTPPCl	$PhCMe_2OOH$	Cyclohexane	Cyclohexanol Cyclohexanone (2:1)	60[a]	12
$RuTMP(O)_2$, HCl added	2,6$(Cl)_2$ pyridine N-oxide	Adamantane	Adamantanol-1 Diol-1,3 One-2	94[a]	28

[a] With respect to the initial substrate.
[b] With respect to the oxidant consumed.

TABLE II.10
Turnover Rates in Alkane Oxidation by Systems with Active Oxygen Donors[83a]

Catalyst	Oxidant	Substrate	Turnover rate, cycles·min^{-1}
MnTPPCl	PhIO	Cyclohexane	0.03
MnTMPCl	NaOCl	Cyclohexane	0.08
MnTMPCl	$KHSO_5$	Adamantane	9
$MnBr_8TMPCl$	$KHSO_5$	Adamantane	29
$MnCl_{12}TMPCl$	$KHSO_5$	Adamantane	67
FeTMPCl	PhIO	Cyclohexane	0.16
FeTDCPPCl	C_6F_6IO	Cyclohexane	1–7[a]

[a] Depending on experimental conditions.

TABLE II.11
Alkane Oxidation in the Systems with Dioxygen and a Reductant

Catalyst	System	Substrate	Products, ratio	Rate, cycles·min^{-1}
MnTPPCl + N-methylimidazol	H_2/Pt–HCl–benzoic anhydride (toluene–ethanol)	Adamantane	ol-1 ol-2 one-2 15:1.7:1	0.004
MnTPPCl	Sodium ascorbate (benzene)	Cyclohexane	ol one 1:7.5	0.006
MnTPPCl + N-methylimidazol	Zn powder–AcOH (CH_3CN-CH_2Cl_2)	Cyclooctane	ol one 1.8:1	0.5
		Adamantane	ol-1 ol-2 one 20:2:1	0.5
FeTpivPPCl	Zn powder–AcOH (CH_3CN)	Cyclohexane	ol one 1.7:1	2
FeTpivPPCl	Zn powder–AcOH–MV (CH_3CN)	Cyclohexane	ol one 3.3:1	50
FeTPPCl	Zn powder–AcOH–MV (CH_3CN)	Cyclohexane	ol one 2.0:1	340
FeTPPCl	Zn powder–benzoic acid–MV (CH_3CN)	Cyclohexane	ol one 1.5:1	340

amalgam), were used later. Proton donors like mineral or carboxylic acids and water are also needed.[89-92] In a communication[91] acetic anhydride was reported to be an effector in the system containing dioxygen and zinc amalgam in acetonitrile, but later acetic acid was found to be necessary for cyclohexane oxidation.[92]

Similarly to the systems with active oxygen donors, the oxidation of alkanes with O_2 and electron (and proton) donors proceeds at ambient temperatures and pressures. Selectivity and reaction rate are determined by the nature of metalloporphyrins and also of the electron and proton donors (Table II.11).[83b]

The product yields per reductant are usually rather small: the highest yield achieved is 37% with Zn as a reductant in the oxidation of hydrocarbons with low C–H bond energies (indane and tetralin), but the rates are close to those of enzymatic oxidations. High yields per reductant in nonenzymatic systems are difficult to achieve because the active particles, being strong oxidants, usually react readily with the reductant directly, thus the latter competes with a substrate for an active oxygen.

The systems based on dioxygen with zinc powder or amalgam are usually less stereoselective than those based on PhIO, NaOCl, or $KHSO_5$ (Table II.12).[83b]

TABLE II.12
Stereoselectivity in Alkane Oxidation Catalyzed by Metal Porphyrins

Catalyst	System	Substrate	Retention of config. (%)
MnTPPCl + 2-methylimidazol	O_2–AcOH–Zn powder	*cis*-1,2-Dimethylcyclohexane	25
MnTDCPPCl + 4-*tert*-Bu-Py	$KHSO_5$	*cis*-Decalin	90
FeTPPCl + 2-methylimidazol	O_2–AcOH–Zn powder	*cis*-1,2-Dimethylcyclohexane	0
FeTPPCl	PhIO	*cis*-Decalin	80
FeTFPPCl	NaOCl	*cis*-1,2-Dimethylcyclohexane	92

II.4.1.1.3 With Dioxygen without a Reductant

Ellis and Lyons have reported that *meso*-tetraphenylporphyrins of manganese, iron or chromium with azide as an axial ligand catalyze the autooxidation of isobutane to *tert*-butyl alcohol at 80°C with a selectivity of 88 to 92%.[93] Acetone is a side product. Ellis et al. found later that when hydrogen in tetraphenylporphyrin is replaced by halogen atoms the activity strongly increases (Table II.13).[94]

The redox potential of the pair Fe^{III}/Fe^{II} increases from −0.29 to +0.31 V when FeTPP is replaced by $FeTPPF_{20}\beta\text{-}Br_8$. Hence there is a relation between the ease of Fe^{III}/Fe^{II} reduction and the iron porphyrin catalytic activity. When catalyzed by iron $Fe(TPPF_{20}\beta\text{-}Br_8)Cl$ the oxidation of isobutane proceeds at room temperature with a selectivity of 92%, the turnovers in 70 h being more than 13,000. The catalytic activity is reported to remain constant during 74 h.

II.4.1.2 Photocatalytic Systems with Dioxygen

Mansuy et al.[95] have described systems of cycloalkane oxidation by dioxygen at 20°C and 200 torr O_2 in the presence of halogenated iron tetraphenylporphyrin complexes, the systems being irradiated with light of wavelength 350 and 450 nm (Table II.14). Cycloalkanes are oxidized in this system to corresponding alcohols and ketones. The number of turnovers does not exceed 230 because of the catalyst destruction in the process of the oxidation.

TABLE II.13
Oxidation of *iso*-Butane in Ellis–Lyons Systems (Time: 6 h)

Catalyst	O_2 uptake (mmoles)	Turnover number (moles O_2 consumed per mole catalyst used)	Selectivity to *t*-Bu alcohol (%)
FeTPPCl	0.0	0	—
FeTDCPPCl	5.0	260	89
$Fe(TPPF_{20}\beta\text{-}Br_8)Cl$	40.2	3090	89
$FeTPPN_3$	1.7	130	92
$FeTDCPPN_3$	15.0	650	80
$FeTF_5PPN_3$	33.0	2060	89

TABLE II.14
Oxidation of Cycloalkanes by O_2 with Photoirradiated Iron Porphyrins

Substrate	Catalyst	One/ol	Initial rate (mol/cat. min)	Turnover
Cyclohexane	FeTDCPPCl	0.45	0.06	10
	FeTDCPPOEt	0.60	0.09	28
	FeTDCPPOH	>17	0.14	82
Cyclooctane	FeTDCPPCl	2.9	0.11	230
	FeTDCPPOH	6.3	0.63	175

Thus the systems described above and based on metalloporphyrins in organic solvents perform the oxidation of alkanes with the process characteristics (temperature and reaction rates) close to those of enzymatic oxidation. The main difference is the stability of the catalyst and the process selectivity, which are usually much poorer compared to the natural systems. To increase the stability and the selectivity metalloporphyrins containing bulky groups were designed with increased steric hindrances of the active center. This has led to considerable enhancement of the primary CH_3 bond hydroxylation in linear alkanes approaching the ω-hydroxylation that is achieved in natural systems.[81c,96-99]

Another way to achieve higher selectivity and catalyst stability is to model the microenvironment of the catalyst active center.

II.4.2 METALLOPORPHYRINS IN MICROHETEROGENEOUS SURROUNDING AND ON THE SUPPORT

II.4.2.1 Micelles

Oxidation of cyclohexane by the system PhIO–FeTHDOPPCl in microheterogeneous media has some special features.[100] The system increases its selectivity compared with the homogeneous one (e.g., the only product is alcohol with no ketone formed), but it is less active. Regioselectivity in the oxidation of linear alkanes is also increased in microheterogeneous media. The selectivity index, which may be defined as the ratio of the primary alcohol yield to that of secondary alcohols, taking into account the number of C–H bonds, shows (Table II.15)[83b] that the regioselectivity in hexane oxidation with respect to the oxidation of the terminal methyl group approximates the regioselectivity for certain forms of cytochrome P-450.

II.4.2.2 Metalloporphyrins in Phospholipid Bilayer

The highest achievement in organization of the microenvironment for the catalyst active center is the system of Groves and Neumann.[101] Membrane-bound tetrakis(*o*-cholenylamidophenyl)porphyrin-H_2(ChP) was synthesized by addition of 3β-hydroxy-5-cholenic acid to the α,β,α,β atropoisomer of *meso*-tetrakis(*o*-aminophenyl)porphyrin in the presence of 4-(dimethylamino)pyridine via the formation of amido bondings. The mixture of the metallated steroid porphyrin with dimyristoylphosphocholine in phosphate buffer produced phospholipid vesicles by means

TABLE II.15
Regioselectivity of *n*-Hexane Oxidation by Metalloporphyrin Chemical and Enzymatic Systems

Catalyst	Oxidant	Medium	Selectivity index
FeTHDOPPCl	PhIO	C_6H_6	0.003
		H_2O–triton X-100	0.021
		H_2O–DMPCh	0.050
		H_2O–CD	0.211
		H_2O–MeCD	0.234
MnTPPOAc	PhIO	C_6H_6	0.027
MnTTMPPOAc			0.041
MnTTPPPOAc			0.310
Cytochrome P-450	O_2	H_2O	0.030
Cytochrome P-450 LM_2	O_2	H_2O	0.160

TABLE II.16
Oxidation of Alkanes and Cholesterol Catalyzed by Membrane-Bound Mn(ChP)Cl

Substrate	System	Time, h	Products (ratio)	Yield, % mol product/mol substr
Cyclooctane	O_2–ascorbic acid	15	Cyclooctanone	8.5
Cyclooctane	$NaIO_4$	3	Cyclooctanol Cyclooctanone (12:1)	15.5
Adamantane	O_2–ascorbic acid	3	Adamantanol-1	4.0
Ethylbenzene	O_2–ascorbic acid	15	Acetophenone	15.0
Ethylbenzene	$NaIO_4$	3	1-Hydroxyethylbenzene acetophenone (8:1)	22.0
Cholesterol	O_2–ascorbic acid	15	25-Hydroxycholesterol	2.0

of ultrasonication. The porphyrin ring has been determined by physical methods to be situated in the center of the bilayer and parallel to the aqueous interface. This membrane-bound manganese porphyrin, Mn^{III} (ChP)Cl, is a catalyst of alkane hydroxylation. Cholesterol is selectively hydroxylated in this system at the C-25 position by dioxygen together with a reductant (ascorbic acid) or by sodium periodate as an oxidant (Table II.16). Similarly to the systems with metalloporphyrins in organic solvents, the selectivity of the process depends on the oxidant.

The oxidation proceeds very slowly, but the catalyst remains stable for a long time. The most striking feature of the catalytic system is selective formation of 25-hydroxo-derivative of cholesterol without affecting its highly reactive Δ^5 double bond. The authors[101] explain this effect by the proximity of the C-25 tertiary carbon to the metal center.

II.4.2.3 Metalloporphyrins on the Support

A serious shortcoming of the systems based on metalloporphyrins in solution, except for the Ellis–Lyons system, is the loss of system activity during the process,

TABLE II.17
Oxidation of Alkanes Catalyzed by Supported Metalloporphyrins

System	Substrate	Catalyst	ol/one ratio	Turnover number
PhIO	*n*-Octane	$MnCl_{12}TSMPOAc$		
		In solution	1.8	42
		On PVP	2.4	24
		On $[PVPMe^+][TSO^-]$	11.7	76
PhIO	Adamantane	$MnCl_{12}TMSPOAc$		
		On PVP	21	238
		On $[PVPMe^+][TSO^-]$	53	322
		On amberit-IRA-900	16	50
PhIO	Adamantane	Mn(T4MPyP)Cl		
		In solution	14	1.5
		On SiO_2	37	7.6
		On clay (5%)	28	11.8
PhIO	Cyclooctane	Mn(T4MPyP)Cl		
		In solution	0.8	2.2
		On SiO_2	5.2	8.1
		On clay (5%)	8.8	11.8
PhIO	Cyclohexane	Mn(T4MPyP)Cl		
		In solution	2.0	1.2
		On SiO_2	6.5	6.0
		On clay (5%)	8.8	7.8
O_2 + Zn + AcOH + MV	Cyclohexane	FeTHDOPPCl		
		In solution	2	10
		On porous glass	5	1800

resulting in a decrease in the number of turnovers. The situation is much improved in the Groves bilayer-bound system but the latter is very complicated. The stability can be increased in simpler systems using various support materials for metalloporphyrins. The examples of the systems with supported metalloporphyrins are given in Table II.17,[83b] which illustrates that the number of turnovers for catalysts on polyvinylpyridine or porous glass are substantially increased compared with the same catalysts in homogeneous solution. Selectivity with respect to alcohol formation is also increased.

As could be expected, the reaction rates on the supported catalysts are lower than in solution.

II.4.3 THE NATURE OF ACTIVE SPECIES AND MECHANISM OF THEIR FORMATION

Oxoiron(IV) porphyrin π-cation radical complex similarly to compound I of HRP is generally accepted to be the most probable candidate for the active species in the catalytic cycle of P-450.

Systems based on synthetic metalloporphyrins with various oxygen donors are good functional models of P-450 (see Tables II.9 to II.17) because they oxidize alkanes under mild conditions with rates and selectivities close to those in the enzymatic oxidations.

The mechanism of involvement of alkanes in various oxidation reactions may include their interaction with free radicals or with high-valence states of metal compounds. The nature of the active particles interacting with alkanes can be determined by physical methods or (particularly when the concentration of the active particles is small and their reactivity is high) by applying different tests based on product sensitivity, stereoselectivity, regioselectivity (selectivity for the site of the attack), kinetic isotope effect (KIE), or kinetic equation of the process (see Section II.2.3). Classical radical-chain oxidations usually require high temperatures and pressures and are characterized by poor selectivity. At the same time, reactions of alkanes in the presence of metal compounds involving such active intermediates as hydroxyl or alkoxyl free radicals may be selective processes, e.g., as was shown for the cyclohexane-coupled oxidation with $SnCl_2$ in an acetonitrile–water mixture with a free radical branching chain mechanism[102,103] (*vide infra*). Therefore a careful approach is needed to make a definite conclusion, and many suggested schemes remain speculative. Only combined application of various tests helps to understand the nature of an active particle and the form of a transition state in the interaction with alkane molecules.

II.4.3.1 Active Species in the Systems Based on Oxygen Donors

Groves et al.[104] have identified a green intermediate in the oxidation of FeTMPCl by *m*-perbenzoic acid in methylene chloride–methanol mixture at –78°C. It was found to be able to oxidize hydrocarbons and was characterized as oxo-Fe^{IV} porphyrin π-cation radical ($P^{+\cdot}Fe^{IV}O$), similar to compound I of HRP. Later Groves and Watanabe investigated detailed mechanisms of iron porphyrin interaction with peracids at –46°C using low-temperature spectroscopy, including the previously elusive step of O–O bond cleavage.[105] The first step of the process turned out to be the exchange of the ligands, producing an intermediate iron(III) alkylperoxo complex. The next step, O–O bond cleavage, depends on the properties of the medium: in methylene chloride acid, catalyzed heterolysis takes place forming $P^{+\cdot}Fe^{IV}O$ whereas in toluene a homolytic cleavage proceeds forming iron porphyrin-N-oxide. In polar media $P^{+\cdot}Fe^{IV}O$ is an active intermediate in the hydrocarbon oxidation while in nonpolar media an unidentified intermediate is an active species that is further transformed into N-oxide of iron porphyrin in the absence of a hydrocarbon.

The results, supporting those mentioned and disclosing the nature of active particles, were obtained by Khenkin and Shteinman in the system PFe^{III}–KO_2–Ac_2O–AcOH.[106-108] At –70°C an intermediate was observed that was characterized by its spectrum as $P^{+\cdot}Fe^{IV}O$. High-valent metal porphyrin complexes have also been found in the systems with other donors of active oxygen: NaOCl, PhIO, and $KHSO_5$. At –78°C iron(III) porphyrin reacts with PhIO to produce a red intermediate[104] characterized as $TMPFe^{IV}O$ (model of HRP compound II), which may be oxidized by chlorine or bromine[109,110] to produce the green intermediate $TMP^{+\cdot}Fe^{IV}O$ with the structure of the porphyrin π-cation radical, the analogue of peroxidase compound I. In the system TMPMnCl–NaOCl the bond length of Mn–O in the active intermediate estimated using EXAFS was found by Meunier et al. to be 1.84 ± 0.02 Å, which is too long for Mn=O and closer to Mn–OCl.[111] Manganese was found to correspond to the Mn^{IV} state, but in spite of a number of physical methods applied (I.R., PMR, magnetic susceptibility, electrochemistry, and EXAFS), the authors were unable to confirm the presence of a chlorine atom in the Mn coordination sphere and to determine the structure of the active intermediate. In a similar system, but based on iron porphyrin, NaOCl–TPPFeCl in benzene or methylene chloride at 6°C, a red compound is formed also with adsorption spectrum characteristic of iron(IV) porphyrin complexes and different from that of $P^{+\cdot}Fe^{IV}O$.[100] In aqueous solution with pH 9.2, potassium monopersulfate generates a water-soluble high-valent iron-oxo $PFe^{IV}O$ complex stable at room temperature.[112] Other oxidants are less suitable for preparation of high-valent metal porphyrin oxo complexes, e.g., water-free hydrogen peroxide destroys iron complexes even at –30°C;[113] *tert*-butylhydroxyperoxide generates cationic *iso*-porphyrin compounds with an ROO group attached to one of the meso positions.[114]

Thus, the active intermediate $P^{+\cdot}Fe^{IV}O$, an analogue of the HRP compound I, is identified by physicochemical means only in the system with peracid. Recently an example of Mn^{V} oxo-complex was reported containing macrocyclic tetramide as a ligand.[115] In most other systems with donors of active oxygen either $PM^{IV}O$ (chemical analogue of the HRP compound II) or PM^{III}–OL has been identified, but in the latter compound the presence of L in the coordination sphere has not been proven. Because $PM^{IV}{=}O$ complexes are able to oxidize alkenes but are not sufficiently active to hydroxylate alkanes,[116,117] the ability to react with alkanes is explained by formation of a small concentration of more active intermediates. Some authors believe that in all the systems based on metal porphyrins with donors of active oxygen, $P^{+\cdot}M^{IV}O$ is the active intermediate reacting with alkanes,[112,118] and it could be formed, for example, in the process of disproportionation of two ferryl ($PFe^{IV}O$) complexes. Others[81d] are inclined to think that active complexes reacting with alkanes could be different, depending on the structure of the initial compound and the reaction conditions. Differences in stereoselectivity (Table II.13) and KIE values (Table II.18) in the systems with different oxygen donors are indicative of different active species present.[83b]

Oxoiron(IV) porphyrins $(TMP)Fe^{IV}{=}O$ prepared by chemical and electrochemical oxidation of iron(III) porphyrins were found[118a] to react with substituted styrenes with quite different selectivity from that observed for reaction of $(TMP^{+})Fe^{IV}{=}O$. Whereas the latter readily reacts with *cis*-β-methylstyrene affording mainly *cis*-epoxide (cis-to-trans ratio = 11.3), the former ($(TMP)Fe^{IV}{=}O$) gives a cis-to-trans ratio of only 1.0 to 1.3. It is interesting that although the reactivity of $(TMP)Fe^{IV}{=}O$ towards olefins is comparatively low, it is highly reactive towards saturated molecules of methylene chloride (used as one of the solvents), producing $(TMP)Fe^{III}Cl$ even at low temperatures. We shall discuss possible implications later.

It may be noted that $P^{+\cdot}M^{IV}O$ must be positively charged, and, depending on the medium and the nature of a counterion, there can be different particles in solution starting from free ions $[P^{+\cdot}M^{IV}O]^{+}$ in polar media to ion pairs $[P^{+\cdot}M^{IV}O]^{+}L^{-}$ and covalently bonded $[PM^{III}OL]$ in less polar and nonpolar media. Naturally, their activity towards alkanes will be different. Nucleophilic and electrophilic assistance are of importance in producing more active $[P^{+\cdot}M^{IV}O]^{+}$ particles from presumably

TABLE II.18
Kinetic Isotope Effects (KIEs) in Cyclohexane Oxidation (C_6H_{12} vs. C_6D_{12}) with Different Oxidants and Catalysts

Oxidant	Medium	Catalyst	KIE
PhIO	CH_2Cl_2	MnTDCPPCl	2.6 ± 0.4
		MnTpyPP($O_2C_8F_{15}$)	5.7
		MnTpyPP(OAc)	11.8
		FeTDCPPCl	9 ± 3
NaOCl	C_6H_6–H_2O	MnTMPCl	7.0 ± 0.4
		FeTFPPCl	10.9 ± 0.9
		FeTMPCl	21.9 ± 1.9
	PhCN–H_2O	FeTFPPCl	6.0 ± 0.6
		FeTMPCl	10.2 ± 0.6
	C_6H_6–H_2O + 15 *M* NaOH	FeTMPCl	11.2 ± 0.5
NaOBr	C_6H_6–H_2O	FeTMPCl	8.5 ± 0.9
$KHSO_5$	CH_2Cl_2–phosphate	MnTDCPPCl	2.1 ± 0.2
		MnTMPCl	3.1
n-Bu_4NHSO_5	CH_2Cl_2–H_2O	MnTDCPPCl	2.8 ± 0.6
		FeTDCPPCl	2.3 ± 1.2
		FeTPPCl	2.3 ± 0.3
O_2, Zn powder	CH_3CN–0.2 *M* AcOH	FeTFPPCl	2.6 ± 0.3
		MnTPPCl	2.8

less active [$PM^{III}OL$] ones. Axial ligands, such as imidazol, will play the role of a nucleophile and protic acid may facilitate heterolytic cleavage of the O–L bond (*vide infra*).

To make a reliable interpretation the concentration of active species in each system has to be determined. This problem cannot be solved on the sole basis of kinetic data. On the one hand, the constant ratio of products may be observed along the reaction path with two or even more kinds of active particles present, particularly when those are produced from one precursor with some definite ratio of their formation rate constants. On the other hand, the ratio of the reaction products may change in the process of the reaction even with one kind of active species, particularly when a consecutive oxidation is involved. Comparison of the reactivities of alkanes and alkenes is not sufficiently informative because different species can be responsible for these two processes.

The systems based on ROOH or H_2O_2 may generate several kinds of active species: hydroxyl or alkoxyl radicals and porphyrin oxo complexes. In the case of homolytic cleavage of the O–O bond of hydrogen peroxide or alkyl hydroxyperoxide coordinated to metal porphyrin, $PM^{IV}{=}O$ is produced together with OH or OR free radicals. Heterolytic cleavage results in formation of $P^{+}M^{IV}O$ and water and the corresponding alcohol. It may be thought that in the absence of an axial ligand on the metal porphyrin mostly homolytic O–O bond cleavage proceeds in alkyl hydroperoxides.[119-121] However, Traylor and Ciccone suggest formation of alkylperoxo complex of iron(III) with subsequent heterolytic splitting of the O–O bond, which is catalyzed by acid.[122]

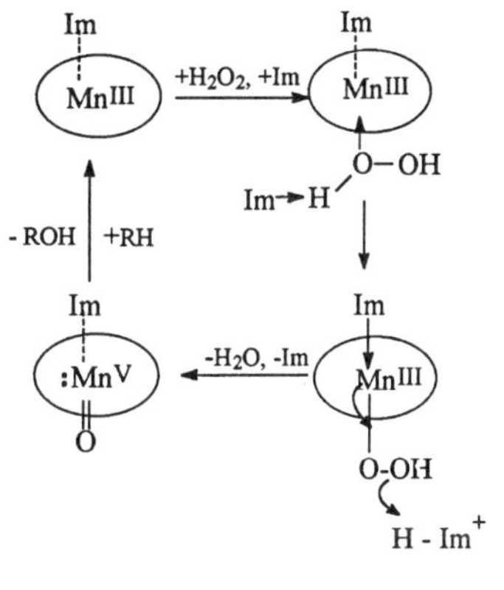

FIGURE II.17 Possible roles of imidazol in a substrate hydroxylation with H_2O_2 catalyzed by manganese porphyrin. (Adapted from Battioni, P. et al., *J. Am. Chem. Soc.*, 110, 8462, 1988.)

NMR investigation has shown[123] that upon the activation of *tert*-BuOOH by Fe^{II}(TTP), alkylperoxo iron(III) complex is formed, which produces (N-MeIm)Fe^{IV}(TTP)O. In the systems based on cumyl hydroxyperoxide and *tert*-butyl hydroperoxide with iron and manganese porphyrins, kinetic characteristics and reaction selectivities turned out to be independent of the metal porphyrin nature and this has led the authors to the conclusion that free RO and ROO radicals are the active species.[124-126] Recently several researchers found that heterolytic cleavage of ROOH by iron or manganese porphyrins at least is not the main route of the process.[120,121,126] From the identity of kinetic characteristics and product composition in the presence of H_2O_2 or PhIO with TDCPPMnCl and imidazol Mansuy et al. came to the conclusion about the identity of active particles produced by both these oxygen donors: high-valent $PMn^{V}O$ (or $P^{+\cdot}M^{IV}O$) intermediates were formed upon the heterolytic cleavage of the O–O bond in coordinated hydrogen peroxide or O atom transfer from iodosylbenzene to metal porphyrin.[127] The authors explain the high activity of the system with H_2O_2 in the presence of imidazol by the latter two functions: as an axial ligand with nucleophilic assistance of heterolytic cleavage of the O–O bond as well as an acid giving electrophylic assistance (Figure II.17).

Panicucci and Bruice[128] and Traylor and Xu[117] have found the rate-limiting stage of metal porphyrin reactions with H_2O_2 to be the O–O bond and O–H bonds cleavage in the hydrogen peroxide molecule coordinated on the metal porphyrin.

II.4.3.2 Active Species in the Systems with Dioxygen and Electron Donors

Similarly to the catalytic cycle of cytochrome P-450, the formation of active intermediates in model systems based on dioxygen and electron donors proceeds in several consecutive steps. The first, usually rate-controlling step, is a metal porphyrin reduction. Reduced metalloporphyrin adds dioxygen with a high rate constant (on the order of 10^7 $M^{-1}s^{-1}$)[129] to form a dioxygen complex. This reaction is reversible, but in the case of sterically hindered FeTpivPPCl with N-methylimidazol or 1-tritylimidazol[130] the complex TpivPP$Fe^{II}(O_2)$(Im) can be isolated. Tabushi et al.[131]

have made a detailed investigation of this complex decomposition. In ethanol–toluene solution at 25°C the complex decomposes very slowly (pseudo-first-order constant of $1.9 \times 10^{-4} s^{-1}$) but the addition of a reducing agent (H_2/Pt) considerably increases it decomposition rate (rate constant of 1.8×10^{9} $M^{-1}s^{-1}$). A similar rate constant was found for the reduction of $TpivPPFe^{II}(O_2)$(Tr-Im) by the radical cation of methyl viologen: $k = (2.2 \pm 0.4) \times 10^{9}$ $M^{-1}s^{-1}$.[132] Thus in the presence of a reductant the step following O_2 addition is the reduction of the dioxygen complex:

$$PM^{III} \xrightarrow{e} PM^{II} \overset{O_2}{\rightleftharpoons} PM^{II}(O_2) \xrightarrow{e} PM^{III}(O_2)^{2-} \tag{59}$$

Protonation of the oxygen atom in the peroxo group leads to a homolytic or heterolytic cleavage of the O–O bond (similarly to the systems on the base of H_2O_2). The mechanism of O–O cleavage depends on a solvent.[133] In tetrahydrofurane the reaction of peroxoiron(III) tetra(*o*-pivaloylamidophenyl)porphyrin with $HBF_4{\cdot}Et_2O$ at –53°C produces red oxoiron(IV) derivative $THFPFe^{IV}{=}O$, which is inert even towards cyclohexene, and at room temperature produces initial iron porphyrin. The peroxo complex reaction with $HBF_4{\cdot}Et_2O$ at –83°C in toluene produces the green intermediate $P^{+\cdot}Fe^{IV}{=}O$; in methylene chloride a mixture of the red and green intermediates are produced. It should be noted that this solvent influence is different from that found by Groves and Wantanabe[105] in the system Fe^{III}TMPX–peracid (X = OH, Cl, or mercaptobenzoate, MCB) for which only $P^{+\cdot}Fe^{IV}O$ is produced in methylene chloride (for X = Cl or OH); $PFe^{IV}O$ is formed in toluene for X = OH and N-oxide of iron porphyrin is observed for X = MCB. Thus, there is no direct correlation for all model systems between solvent polarity and heterolytic or homolytic O–O bond cleavage in metalloporphyrin intermediates. Therefore to establish the nature of the active particles tests are essential.

Kinetics of alkanes oxidation was investigated in details in the system O_2-carboxylic acid–Zn powder in CH_3CN (with and without methyl viologen, MV^{2+}).[92] In the absence of MV^{2+} dioxygen is mainly consumed in the reaction with reduced iron porphyrin. The use of stronger acids, e.g., sulfuric instead of acetic, leads to smaller oxidation rates in correspondence with competitive reduction of dioxygen complex of metalloporphyrin:

$$PFeO_2 + e \rightarrow P(FeO_2)^- \tag{60}$$

and its decomposition:

$$PFeO_2 + H^+ \rightarrow PFe^+ + HO_2{\cdot} \tag{61}$$

These experimental findings show that dioxygen is activated directly on the porphyrin complex. The introduction of MV^{2+} leads to the increase of O_2 consumption rate and at sufficiently high MV^{2+} concentration the rate ceases to depend on metalloporphyrin concentration in the reaction mixture. Noticeable concentrations of H_2O_2 are detected in the reaction solution under these conditions (about 10^{-2} M)

and the nature of an acid exerts no more influence on the rate of hydrocarbon oxidation. The role of metalloporphyrin is to form an active particle with hydrogen peroxide produced in reaction with reduced methyl viologen in the presence of an acid. The formation of HO_2 radicals and H_2O_2 may lead to the suggestion that several active species, hydroxyl and alkoxyl radicals along with active metalloporphyrin complexes, interact with hydrocarbon molecules with comparable rates. To verify this suggestion tests were made of stereoselectivity, KIE values, and substrate selectivity. Racemization was observed in the oxidation of *cis-* and *trans-*1,2-dimethylcyclohexane (Table II.12). Relatively low KIEs for the oxidation of cyclohexane and cyclohexane-d_6 were also observed (Table II.18). This may be explained by both free radicals and iron porphyrin complexes as active species.

Substrate selectivity measurements turned out to be more informative. These were made for competitive oxidation of cyclopentane and cyclohexane (so-called parameter 5/6 was measured, which is the ratio of respective rate constants corrected for the difference in the number of C–H bonds in C_5H_{10} and in C_6H_{12}). Because both alcohols and ketones are produced in the system in parallel (this was established in kinetic experiments) calculation of the 5/6 parameter was made by three ways: for initial rates of the alcohols formation, $(5/6)_1$, the same for the ketones, $(5/6)_2$, and for the sum of both the alcohols and ketones, $(5/6)_3$. The results are presented in Table II.19.[83b] It can be seen that for each metalloporphyrin the introduction of MV^{2+} or hydroquinone leads to considerable changes in $(5/6)_1$ and $(5/6)_2$ parameters, while $(5/6)_3$ remains constant within the limits of experimental errors, depending only on the nature of metalloporphyrin. A simple analysis shows[92] that this corresponds to the presence in the system of a single kind of active particles producing the major part of both the alcohols and ketones (with their ratio depending on added reagents). Other active species if present do not make any significant contribution to the reaction products. This also shows that no other products besides alcohols and ketones are produced. This approach is particularly effective at the initial stages of the reaction when balance experiments are difficult to perform. It should be noted that additional reagents that change the 5/6 parameters must not affect the properties of the active particles, e.g., forming complexes with the latter, such as tetrahydrofurane, dimethylsulfoxide, etc.

Thus, the true substrate selectivity for the active particle is reflected in the parameter $(5/6)_3$. The absence of the influence of methyl viologen on this parameter and its dependence on the nature of the metalloporphyrin shows that MV^{2+} does not change the direct reagent and that metalloporphyrin constitutes an important part of it. A $(5/6)_3$ numerical value of 0.38 to 0.50 is characteristic for electrophilic attack on alkanes by positively charged reagents[134] and is essentially different from the value of this parameter in the absence of the catalyst (0.71 ± 0.07). To identify the active particles in the system O_2–AcOH–Zn–MV^{2+} generating H_2O_2 and those based on H_2O_2 itself the parameter $(5/6)_3$ was determined in the system with H_2O_2 in the presence of FeTDCPPCl. The results show that in the latter system in the presence of AcOH and (or) imidazol the parameter $(5/6)_3$ does not depend on the presence or absence of hydroquinone and coincides with that in the system based on O_2 and proton and electron donors. In the absence of the acid and imidazol in the system with H_2O_2, $(5/6)_3 = 0.67 \pm 0.03$, i.e., coincides with this parameter in the system

TABLE II.19
Substrate Selectivity Parameters in Cyclohexane Oxidation in Acetonitrile with Different Metalloporphyrins[a]

1. In the System with O_2 + Zn Powder, [AcOH] = 0.2 *M*, [Im] = 0.05 *M*

Catalyst	[MV], $M \times 10^3$	[Hydroquinone], M	$(5/6)_1$	$(5/6)_2$	$(5/6)_3$
None	7.5	0	0.67	1.00	0.71
FeTPPCl	0	0	0.23	0.77	0.45
	7.5	0	0.24	1.05	0.45
FeTFPPCl	0	0	0.18	0.75	0.38
	7.5	0	0.26	0.77	0.38
	7.5	0.04	0.24	1.25	0.37
FeTDCPPCl	7.5	0	0.38	0.77	0.50
	7.5	0.04	0.40	1.25	0.49
MnTPPCl	0	0	0.4	1.6	0.7
	7.5	0	0.7	—	0.7

2. In the System with H_2O_2 as an Oxidant (Catalyst: FeTDCPPCl, [MV] = 0)

[AcOH], *M*	[Im], *M*	[Hydroquinone], *M*	$(5/6)_1$	$(5/6)_2$	$(5/6)_3$
0	0	0	0.54	0.85	0.67
0	0	0.09	0.33	0.73	0.49
0.1	0	0	0.48	0.52	0.50
0	0.05	0	0.35	0.65	0.49
0.1	0.05	0	0.44	0.55	0.50
0	0.05	0.09	0.32	0.73	0.49

[a] Calculated as the ratios of the initial rates of the formation of alcohols $(5/6)_1$, ketones $(5/6)_2$, and the total rates ratio (alcohols + ketones) $(5/6)_3$.

O_2–AcOH–Zn–MV^{2+} in the absence of metalloporphyrin. Therefore in the system under consideration with O_2 (with or without MV^{2+} and Im) and in the system with H_2O_2 with at least one of the two reagents (Im or AcOH) the reagent attacking the C–H bond is PFeOOH or $(PFeO)^+$. In the system with H_2O_2 (but without Im and AcOH) catalyzed by iron porphyrin, and in the system O_2–AcOH–Zn–MV^{2+} without a catalyst the direct reagents are presumably cycloalkoxyl radicals formed in a radical-chain process developing after initial H atom abstraction by a hydroxyl radical $(5/6)_3$ for the OH radical is known to be 1.1).[20] An OH radical can be formed in a homolytic cleavage of H_2O_2 (free or coordinated on iron porphyrin).

In the reaction catalyzed by MnTPPCl, distinct from the catalysis by iron porphyrins, the value of $(5/6)_3$ is very close to that observed for the system in the absence of metalloporphyrin. However, partial retention of configuration for this case (Table II.12) shows that in the presence of MnTPPCl the reaction is at least in part non free radical. Mansuy et al.[90] demonstrated the characteristics for the oxidation of olefins and of *n*-heptane in the systems MnTPPCl–O_2–AcOH–Zn–MeIm and PhIO–MnTPPCl–MeIm. On the basis of all the results obtained so far it may be concluded that the same mechanism of active particles formation operates in the presence of iron and manganese porphyrins in the systems with electron and proton donors. In these systems the active intermediates seem to be $(PMO)^+$ or PMOL formed as a result of reductive activation of O_2 in the presence of a proton donor either directly on a metalloporphyrin or via intermediate formation of H_2O_2 in a catalytic cycle similar to that of cytochrome P-450. The active particles in these systems as well as in the systems based on various donors of active oxygen are apparently somewhat different for different reagents. This follows from different stereoselectivities (Table II.12) and KIE values (Table II.18) for different systems generating active metalloporphyrin complexes. As mentioned above, these differences may be connected with different solvating properties of media and the nature of counterions in the active metalloporphyrin complexes.

II.4.3.3 Active Species in the Systems with Dioxygen without a Reductant

The high selectivity of Ellis and Lyons systems (Table II.13) and the absence of a reducing agent attracts much interest to these systems. Absorption spectra obtained at room temperature show that in the system $FeTF_5PPCl–O_2–C_6H_6$–*iso*-butane, $Fe^{II}TF_5PPCl$ and Fe-μ-oxodimer, $(FeTF_5PP)_2O$, are formed, the latter isolated from the solution.[94] Nonhalogenated iron porphyrins FeTPP form inert μ-oxo complexes unable to catalyze alkane oxidation. As a result of the halogenation the electron density is shifted from porphyrin rings. Compared with nonhalogenated porphyrins, iron reduction is believed to be facilitated; iron(II) formed reacts with dioxygen and then with another iron(II) complex, producing peroxodimer, which can decompose into two $PFe^{IV}{=}O$ particles considered to be active intermediates in the system. However, this scheme is apparently oversimplified, and in fact free radical-chain mechanisms have been shown[134] to be operating in the system. The oxidation is strongly inhibited by a radical trap, 2,6-di-*tert*-butyl-4-methylphenol. The radicals are suggested to be generated by oxidation and reduction of alkyl hydroperoxides.

II.4.3.4 Active Species in the Photocatalytic System with Dioxygen

A free radical oxidation mechanism is proposed in this system on the basis of the process selectivity: no epoxides are produced from olefins and high yields of alcohols and ketones are detected (5000 and 4500 turnovers for cyclohexanol and cyclohexanone, respectively).[95] Active species are suggested to be hydroxyl radicals originated at the photoactivation of $PFe^{III}–OH$ and peroxoradicals ROO produced in alkyl radicals reaction with dioxygen.

II.4.3.5 Active Species in the Systems with Metalloporphyrins on Support

These systems were investigated mainly with the purpose of increasing the product yields and selectivity (Table II.17) on the basis of the substantially increased catalyst stability demonstrated as compared with similar systems in solution. It is generally accepted that the active intermediates in metalloporphyrin systems on support are identical to those in homogeneous media. In fact, this is probably often the case, but it has to be taken into account that the binding of the catalyst with a support may change the former coordination properties and the ability to form the active intermediates. Consideration of the published results shows that free radicals may be active intermediates at least in some of the systems. An example may be the system based on PhIO with Mn^{III} (T4PyP)Cl supported on SiO_2.[135] The results demonstrate strong influence of competing substrates introduced in the system on the pentane oxidation selectivity: the pentanol-to-pentanone ratios were 4, 16, and 1.6 in the presence of adamantane, cyclohexane, and cyclooctane, respectively. The mechanism proposed by the authors[135] involves metalloporphyrin oxo complex attack on an alkane without free radicals leaving the solvent cage. The differences in selectivity seem to be more easily explained by involvement of a radical chain mechanism (the interaction of the chains of different alkanes oxidations may change the product ratio). The decrease of the substrate selectivity observed in this system compared with that of the soluble metalloporphyrin system can be explained by involvement of a free radical process. Thus, a superposition appears to be probable in this system of nonradical and free radical mechanisms.

II.4.4 MECHANISMS OF ALKANE OXIDATION

The mechanism of hydroxylation of nonactivated C–H bonds in alkanes catalyzed by metalloporphyrins definitely can be different in different systems. Free radicals can be formed and can initiate a radical-chain mechanism, although it is often difficult to determine that this is the case. The radical character of the intermediates formed in the reaction is confirmed by the observation that such reactions carried out using synthetic metalloporphyrins and single oxygen atom donors in the presence of CCl_3Br give substantial amounts of alkyl bromides.[136a] Ellis–Lyons systems presumably involve a radical-chain mechanism, porphyrin complexes facilitating the decomposition of peroxides formed intermediately. Photochemical oxidation reactions, as mentioned above, were described as free radical-chain processes.

Reactions in the presence of reductants, dioxygen, and electron transfer agents may lead to H_2O_2 formation in the reaction

$$O_2 \xrightarrow{e,H^+} HO_2\cdot \xrightarrow{e,H^+} H_2O_2 \qquad (62)$$

and its further decomposition with participation of reduced iron porphyrin will produce hydroxyl radicals, reacting further with hydrocarbons to afford alkyl radicals.

At the same time, in a number of systems the reaction of porphyrin iron complexes with C–H bonds does not seem to involve radicals, at least free radicals in solution. For example, this was established in the systems transferring oxygen atoms to metalloporphyrin complexes to produce a high-valent iron porphyrin complex, presumably $P^{\dagger}Fe^{IV}O$, or its close analogue. High selectivity, in particular stereoselectivity, e.g., in the systems with PhIO or NaOCl and $NaHSO_5$ and the absence of the influence of inhibitors of radical-chain reactions, no doubt, require some form of direct oxygen atom transfer to form an alcohol molecule. It may be thought that these systems are real models of cytochrome P-450, which definitely must avoid free radicals. We have mentioned above that the oxygen rebound mechanism, with a free radical formed and immediately rebound with an oxygen atom on iron, is particularly widely used in the literature to explain the observed results for the natural systems, and the same mechanism is generally proposed for the chemical model systems. Similarly to enzymatic systems, direct insertion of the oxygen atom into a

C–H bond in chemical model systems is considered unlikely mostly because of the high KIE and also isomerization, which is often observed in the products of hydroxylation. It is expected that the isotope effect should be low for oxygen atom insertion and no isomerization in the products should be observed. However, we have seen in the preceding section that high isotope effects are sometimes likely due to proton tunneling and therefore do not provide direct evidence for the linear transition state expected for initial hydrogen abstraction from a hydrocarbon molecule. As to the isomerization, it can be reconciled with oxygen atom insertion if this step is preceded by a *five-coordinate carbon intermediate.* Model systems provide additional support for these arguments, although free radicals are not excluded in certain reactions with high-valent iron porphyrin complexes, particularly for hydrocarbons and their derivatives with weak C–H bonds. In some of the model systems proton tunneling is particularly likely, e.g., very high isotope effects obtained for NaOCl–iron porphyrin systems and their temperature dependence fully correspond to the tunneling involvement. For example, k_H/k_D values for the oxidation of cyclohexane in the system TMPFeCl–NaOCl reach 21.9 at 20°C (Table II.18). Moreover, Arrhenius temperature dependence of KIE shows that the ratio of preexponential factors $A_H/A_D = 0.01$ and $E_H - E_D = 4.5 \pm 0.5$ kcal/mol.[136b] For isotope effects based only on zero-energy difference the lower limit of A_H/A_D is about 0.7 to 0.9[18] and therefore a strong contribution of proton tunneling in this reaction appears to be certain. Intramolecular isotope effect obtained for specially prepared substrate containing deuterium confirmed the tunneling mechanism for iron porphyrin complexes.[136c] It is natural to accept that in other reactions with a similar mechanism, as well as in hydroxylations catalyzed by P-450 itself, high isotope effects that are difficult to reconcile with "classic" interpretation involve some tunneling contribution, although k_H/k_D and A_H/A_D values may be close to the extreme theoretical values found for the "classic" approach (k_H/k_D of about 8 at room temperature). Indeed, it is more logical to accept that the tunneling contribution is always present and differs quantitatively for reactions of similar mechanism than to believe that there is a strong effect of tunneling in certain cases while in others it disappears completely.

A question then arises of why KIEs are so different if similar mechanisms take part in hydroxylations. In some of the systems this could be the result of a complex reaction mechanism with a rate-controlling step other than C–H cleavage. However, even when the hydroxylation determines the rate strong differences are often observed. It should be noted that for the systems with high KIE ($k_H/k_D > 6$) the increase of the KIE is accompanied by the increase in stereoselectivity. For example, for FeTMPCl with NaOCl a record retention of configuration is observed (92%) for *cis*- and *trans*-1,2-dimethylcyclohexane hydroxylation.[100] At the same time, for the systems with relatively low KIE (about 2) the stereoselectivity varies greatly for different systems, e.g., from 0% for the system FeTPP–O_2–Zn–AcOH to 90% for the system MnTDCPPCl–$KHSO_5$ (Table II.12).

Tunneling must be caused by high and narrow potential barriers. For metalloporphyrins the barriers may be due to steric influences of the anion Y in PMOY formed in the reaction of the metalloporphyrin PM with an oxygen donor YO. As was mentioned, the extent of O–Y bond ionization may be different depending on the media polarity, acidity, and the presence of a base X that may function as an axial ligand at the metalloporphyrin oxo complexes. The decrease of KIE and the stereoselectivity and the increase of the reaction rate may be connected with the transition to the free ion $[P^{\ddagger}Fe^{IV}O]^+$ from ion pairs $[P^{\ddagger}Fe^{IV}O]^+Y^-$ and from covalently bound $PFe^{III}OY$. It may be remembered that the axial cysteine $-S^-$ ligand seems to facilitate the O–O bond heterolytic cleavage in natural cytochrome P-450 systems.

Addition of alkali increases the rate and decreases the KIE in the system TMP-FeCl–NaOCl in a benzene–water biphasic system, OH^- possibly functioning as an axial ligand in the metalloporphyrin:

$$HO^- + \underset{P}{Fe} - O - Cl \rightarrow HO - \underset{P}{Fe} = O + Cl^- \qquad (63)$$

A similar role may be played by imidazol and an acid in the systems involving metalloporphyrins and H_2O_2, e.g., in the system $PFe–O_2–Zn–MV^{2+}$, which was described earlier. The acid may facilitate the ionization of the O–O bond protonating an oxygen atom:

$$PFe^{III}OOH + HA \rightarrow P^{\ddagger}Fe^{IV}O + H_2O + A^- \qquad (64)$$

The same effect can be produced by increasing polarity: it will increase the extent of ionization, the potential barrier will be decreased, and therefore KIE will be decreased as well. Another explanation for smaller isotope effects in more polar media may be found taking into account the effect of the proton tunneling: in more polar media the proton is "heavier" because it is solvated by the solvent molecules, therefore the mass ratio of solvated deuteron and proton is less than that of the free entities.

These considerations may explain the finding that although the iron porphyrin oxo complex is the active particle in the systems involving hydrogen peroxide, the KIE values are small and no stereoselectivity is observed. The reaction rate may be close to that of P-450 hydroxylation in which high KIEs and selectivity may be due to the protein surrounding. At the same time, the substrate selectivity 5/6 < 0.5 for the model system and this value corresponds more to the bound oxygen attack on carbon atom (perhaps via a five-coordinate carbon intermediate) than to a radical abstraction of the H atom for which 5/6 values are close to unity. It may be suggested that for cyclopentane and cyclohexane the reaction proceeds via the insertion of the electrophilic O atom into the C–H bond or even with the intermediate formation of the iron–carbon bond and subsequent reductive elimination of ROH from Fe(R)OH formed. Although it is difficult to reconcile this mechanism with the absence of stereoselectivity in the cases of *cis*- and *trans*-1,2-dimethylcyclohexanes, it is not excluded that for different substrates different mechanisms operate, and whereas for hydroxylation of tertiary C–H bonds free radicals are formed, a concerted mechanism prevails for secondary C–H bonds in cycloalkanes. In all cases an alcohol will be produced independently of the reaction mechanism.

Ketones are often formed together with alcohols and kinetic data show that at the early stages of the reaction they are formed in parallel. Usually it is explained by a free radical mechanism. The alcohol and ketone may be produced upon the disproportionation of two hydroperoxide radicals:

$$2RO_2\cdot \rightarrow ROH + R'_2C{=}O + O_2 \tag{65}$$

and ketone can be also the product of a hydroperoxide rearrangement. However, there must be another mechanism of ketone formation not involving free alkyl radicals.

Thus in the system O_2–PM–Zn–MV^{2+}, described above, there was much evidence against a radical-chain mechanism and at least a part of the products (alcohols and ketones) did not come from a hydroperoxide. The formation of ketones in parallel with alcohols may be explained by suggesting that an iron alcoholate is formed initially after oxygen atom addition to carbon:

$$PFe^{V}O + H-\overset{|}{\underset{|}{C}}-H \rightarrow PFe^{IV}-\overset{H}{\underset{+}{O}}-\overset{|}{\underset{|}{C}}-H = PFe^{III}-O-\overset{|}{\underset{|}{C}}-H + H^+ \tag{66}$$

A subsequent reaction with a base (to abstract H^+) or with dioxygen (to abstract H atom) will lead to the ketone formation:

$$PFe^{III}-O-\overset{|}{\underset{|}{C}}-H + O_2 \rightarrow PFe^{II} + O{=}\overset{|}{\underset{|}{C}} + HO_2\cdot \tag{67}$$

Accordingly, relative yields of the ketone are increased when dioxygen is present. The proposed mechanism explains the influence of an acid, reducing agent, and O_2 on the alcohol/ketone ratio.[92]

We have seen that some of the isomerization results for camphor hydroxylation in a natural P-450 system do not agree with the oxygen rebound mechanism but are better explained by an oxygen atom addition to a four-coordinate carbon, leading to the partial isomerization. A similar mechanism may be suggested for model systems and some data for model systems as well as for cytochrome P-450 are also better explained by this nonradical mechanism.

Let us consider, for example, the results presented in one paper[137] for the hydroxylation of norbornane and *exo,exo,exo,exo*-2,3,5,6-tetradeuterionorbornane catalyzed by halogenated iron(III) porphyrins such as iron(III)tetrakis(2,6-dichlorophenyl)octabromoporphyrin with pentafluoroiodosylbenzene as an oxidant. The reaction yields the products which consist of *exo*-norbornan-2-ol, *endo*-norbonan-2-ol, and 2-norbornanone in ratios of 86:13:<1 and 84:16:<1 for both substrates, respectively. The results show that the reaction involves partial loss of stereochemistry and has a large isotope effect. The results of the stereochemistry studies are presented in Table II.20.

The authors[137] explain the data by the oxygen rebound mechanism: formation of a radical in a cage, its partial isomerization (from exo to endo and from endo to exo via rotation), and a collapse of the radical pair formed to yield the alcohol. According

TABLE II.20
The Results of Stereochemistry Studies for Hydroxylation of *exo,exo,exo,exo*-4-D-norbornane

Substrate	Products (%) formed	
	Without epimerization	
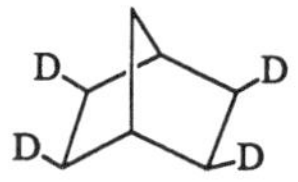	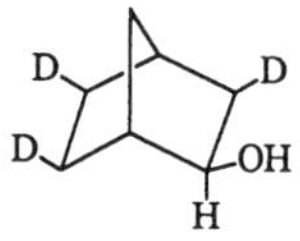 (70)	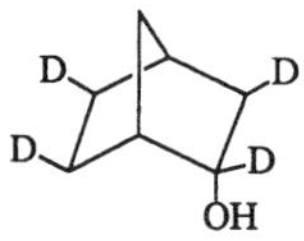 (14)
	With epimerization	
	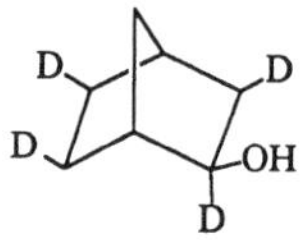 (14)	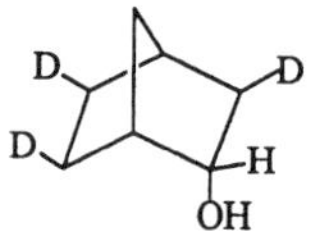 (2)

to this mechanism, D abstraction from the exo position produces a D_3 radical that can isomerize to D_3 *endo*-alcohol in the process of a rotation and subsequent recombination, and D_4 *endo*-radical produces D_4 *exo*-alcohol in a similar process. From this scheme the authors estimate exo–endo leakage to be 4% and endo–exo leakage to be 42%. These results gave them the rate constants ratios $k^c_{ex}/k^r = 8$ and $k^c_{en}/k^r = 0.6$ (here k^c_{ex} and k^c_{en} are the rate constants of the collapse of the radical pairs for exo and endo structures, respectively, and k^r is the rate constant for the rotation to form exo structure from endo and vice versa). Thus, the ratios are more than one order of magnitude different, but, as seen in Table II.20, the resulting percentage of the isomerized and nonisomerized forms are rather close for both *exo-* and *endo-*isomers. It is a sheer coincidence for the accepted mechanism. If, however, we assume that there is a nonradical mechanism of exo–endo isomerization by H–D rotation in the intermediate complex with oxygen bound to carbon in a five-coordinate carbon intermediate *without* the rotation of the radical(s) as a whole with the subsequent O atom insertion into the C–H bond then it seems natural that the ratios are not very different. The difference observed may be explained by the kinetic isotope effect: the hydroxylation involves an insertion of an O atom into a C–H or C–D bond whereas the isomerization in both cases proceeds through H–D and D–H rotation; for C–H bond insertion the ratio of hydroxylation to H–D rotation must be higher. Accordingly, the difference is more pronounced for cytochrome P-450, for which the KIE is much higher than for the model system (11.5 and 5, respectively).

Intermediate formation of a five-coordinate carbon intermediate in reactions of iron–oxoporphyrin complexes with saturated organic compounds may explain the facile reaction of $PFe^{IV}O$ complex with methylene chloride used as a solvent to produce HCl.[118a] The C–H bond in methylene chloride is strong and no H atom abstraction could be expected. However, close contacts between some of the atoms in the intermediate may facilitate the reaction of oxygenation accompanied by HCl elimination:

$$PFe^{IV}O + CH_2Cl_2 \rightarrow \left[PFe\cdots O{=}C\langle^{H}_{Cl}\right] \rightarrow PFe^{II} + O{=}C\langle^{H}_{Cl} + HCl \rightarrow \quad (68)$$

$$PFe^{II} + CO + 2HCl$$

Another mechanism of C–H hydroxylation may involve intermediate Fe–C bond formation, which will correspond to electrophilic substitution followed by reductive elimination of the alcohol molecule.

$$PFeO + RH \rightarrow PFe(R)OH \rightarrow PFe + ROH \quad (69)$$

There is evidence that an iron–carbon bond may be formed at least in some cases, e.g., in the metabolic oxidation of a benzodioxole derivative.[138]

Formation of Fe–C bonds may proceed without free radical formation via the mechanism of electrophilic substitution.

II.4.5 CONCLUSION

The results and considerations presented in this section demonstrate much progress in the development of new and very effective metalloporphyrin-based systems and in understanding the hydroxylation mechanism in natural and model systems. They demonstrate also that both free radical and nonradical hydroxylation mechanisms are possible in model systems, and although the proposed schemes are in many cases still speculative and somewhat contradictory, nonradical mechanisms are very likely in some of the systems and this supports the view that cytochrome P-450 and its closest chemical model systems can directly insert an oxygen atom into the C–H bond of a substrate, possibly via a five-coordinate carbon intermediate.

ABBREVIATIONS USED IN SECTION II.4

TPP	tetraphenylporphyrin dianion
$TPPF_{20}\beta$-Br_8	*meso*-tetrakis(pentafluorophenyl)-β-octabromoporphyrin dianion
TFPP	tetra(*o*-fluorophenyl)porphyrin dianion
TDCPP	*meso*-tetrakis(2,6-dichlorophenyl)porphyrin dianion
$Cl_{12}TMP$	*meso*-tetrakis(3-chloro-2,4,6-trimethylphenyl)-β-octachloroporphyrin dianion
TTP	*meso*-tetrakis(*p*-tolyl)porphyrin dianion
TF_5PP	tetrakis(pentafluorophenyl)porphyrin dianion
TMP	*meso*-tetramesitylporphyrin dianion
Br_8TMP	*meso*-tetramesityl-β-octabromoporphyrin dianion
TpivPP	α,α,α,α,-*meso*-tetrakis(*o*-pivaloylamidophenyl)porphyrin dianion
TDOPP	tetrakis (*p*-hexadecyloxophenyl)porphyrin dianion
TTMPP	*meso*-tetrakis(2,4,6-trimethoxyphenyl)porphyrin dianion
TTPPP	*meso*-tetrakis(2,4,6-triphenylphenyl)porphyrin dianion
$Cl_{12}TSMP$	*meso*-tetrakis(3-chloro-5-sulfonato-2,4,6-trimethylphenyl)-β-octachloroporphyrin dianion

T4MPyP	*meso*-tetrakis(4-N-methylpyridinium)porphyrin dianion
ChP	tetrakis(*o*-cholenylamidophenyl)porphyrin dianion
PVP	polyvinylpyridine
MV	methyl viologen
Im	imidazole
DMPCh	dimyristoylphosphocholine
CD	cyclodextrine
MeCD	methyl-β-cyclodextrine
PPAA	phenylperoxyacetic acid

II.5 OXIDATION OF ALKANES BY NONPORPHYRIN HIGH-VALENT METAL COMPLEXES, ANALOGUES OF NONHEME MONO- AND DIOXYGENASES

The mechanism of C–H oxidation by nonheme mono- and dioxygenases seems to be essentially similar to that of heme-containing enzymes, particularly of cytochrome P-450. The coupling with the oxidation of an electron donor (NADH or NADPH) helps to create a high-valent iron compound containing monomeric ($Fe^{V}{=}O$ or $Fe^{IV}{=}O$ and dimeric $Fe^{IV}{-}O{-}Fe^{IV}{=}O$ or $Fe^{IV}{<}^{O}_{O}{>}Fe^{IV}$ fragments capable of inserting an O atom into the C–H bond. The special case of ketoglutarate-dependent dioxygenase involves formation of similar $Fe^{IV}{=}O$ fragment by the coupling with the oxidation of the ketoglutarate, accompanied by formation of thermodynamically stable CO_2.

The use of metalloporphyrins has led (as we have seen) to effective models of cytochrome P-450, both structural and functional, which can readily hydroxylate C–H bonds in typical substrates of cytochrome P-450, including alkanes. At the same time, various mechanisms may materialize in enzymatic and particularly in model systems, including free radical formation and concerted insertion of an O atom into the C–H bond. The same situation exists in nonheme model systems, with perhaps an even greater number of possible oxidation mechanisms. In spite of the many efforts and the large number of systems investigated, only few of them can pretend to be real models of enzymatic oxidation, particularly for the most interesting case of methane monooxygenase.

II.5.1 HYDROXYLATION OF HYDROCARBONS COUPLED WITH OXIDATION OF LOW-VALENT METAL IONS AND COMPLEXES

Participation of metal ions in most enzyme systems, which hydroxylate organic compounds, was the reason behind many attempts to create analogous purely chemical systems.

In 1954 Udenfriend et al. proposed a nonenzymatic system as a possible model of monooxygenase.[139] It involves an iron(II) complex with EDTA and ascorbic acid and is capable of hydroxylating aromatic compounds. Later Hamilton et al. found that this system was able to hydroxylate cyclohexane (although with very low yields) and epoxidize cyclohexene.[140a] Hamilton proposed the so-called oxenoid mechanism of direct oxygen atom insertion into the C–H bond.[140b] Subsequently a number of

TABLE II.21
Systems Involving Ions or Metal Complexes and Dioxygen

System	Substrate	Products *(o:m:p)*
Fe^{2+} + ascorbate + EDTA	Anisole	Methoxyphenols (43:18:39)
Fe^{2+} + thiosalicylic acid	Toluene	Cresols (61:13:16)
Fe^{2+} + Tetrahydropterin	Phenylalanine	Tyrosine (2:1:1)
Hemin + thiosalicylic acid	Cyclohexane	Cyclohexanol
Fe^{2+} + EDTA	Anisole	Methoxyphenols (56:8:36)
$TiCl_3$	Toluene	Cresols (61:16:23)
$TiCl_3$	Anisole	Methoxyphenols (54:0:46)
CuCl	Toluene	Cresols (29:29:49)
CuCl	Anisole	Methoxyphenols (75:8:17)
$SnHPO_4$	Toluene	Cresols (46:38:16)
$SnHPO_4$	Anisole	Methoxyphenols (53:20:17)
$MoCl_3$	Cyclohexane	Cyclohexanol

similar systems was proposed. For example, Ullrich found two models that are more effective than the Udenfriend system. One of the includes a stannous phosphate complex in water solution,[141] and the other one a ferrous complex with mercaptobenzoic acid in aqueous acetone solution.[142] Table II.21[143] lists the products of some coupled hydroxylation of hydrocarbons induced by metal ion oxidation by dioxygen.

The reactions are analogous to those catalyzed by monooxygenases: hydroxylation of alkanes and aromatic compounds, and in some cases epoxidation of olefins. At the same time, there are some essential differences between biological oxidation and the processes occurring in the model systems. In the latter there is no NIH shift (migration of a hydrogen atom to the position next to the point of oxygen insertion), which is characteristic of biological systems in the hydroxylation of aromatic compounds. The model systems are mostly less selective in the reactions with aromatic compounds (a large amount of *meta*-substituted product is formed); they do not show stereospecificity in the hydroxylation of alkanes (e.g., a mixture of both possible isomers in the hydroxylation of *cis*- and *trans*-dimethylcyclohexane is produced).

Very simple systems, for example, chlorides of some metallic ions, are effective in coupled oxidation of hydrocarbons in organic solvents. Thus, coupled oxidation of C_1–C_4 alkanes and $SnCl_2$ was found to proceed in acetonitrile at room temperature. The yield of alcohols, which are the oxidation products, reach 7 to 15% per $SnCl_2$. Comparatively high yields of cyclohexane oxidation products were obtained in the presence of oxidizing tin(II) or iron(II) chlorides when the reaction is carried out in acetonitrile as a solvent (Table II.22).[144,145] The yield, if based on the metal salt present, may be increased when MCl_2 concentration is decreased. Under optimum conditions at very small MCl_2 concentrations the yield of products increases to 20% with $SnCl_2$ and to 30% with $FeCl_2$.

However, there is much evidence that the mechanism of the oxidation is different from that of the enzymatic oxidation. In effect a free radical mechanism seems to function in these reactions with hydroxyl radicals as the most likely intermediates. In particular, consideration of the distribution of isopentane hydroxylation products

TABLE II.22
Hydroxylation of Cyclohexane Coupled with the Oxidation of $SnCl_2$ and $FeCl_2$

MCl_2	Medium	$[MCl_2]$, $M \times 10^2$	$[C_6H_{12}]$, $M \times 10^2$	ΔO_2, $M \times 10^2$	$[C_6H_{10}O]$, $M \times 10^3$	$[C_6H_{11}OH]$, $M \times 10^3$	Yield on MCl_2, mol%	Yield on C_6H_{12}, mol%
$SnCl_2$	4 ml CH_3CN, 0.7 ml H_2O, 0.2 *N* HCl	10.0	19.0	5.5		16.0	16	8.5
$FeCl_2$	4 ml CH_3CN, 0.25 *N* HCl	4.0	23.0	2.1	0.6	3.25	9.6	1.7

in various systems, including photochemical hydroxylation by hydrogen peroxide, where free hydroxyl radicals are known to be formed as intermediates, shows that the ratios of selectivities with respect to the site of the attack (1°:2°:3°) in the systems involving Fe^{2+} and Sn^{2+} ions are very close to each other and to the selectivity observed for the attack by hydroxyl. This leads to the conclusion that the same active species (presumably hydroxyl radicals) participate in all these and similar systems.[146] This is also supported by a low isotope effect (k_H/k_D = 1.2 to 1.3) when comparing reactions of C_6H_{12} and C_6D_{12}. The yield of cyclohexane hydroxylation products depends strongly on the nature of the solvent, again in agreement with hydroxyl radicals as active species of the hydroxylation. The radicals react with both the substrate and the solvent and the yield is determined largely by this competition. The rate constants of the hydroxyl radical reactions with cyclohexane are $5 \times 10^9\ M^{-1}s^{-1}$; with acetonitrile, $2.1 \times 10^6\ M^{-1}s^{-1}$; with acetone, $8.1 \times 10^7\ M^{-1}s^{-1}$; and with methanol, $7 \times 10^8\ M^{-1}s^{-1}$.[147] Therefore, only in acetonitrile when $[C_6H_{12}] > 10^{-2}\ M$ may the reaction of OH radicals with the solvent be neglected, whereas in other solvents, such as acetone or methanol, it competes with that of the substrate. Accordingly, in acetonitrile the yield is the highest. In the case of iron salts the accepted mechanism involves the formation of hydroxyl radicals:

$$Fe^{2+} + O_2 + H^+ \rightarrow Fe^{3+} + HO_2^{\cdot} \quad (70)$$

$$Fe^{2+} + HO_2^{\cdot} + H^+ \rightarrow Fe^{3+} + H_2O_2 \quad (71)$$

$$Fe^{2+} + H_2O_2 \rightarrow Fe^{3+} + {}^{\cdot}OH + OH^- \quad (72)$$

$$Fe^{2+} + {}^{\cdot}OH \rightarrow Fe^{3+} + OH^- \quad (73)$$

$$2H^+ + 2OH^- \rightarrow 2H_2O \quad (74)$$

The total reaction:

$$4Fe^{2+} + O_2 + 4H^+ \rightarrow 4Fe^{3+} + 2H_2O \quad (75)$$

For stannous chloride a detailed study of its oxidation showed that it proceeds according to a radical chain mechanism with branching involving hydroxyl radicals.[2]

Initiation

$$Sn^{2+} + O_2 \xrightarrow{H_2O} Sn^{3+} + HO_2^{\cdot} \tag{76}$$

$$Sn^{3+} + O_2 \xrightarrow{H_2O} Sn^{4+} + HO_2^{\cdot} \tag{77}$$

Branching

$$HO_2^{\cdot} + Sn^{2+} \begin{cases} \xrightarrow{\beta} H_2O_2 + Sn^{3+} \\ \xrightarrow{1-\beta} {}^{\bullet}OH + Sn^{4+} \end{cases} \tag{78}$$

$$H_2O_2 + Sn^{2+} \begin{cases} \xrightarrow{\alpha} {}^{\bullet}OH + Sn^{3+} \\ \xrightarrow{1-\alpha} H_2O + Sn^{4+} \end{cases} \tag{79}$$

Termination

$${}^{\bullet}OH + Sn^{2+} \rightarrow Sn^{3+} \tag{80}$$

$$Sn^{3+} + Sn^{3+} \rightarrow Sn^{2+} + Sn^{4+} \tag{81}$$

Here α and β are the probabilities of one-electron reactions of Sn^{2+} with H_2O_2 and $HO_2^{\cdot}$, respectively. In steady-state conditions

$$\alpha\beta k_1[Sn^{3+}][O_2] = k_5[Sn^{3+}]^2 \tag{82}$$

The reaction rate is practically equal to that of chain propagation

$$-d[O_2]/dt = k_1[Sn^{3+}][O_2] = \alpha\beta k_1^2/k_5[O_2]^2 \tag{83}$$

and, in agreement with the experimental data, depends only on dioxygen concentration.

The branching chain mechanism is confirmed by the sensitivity of the reaction to the inhibitor concentration. There exists some limiting inhibitor concentration that practically stops the reaction (the induction period increases to infinity on the addition of a definite amount of an inhibitor). Hydrogen peroxide was detected in the reaction mixture in the course of the oxidation of stannous chloride. Using the reaction scheme, it was possible to estimate the rate constant of the reaction of H_2O_2 with Sn^{2+} from the dependence of H_2O_2 concentration on time. The rate constant found coincided with the value obtained for the reaction of H_2O_2 with Sn^{2+} determined in separate experiments, thus confirming the proposed mechanism.

Although the probability (α) of a one-electron reaction of hydrogen peroxide with Sn^{2+} to give a hydroxyl radical is not very high ($\alpha \approx 5 \times 10^{-4}$ for the reaction in acetonitrile), it is sufficient to ensure the high rate of radical formation in the branching chain reaction under steady-state conditions. Thus, the autooxidation of stannous chloride occurs by a branching chain mechanism with the participation of OH and HO_2 radicals. Reactions of radicals with hydrocarbons should naturally be taken into account in the process of their oxidation coupled with the autooxidation of tin(II) salts.

Upon the introduction of cyclohexane into aqueous acetonitrile solution of tin(II) chloride, the rate of dioxygen consumption increases and the induction period decreases. An increase of cyclohexane concentration decreases the induction period and then causes its complete disappearance. The rate of hydrogen peroxide production is also increased in the presence of cyclohexane. These data can be understood on the basis of the above chain mechanism of tin(II) oxidation, where free hydroxyl radicals intermediately formed react with cyclohexane.

$$^{\bullet}OH + C_6H_{12} \rightarrow H_2O + {}^{\bullet}C_6H_{11} \qquad (84)$$

Taking into account the value of this rate constant it is easy to see that with an appropriate hydrocarbon concentration all the hydroxyl radicals will interact with the latter to produce cyclohexyl radicals ($R^{\bullet}$). These very readily enter into the addition reaction in the presence of O_2:

$$R^{\bullet} + O_2 \rightarrow RO_2^{\bullet} \qquad (85)$$

In hydrocarbon autooxidation reactions, the hydroperoxo radicals form hydroperoxide or disappear in a biomolecular disproportionation process:

$$2RO_2^{\bullet} \rightarrow R'COR'' + ROH + O_2 \qquad (86)$$

However, in the presence of a high concentration of tin(II), practically all hydroperoxide radicals interact with Sn^{2+} in a manner similar to HO_2 radicals:

$$RO_2^{\cdot} + Sn^{2+} \xrightarrow{H_2O} \begin{cases} \xrightarrow{\beta} RO_2H + Sn^{3+} \\ \xrightarrow{1-\beta} RO^{\bullet} + Sn^{4+} \end{cases} \qquad (87)$$

The interaction of hydroperoxide with tin(II) is similar to that of hydrogen peroxide

$$RO_2H + Sn^{2+} \begin{cases} \xrightarrow{\alpha} RO^{\bullet} + Sn^{3+} \\ \xrightarrow{1-\alpha} ROH + Sn^{4+} \end{cases} \quad (88)$$

RO radicals react with RH and Sn^{2+}

$$RO^{\bullet} + RH \rightarrow ROH + R^{\bullet} \quad (89)$$

$$RO^{\bullet} + Sn^{2+} \xrightarrow{H_2O} ROH + Sn^{3+} \quad (90)$$

leading to the formation of the alcohol, which is practically the only product.

Thus, the participation of Sn^{2+} (as well as Fe^{2+} in the case of iron salts) ensures high selectivity of the reaction to produce cyclohexanol as virtually the sole product. It is important that the situation changes if the stannous chloride concentration is low; for example, if hydrogen peroxide is added in excess. In this case Sn^{2+} quickly disappears and the products formed are different from those observed in the oxidation by dioxygen. If we disregard the differences in reactions with high and low Sn^{2+} concentration, it may lead us to an incorrect conclusion about the nonradical character of hydrocarbon oxidation coupled with stannous chloride oxidation by dioxygen.

It is interesting to note that if all the RO radicals disappear in the reaction with hydrocarbon (when its concentration is sufficient), the stoichiometry of the reaction may correspond to the ratio $Sn^{2+}:O_2:ROH = 1:1:1$, that is, the same as the stoichiometry of oxidation in the presence of monooxygenase.

Thus, in systems with an excess of reducing agent the reaction mechanism can be explained by the participation of the hydroxyl radicals. Obviously this mechanism was unacceptable for living organisms, at least for the reason of nonselective side reactions, which would undoubtedly destroy the enzyme. Naturally, when an external reductant is added the mechanism remains the same although the yield may be greatly increased. For example, if the coupled oxidation of $SnCl_2$ and C_6H_{12} is carried out in the presence of metallic tin, then the yield of cyclohexanol can be increased several times. For $FeCl_2$-containing systems, metallic mercury may be used as a reducing agent. This also increases the yield of hydroxylation products. In the systems involving, besides metal compounds, such donors of electrons as hydrazobenzenes or *o*-phenylenediamine, the yield of hydroxylation products may exceed the amount of the metal compound taken. For example, cyclohexane produces cyclohexanol and cyclohexanone in the presence of such salts as $FeCl_2$, $FeBr_2$, FeI_2, $FeClO_4$, $MnCl_2$, $MnBr_2$, VCl_2, $CuCl_2$, and $NiCl_2$, the yields being not very different from each other.[148] In the case of cyclohexene, cyclohex-1-ene-3-ol is mainly produced (without epoxide formation). Hydroxylation of 2-methylbutane produces isomeric alcohols, the selectivity with respect to the site of the attack following the "normal" order: 3°>2°>1°. In toluene hydroxylation, the formation of both benzyl alcohol and isomeric cresols is observed, *o*- and *p*-isomers prevailing. An oxenoid (i.e., with O atom insertion) mechanism of hydroxylation has been suggested,[148] but in view of all the considerations presented above the mechanism via OH radicals seems more plausible: it is just nonselective reactions of the OH radicals that make the reaction look like selective hydroxylation, because the radicals readily attack mainly hydrocarbon molecules when the latter are in excess. Another example of such reactions is hydroxylation of cyclohexane in the presence of $CuCl_2$, with phenylhydrazine as a reducing agent and dioxygen as an oxidant.[149] The yield of products (cyclohexanol and cyclohexanone) may correspond to 40 redox cycles per mole of Cu salt with one portion of phenyl hydrazine, and even several hundred cycles, if new portions of the reducing agent are added after the previous phenyl hydrazine has been oxidized. In all these cases a free radical mechanism, probably including OH radicals, is very likely.

II.5.1.1 Reactions with an Excess of Oxidizing Agent. Oxygen Transfer from Peroxo Metal Complexes

The Fenton-like reaction

$$M^{2+} + H_2O_2 \rightarrow M^{3+} + {}^{\bullet}OH + OH^- \quad (91)$$

is presumably responsible for the free radical mechanism of hydrocarbon hydroxylation in coupled reactions with the oxidation of low-valent metal ions, particularly in aqueous solutions. As we have seen in the case of metalloporphyrins in the presence of an oxidant, three-valent iron and manganese complexes produced high-valent compounds capable of reacting with alkanes in the nonradical mechanism, avoiding dangerous OH radicals. We may apply the same approach to nonheme iron and other metal complexes. The difference between porphyrin complexes and nonheme iron mononuclear complexes is the easy electron transfer from the porphyrin ring to form a radical cation, facilitating two-electron oxidation and avoiding a radical path. A homolytic cleavage of the O–O bond, e.g., in peroxo complexes, may lead to the formation of free OH or OR radical together with Fe^{IV}=O, which is the presumed active species in the biological oxidation. In mononuclear iron complex of the ketoglutarate-dependent dioxygenase, CO_2 formation helps in the use of dioxygen as a double-two electron oxidant, yielding Fe^{IV}=O and succinate from Fe^{II} and ketoglutarate, respectively. In ordinary cases mononuclear complexes will produce Fe^{V}=O together with OH^- or H_2O in the case of the heterolytic O–O cleavage: Fe^{III}–OOH → Fe^{V}=O + OH^-. Perhaps thermodynamically it is easier to form Fe^{IV}=O and OH radicals in many cases and therefore several different active particles might be expected in the case of nonheme iron complexes in the presence of oxygen atom transfer agents, e.g., peroxides. This is probably the reason for much of the controversy that still exists in a number of publications in this field. There is some evidence that a nonradical or at least a nonhydroxyl radical mechanism operates in such systems, while there is also much evidence that free alkyl radicals are formed in similar systems.

Stereoselectivity is observed in the oxidation of tin(II) organic derivatives (e.g., diphenyltin) in aprotic media.[150] Cyclohexane oxidation produces cyclohexanol; for the isopentane oxidation the ratio of selectivities (1°:2°:3° = 1:16:107) differs substantially

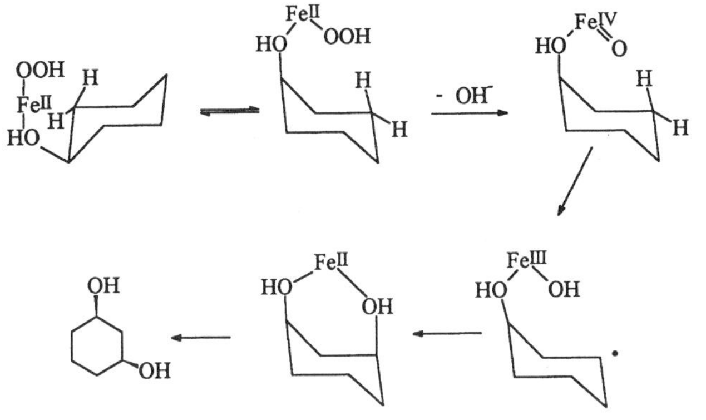

SCHEME II.5 Mechanism of hydroxylation of cyclohexanol by H_2O_2 in the presence of ferrous perchlorate.

from that for hydroxyl radicals. The hydroxylation of *cis*- and *trans*-dimethylcyclohexane proceeds with a partial retention of configuration of the alcohol formed (about 20%). In the absence of the hydrocarbon, for example, in CCL_4 solution, dioxygen forms an adduct with diphenyltin, presumably a peroxide, which is, thus, capable of hydrocarbon hydroxylation.

A study of cyclohexane co-oxidation with ferrous chloride by dioxygen in ethanol[151] led to the conclusion that, depending on the conditions, the hydroxylating species may be both OH and HO_2 radicals and such species as FeO_2^{2+}, as well as $FeOOH^{2+}$ and FeO^{2+}.

Hydroxylation by species other than ·OH and HO_2· radicals has been demonstrated in one paper,[152] where cyclohexanol was hydroxylated by hydrogen peroxide in the presence of ferrous perchlorate in acetonitrile. In this case the product is mainly *cis*-1,3-cyclohexanediol (with more than 70% yield). Stereoselective oxidation was observed to occur in this system also for 7-hydroxynorbornane. Selective oxidation of alkanes may be explained by assuming that iron(IV) oxo complex is formed with cyclohexanol as one of the ligands. Therefore a cyclic transition state is formed, which results in formation of the final product (Scheme II.5).

The same system was later used to oxidize simple alkanes.[153a] Methane is found to be oxidized with k_H/k_D equal to 5, which is very different from the isotope effect in the case of OH radicals. The ethanol yield in ethane oxidation with respect to H_2O_2 approaches 15% and the reaction is catalytic with respect to the iron complex. The oxidation of methylcyclopropane affords mainly cyclopropylmethanol and dimethylsulfoxide is oxidized preferentially to dimethylsulfone, which is difficult to reconcile with the radical mechanism of the oxidation. The addition of water to this system up to 1.5 *M* concentration led to the increase in the rates of ethanol and phenol formation and did not change the selectivity in dimethylsulfoxide oxidation. At higher H_2O concentrations the yields of sulfone and ethanol diminished, however, that of phenol increased. The data confirm the suggestion that at high water concentrations the system is transformed into an ordinary Fenton system generating free OH radicals. At lower water concentrations iron(IV) oxo complex may be the reactive species and its formation from the iron peroxo complex may require some water presence.

A more recent work[153b] demonstrated that for several iron mono-, di-, and tetranuclear complexes, including $Fe(ClO_4)_3{\cdot}6H_2O$, in acetonitrile using H_2O_2/O_2 as the oxidant, cyclohexyl hydroperoxide was formed as an intermediate in the cyclohexane oxidation. The final products were cyclohexanol and cyclohexanone. The reaction was inhibited by 2,4,6-tri-*tert*-butylphenol. The results were explained by a free radical-chain mechanism in which cyclohexyl radical reacted with dioxygen forming cyclohexylperoxyl radical, which abstracts a hydrogen atom to give hydroperoxide. The results are convincing, but they do not prove that a free radical mechanism is the *only* path of the reaction.

Different mechanisms have to be often suggested even for reactions with the same catalysts acting under similar conditions. For example, catalytic oxidation of alkanes with hydrogen peroxide in the presence of *cis*-$[Ru(dmp)_2(S)_2(PF_6)_2$ (S = H_2O, CH_3CN; dmp = 2,9-dimethyl-1,10-phenantroline) was found to involve free radicals, whereas oxidation of alkenes with dioxygen catalyzed by the same compound results in high-selectivity epoxidation (94%) and is better explained by a nonradical mechanism.[154a] Allyl rearrangement was not observed upon oxidation of 3,3,6,6-d_4-cyclohexene by the high-valent $Ru^{IV}{=}O$ complex and the KIE was found to be k_H/k_D = 18 at 25°C.[154b] These data are consistent with nonradical oxygen insertion, although from them alone an oxygen rebound mechanism may be also suggested.

Olefins can be oxidized selectively by alkyl hydroperoxides, e.g., *t*-butyl hydroperoxide, in the presence of metal complexes, those of molybdenum, vanadium, or titanium, producing high yields of epoxides.[154c] The mechanism involves nonradical oxygen atom transfer from alkyl peroxo ligand in the coordination sphere of the metal.

There is certain evidence that metal peroxo complexes can oxidize alkanes also by nonradical mechanisms. Peroxo complexes formed in hydrogen peroxide reaction with high-valent metal compounds in acid solutions can react with alkanes, e.g., cyclohexane, and produce alcohols and ketones in a reaction that does not seem to involve free radicals, at least in the main part.[2] Particularly convincing results were obtained recently by Moiseev et al.[154d] Thus peroxovanadium complex produced in reaction of $VO(acac)_2$ or KVO_3 with H_2O_2 reacts with C_6H_{12} to form cyclohexanol and cyclohexanone in acetic acid. All the evidence is against a free radical mechanism, e.g., the reaction is not inhibited by typical inhibitors of chain reactions, and the reaction rate and the yield of cyclohexanol is not influenced in the presence of *iso*-propyl alcohol and carbon tetrachloride, which would be easily attacked by free radicals if they were present. The reaction is much faster in trifluoroacetic acid, producing cyclohexanol and cyclohexyl trifluoroacetate but no cyclohexanone. The conversion of the hydrocarbon reaches 5 to 8% in acetic acid in tenfold excess of hydrogen peroxide and 95 to 98% in trifluoroacetic acid with only four- to fivefold excess of H_2O_2. When the H_2O_2/C_6H_{12} ratio is diminished, 100% selectivity for reacted cyclohexane is reached.

TABLE II.23
The Nomenclature of Gif Chemistry[a]

System	Catalyst	Electron source		Oxidant
Gif^{III}	Fe(II)	Fe^0		O_2
Gif^{IV}	Fe(II)	Zn^0		O_2
GO	Fe(II)	Cathode		O_2
$GoAgg^{I}$	Fe(II)		KO_2/Ar	
$GoAgg^{II}$	Fe(III)		H_2O_2	
$GoAgg^{III}$	Fe(III)/picolinic acid		H_2O_2	
GoChAgg	Cu(II)		H_2O_2	
Cu^0/O_2	Cu(I)?	Cu^0		O_2

[a] The nomenclature of the Gif systems is geographically based: G stands for Gif-sur-Yvette, O is for Orsay, Agg is for Aggieland, Texas A&M, and Ch is for Chernogolovka, Russia.

In the absence of a substrate *ozone* is detected in the products of the hydrogen peroxide decomposition in the presence of acid: about 1% in acetic acid and *10 to 15%* in CF_3COOH of all the gaseous products formed. Formation of ozone shows that the oxygen atom is easily transferred from the peroxo ligand on vanadium(V) in acidic solutions. Apparently O_3^{2-} ligand is formed initially on vanadium. It is this species or that preceding it that might be the source of ozone and also of the oxygen atom attacking the alkane molecule. It should be noted that ozone reacts with alkanes in superacid media apparently also in a nonradical fashion, thus the complex of vanadium(V) with O_3^{2-} ligand (producing ozone) can be transformed into a complex of vanadium(V) with ozone. Trifluoroacetic acid as the medium and also probably providing some ligands to vanadium makes it and oxygen-containing ligands strong electrophiles capable, in particular, of reacting with alkanes. Peroxoacids are known to react with alkenes to produce epoxides and pertrifluoroacetic acid is able to hydroxylate alkanes, mostly tertiary C–H bonds.

II.5.1.2 Gif Chemistry

In 1983 Barton et al. published a paper[155] describing the oxidation of adamantane by molecular oxygen with unusual selectivity. The catalytic system was based on iron powder and a solvent containing acetic acid in pyridine and also H_2S to increase the solubility of iron. The secondary C–H bonds were oxidized preferentially, the main product being ketone. For the following 10 years the system was intensely developed, and several varieties of the system were discovered, which are presented in Table II.23 (see Reference 156a).

All the systems united by the name Gif chemistry are characterized by unusual regioselectivity: $2° > 3° \approx 1°$. (In the oxidation of adamantane, C_2/C_3 reached the value of 22; for *trans*-decaline it is 36.) The ketone/alcohol ratio in the oxidation products varies from 15:1 to 4:1 for different systems. It is seen from Table II.23 that all the variants of the Gif systems have similar composition: a catalyst (metal ion) and an oxidant (O_2 in the case of electron donors Fe°, Zn°, and cathode), or

KO_2 + Fe^{II} or H_2O_2. The important general feature of all the Gif systems is the use of a mixture of pyridine and acetic acid as a solvent in different proportions, but as a rule 10:1.

In the majority of the Gif systems the catalyst is an iron complex with various ligands that strongly influence the reaction rate. Thus, the addition of pycolinic acid to $FeCl_3{\cdot}6H_2O$ greatly accelerates the rate, whereas pyridine-2 phosphoric acid completely inhibits the reaction. Interestingly, copper can replace iron in Gif systems and preserve the selectivity that is characteristic of the systems: preferential oxidation of the secondary C–H bonds and production of ketones as the main products of the oxidation.[156b] It was found that the regioselectivity (2°:3° ratio in the adamantane oxidation) can be increased when O_2 concentration is diminished.

It was later revealed that free radicals, at least tertiary free radicals, do function in the Gif system after all. The analysis of the products showed that tertiary adamantyl radicals produce the adamantylpyridines when reacting with pyridine; the yield of these products increases at the expense of tertirary alcohol upon the decrease of dioxygen concentration. Secondary adamantyl does not produce substitution products with pyridine even at low dioxygen concentrations and the true 2°:3° ratio, taking into account all the products of the reaction, remains constant and equal to 1.0:1.2. An important step in the understanding of the nature of the active species in the oxidation was the discovery that cyclohexyl hydroperoxide is the intermediate in the cyclohexane oxidation. When the latter is added to the reaction mixture (in the absence of the oxidant) it produces mainly cyclohexanone (the ketone:alcohol ratio reaches 10). Thus, the preferential production of ketones in the Gif systems is evidently due to the decomposition of the hydroperoxide: the solvent consists of an acid and a base and presumably performs the acid–base catalytic dehydration of the hydroperoxide to produce ketone. Alkyl hydroperoxide turned out to be the intermediate also in the systems containing copper. The experiments with $^{18}O_2$ showed that dioxygen (and not hydrogen peroxide) is the source of oxygen atoms in the hydroperoxide. It is natural to suggest that hydroperoxide is originated from alkyl radicals. It corresponds to competitive reaction of tertiary adamantyl radicals with dioxygen and pyridine and complete loss of the configuration in the oxidation of *trans*-decaline. Intermediate formation of secondary alkyl radicals is still uncertain. The experiments with the radical traps seem to confirm the existence of both secondary and tertiary adamantyl radicals. Thus the paramagnetic radical trap TEMPO (2,2,6,6-tetramethylpyridine-N-oxide) produces corresponding derivatives with secondary and tertiary adamantyls, the yield and selectivity factor (2°:3°) being the same as for the oxidation products at normal pressure. At the same time, the absence of secondary adamantylpyridine derivatives at low dioxygen pressure is a serious argument against the formation of free secondary adamantyl radicals. Future investigations have to clarify this controversy. Perhaps there are some products of secondary adamantyl radicals reactions with pyridine still unidentified in the reaction products. For example, it could be suggested that the intermediate radical produced upon the addition of sec-adamantyl to pyridine reacts with another sec-adamantyl radical at the 2nd or the 4th pyridine atom and forms heavy condensation products that have not yet been isolated or identified.

Barton and Doller propose the mechanism in which Fe^{V}-alkyl is formed and, reacting with O_2, produces alkyl hydroperoxide.[156a] The authors realize that there are still unanswered questions and consider the results a "Gif paradox". Moreover, they propose a somewhat romantic explanation for the results obtained that deserves to be presented in full.[156b]

> The only way that we can explain these results is by a hypothesis that the reagent that oxidizes the hydrocarbon is present in a dormant form (Sleeping Beauty) until it collides with the saturated hydrocarbon (the Prince) and reacts with the saturated C–H bond (the kiss) to form the real reagent, which immediately gives the iron–carbon bond (intermediate A). So, the hydrocarbon on contact with the iron species activates and reacts with the activate diron species without separation.

It is difficult to reconcile the proposed mechanism with all the results. It was already mentioned that high-valent metal complexes are unlikely to produce alkyl metal derivatives *from alkanes* because, being hard acids, the complexes would clearly prefer to react with stronger hard bases present. Hydroperoxide formation, no noticeable dependence on the nature of the metal (e.g., iron or copper), and the selectivity observed with different alkanes all point to free radicals as the most likely intermediates. Perhaps several different active species are present, including also $Fe^{IV}{=}O$, which make the systems different from the usual systems involving $\cdot OH$. $Fe^{IV}{=}O$ might behave as $Fe^{III}{-}O^{\bullet}$, reacting with RH by a free radical mechanism: $Fe^{III}{-}O^{\bullet} + RH \rightarrow Fe^{III}{-}OH + R^{\bullet}$. Clearly more work is needed to make definite conclusions about the mechanism.

II.5.2 MODELS OF KETOGLUTARATE-DEPENDENT DIOXYGENASE

Coupled oxidation of iron(II) and ketoglutarate permits the use of all four oxidation equivalents of dioxygen in one process. Carbon dioxide formation facilitates Fe^{II} oxidation to $Fe^{IV}{=}O$. It looks attractive for modeling because neither free radicals nor peroxide are intermediates in this oxidation.

To find if a simple mixture of reagents can provide a mechanism used in the enzymatic reaction a coupled oxidation of cyclohexane and ketoacids was investigated in acetonitrile at room temperature.[157] Cyclohexanol and cyclohexanone were detected as the oxidation products together with carbon dioxide. The yield of the products reached only 4% per dioxygen consumed, which is understandable in view of the absence of space organization in the system. The KIE (k_H/k_D) was found to be about 3. Presumably the mechanism at least partly is similar to that of the enzymatic oxidation.

In a more recent work[158] complexes of iron(II) with ketoacids were synthesized of the type $[Fe^{II}L(PhCOCOO)]ClO_4$. (L is tridentate ligand *tris*(2-pyridylmethyl)amine (TPA) or 6-methyl-2-pyridyl analogue (lutidyl) (TLA).) Reactions of both the complexes with O_2 lead to the decarboxylation of the α-ketoacid to produce benzoic acid with quantitative yield. In the case of $^{18}O_2$ one oxygen atom is transferred to benzoic acid. In the presence of 2,4-di-*tert*-Bu-phenol corresponding biphenols are produced, thus the second oxygen atom goes to the substrate molecule.

II.5.3 DINUCLEAR IRON COMPLEXES

Dinuclear iron complexes have clear advantages over mononuclear ones because they allow avoidance of one-electron reactions both in formation of Fe–O–Fe=O or $Fe^{IV}\langle{}^{O}_{O}\rangle Fe^{IV}$ intermediates and their further reaction with alkanes:

$$Fe^{IV} - O - Fe^{IV}{=}O + RH \rightarrow Fe^{III} - O - Fe^{III} + ROH \quad (92)$$

$$Fe^{IV}\langle{}^{O}_{O}\rangle Fe^{IV} + RH \rightarrow Fe^{III} - O - Fe^{III} + ROH \quad (93)$$

This is a possible reason for their presence in methane monooxygenase, because active methyl radicals formed from methane would be dangerous to the organic surrounding of the active center.

The first models of MMO that contained $Fe_2O(OAc)_2L_2$ were prepared by Wieghardt et al.[159] and Armstrong and Lippard.[160]

$L = C_6H_{15}N_2$ = 1,4,7-triazocyclononane[159]

$L = HB(pz)_3$ = hydro-*tris*(1-pyrazolyl)borate[160]

Later the whole series of dinuclear complexes were synthesized with different terminal and bridging ligands. All these complexes are good models for spectral and magnetic properties of hemerithrin and ribonucleotide reductase, but usually they do not function as acting models of methane monooxygenase, presumably because they do not possess a free vacancy for dioxygen binding in the iron coordination sphere.

Early attempts to prepare dinuclear complexes with open coordination spheres usually resulted in the dimerization of the dinuclear complexes formed to produce tetranuclear ones.

$$2Fe^{/}{}^{O}{}^{\backslash}Fe = Fe - O\langle{}^{Fe}_{Fe}\rangle O - Fe \quad (94)$$

Recently new methods were worked out for preparation of dinuclear complexes with labile coordination sites, in particular using self-assemblies from the simple molecules.[161a,b]

$$2FeOH^{2+} + 2L + 2RCOO^{-} \xrightarrow{S} \left\{Fe_2OL_2\left(RCOO^{-}\right)_2 S_2\right\}^{2+} + H_2O \quad (95)$$

where S = solvent molecule. Depending on conditions di- and tetranuclear complexes can be prepared using this reaction.

Dinuclear iron complexes were also produced by the use of a preorganized, cleft-shaped ligand, *m*-xylenediamine-bis(Kemp's triacid)imide. (μ-Oxo)bis(μ-carboxylato)di-iron(III) were synthesized with this ligand in the self-assembly synthesis.[161c]

Catalytic activity of synthetic dinuclear complexes in alkane oxidation is not sufficiently well studied and in known cases is not particularly high. In one of the early papers of this field,[162] an iron dinuclear complex with tripodal heptadentate ligand $Me_4N[Fe_2L(OAc)_2]$ was shown to produce μ-peroxo complex (ν[oo] = 884 cm^{-1}) with H_2O_2 and with excess of hydrogen peroxide epoxidizes olefins to yield epoxides (up to 10% with respect to H_2O_2). *cis*-Stillbene is oxidized producing 95% of *trans*- and 5% of *cis*-epoxide. A similar but mononuclear iron complex does not produce peroxo complex and does not catalyze the olefin epoxidation. No data were presented concerning the alkane oxidation.

Alkanes were found later to be oxidized by *tert*-BuOOH in the presence of $Fe_2O(OAc)_2Cl_2(bpy)_2$ in acetonitrile solution[163] and by O_2 in the presence of a reductant (zinc) and effector (acetic acid) in CH_2Cl_2.[164] The mononuclear complex $Fe[HB(pz)_3]_2^+$ was found to be catalytically inactive. Later the authors improved the system by introducing an additional stabilizing ligand, hexafluoroacetylacetonate, which played the additional role of the effector instead of acetic acid.[165]

The catalytic complexes are often not sufficiently stable and can decompose into mononuclear complexes or undergo destruction by ligand oxidation. Moreover, nonproductive reactions between oxidized and reduced iron complexes, such as $Fe^{IV}O + Fe^{II}$, very likely decrease the activity of the systems. It may be expected that heterogenized iron complexes on a support will be more stable and the nonproductive bimolecular reactions can be avoided.

To achieve this goal a synthesis of iron complexes on silica surface modified by imidazole groups was performed.[166] The data of UV-VIS spectra of the supported complexes as well as the EPR and Mössbauer spectra have led the authors[166] to the conclusion that dinuclear complexes are formed on the surface. In the presence of the supported catalyst, alkanes, including methane, were found to be oxidized by dioxygen to yield alcohols at room temperature in the presence of a reductant (zinc powder), effector (acetic acid), and electron transfer agent (methyl viologen) in acetonitrile. However, the regioselectivity observed shows that hydroxyl radicals are likely to be the active species in the hydroxylation. In another paper[167] the oxidation of cyclohexane was reported in the presence of similar supported iron catalysts processing two μ-oxo bridges and dimethylamino group instead of imidazol as in the previous complex and not containing carboxylate bridges. The complex oxidizes cyclohexane effectively (57 turnovers per mole of the catalyst) but is not active towards methane.

The complex $Fe_2OCl_4(DTBC)_2$ (DTBC = di-*tert*-butylcatechol) was shown[168] to catalyze the oxidation of methane and other alkanes in the presence of hydrazobenzene as a reductant. The complex turned out to be stable in the process of the reduction. A similar mononuclear complex is one order of magnitude less active than the dinuclear one. Regioselectivity difference when the dinuclear and mononuclear complexes are compared for the oxidation of adamantine and hexane led the authors to the conclusion that the dinuclear oxoferryl iron complex reacts with the alkane molecule. However, the structure of the postulated intermediated remains unclear.

Dinuclear high-valent iron complex (**3**) was obtained in the reaction of dinuclear (μ-oxo)di-iron(III) TPA complex (TPA = tris(2-pyridylmethyl)amine) with hydrogen peroxide.[169,170] The precursor of **3** was found to be $[Fe_2O(TPA)_2(OH)(H_2O)](ClO_4)_3$ (**2**), which can be obtained by the treatment of $[Fe_2O(TPA)_2(H_2O)](ClO_4)_4$ (**1**) with a base. Elemental analysis and electrospray ionization mass spectrometry show that **3** is a dinuclear complex; according to EXAFS spectrum it contains an $Fe_2(\mu\text{-}O)_2$ core and from its formula it follows that the complex oxidation state corresponds to $Fe^{III}Fe^{IV}$. This bis(μ-oxo)di-iron complex might be close to the active state of MMO. However, it may be suggested that in reaction with H_2O_2 initially the complex $Fe^{IV}Fe^{IV}$ is produced (which is the full analogue of the MMO active state), but in disproportionation with the initial $Fe^{III}Fe^{III}$ complex more stable (and less active) complex **3** is formed.

The kinetic isotope effect ($k_H/k_D = 5$) was observed for the cyclohexane oxidation in a similar system $[Fe_2O(TPA)Cl_2]$ $(ClO_4)_2$ with *t*-BuOOH.[171] This high value of the isotope effect speaks against free radical attack on the alkane molecule. However, dinuclear as well as mononuclear iron complexes have been shown to produce alkyl hydroperoxides,[153b] and free radicals can be produced in the interaction of ferryl complex with alkanes. The possibility of a parallel radical chain process together with a nonradical O atom insertion into the C–H bond complicates the investigation of model reactions.

Dinuclear iron complexes were shown[172] to be active in peroxidation of unsaturated fatty acids by dioxygen, while similar complexes of other metals as well as mono- and trinuclear complexes turned out to be inactive.

Dinuclear iron(II) complexes with 1,2-bis(2-hydroxibenzamido)benzene were found to be able to transfer O atoms from peracids and iodosylbenzene via intermediate FeO formation.[173] In the absence of a substrate FeO intermediate controlled by its spectrum undergoes the transformation ($\tau_{1/2}$ = 1.5 h at –78°C) to μ-oxo complex:

$$Fe^{II} + Fe^{IV}O{=}O \rightarrow Fe^{III}\text{–}O\text{–}Fe^{III}$$

A dinuclear iron complex $[Fe_2O(bpy)_2(OBz)_2S_2]$ $(ClO_4)_2$ (S = H_2O or MeOH) in acetonitrile catalyzes the oxidation of methane and other alkanes by hydrogen peroxide.[174] The solvent molecule is labile in these complexes and can apparently be substituted by hydrogen peroxide. The kinetic isotope effect ($k_H/k_D = 3.0$) and the selectivity in the oxidation of hexane, cyclopentane, and cyclohexane permits the conclusion that an FeO core is formed, presumably from FeOOH. The complexes with labile coordination spheres were found to be substantially more active in alkane oxidation than the complexes with stable and closed coordination spheres.

Summarizing the example mentioned, one can ascertain that only initial steps have been made in the search for synthetic functional models for methane monooxygenase and the problem is yet to be solved. Only very few iron dinuclear complexes are known so far to be able to oxidize alkanes, their activity is not too high, and the intermediates are not yet well characterized. At the same time, it may be hoped that creation of effective analogues of MMO will be achieved in the near future. Immobilization of the complex on a solid support is likely to be the best way to create most effective systems.

II.5.4 α-FORM OF OXYGEN ON ZEOLITE CONTAINING IRON

Donors of active oxygen, such as iodosylbenzene, peroxides, etc., can replace molecular oxygen as reducing agent in enzymatic and model systems producing active hydroxylating species. One of these donors is nitrous oxide, N_2O, which, being a gas, is particularly convenient for use in the case of supported complexes that can act as heterogeneous catalysts without a solvent. It was already mentioned that in this way nonproductive disproportionation reactions between complexes of different oxidation states can be avoided.

Efficient catalysts, interesting from the point of view of certain similarity with iron-containing enzymes, have been found using the decomposition of nitrous oxide over iron-containing ZSM-5 zeolite.[175a] N_2O is decomposed at 250°C leading to a specific oxygen form referred to as the α-form of oxygen. These zeolites catalyze the oxidation of aromatics at 300 to 350°C with high selectivity.[175b]

The interaction of benzene with this α-form of oxygen proceeds already at room temperature and results in nearly quantitative yield of phenol.[175c] Methane was reported to be oxidized *selectively to methanol also at room and lower temperatures* with the yields up to 100% of methane consumed. Methanol produced remain tightly bound to the iron complex in the zeolite and cannot be removed from the surface even at elevated temperatures. Upon the increase of temperature to 250 to 300°C, carbon monoxide starts to evolve, demonstrating the decomposition of the sample. Nevertheless, methanol can be extracted from the iron complex by acetonitrile. Dinuclear iron complex is suggested to be the active form in the zeolite in reaction with N_2O and further reaction with hydrocarbon.

Although the exact form of the active iron complex (or complexes) is not known and the reason for the activity of this particular zeolite is not clear, the results illustrate the possibility of oxygen atom transfer from the metal complex to the aromatic or aliphatic C–H bond, presumably in the same way as in the enzymatic process in methane monooxygenase. Indeed, a free radical mechanism for this highly selective oxygen atom transfer in its traditional form or in the more modern form of an oxygen rebound mechanism in the absence of a solvent would be very unlikely.

II.5.5 OTHER EVIDENCE OF NONRADICAL OXYGEN ATOM TRANSFER IN OXIDATION OF ALKANES IN CHEMICAL SYSTEMS

As can be seen, there are many indications that biological oxidation of alkanes, at least for the most important reactions involving alkanes with strong C–H bonds, including methane, proceed as a nonradical oxygen atom transfer following formation of a five-coordinate carbon intermediate. Model chemical studies support this conclusion. However, at the same time free radicals often really participate in the oxidation, and this always leaves some doubts about the mechanism: all the evidence for nonradical oxidation, although sometimes fairly convincing, is rather indirect.

Therefore it is desirable to consider other examples of oxygenation even definitely different from the enzymatic processes in which similar nonradical mechanisms can be suggested.

Low C–H bond energy facilitates a radical mechanism of H atom abstraction and isomerization of the olefin when hydroxylation of a C–H bond in an allylic position takes place is considered as evidence for an allyl type radical intermediate. This isomerization was indeed observed in reactions of high-valent iron and chromium porphyrin complexes with deuterated olefins (*vide supra*). However, this is not always the case and depends on the nature of a metal. For example, oxidation of 3,3,6,6-d_4-cyclohexene by dipyridyl oxo ruthenium(IV) complex is not accompanied by the allylic isomerization.[154b] Hence in this case the first step is probably a direct oxygen atom transfer with insertion into the C–H bond.

We have seen that the donors of single oxygen atoms, such as iodosylbenzene, peroxides, and hypochlorite, are often involved in nonradical oxygen transfer to carbon atoms of alkanes through mediation of metal complexes.

Nitrous oxide is one of the oxygen atom donors, though not very active at low temperatures. At elevated temperatures N_2O reacts with metal oxides to form a surface form of oxygen able to oxidize C–H bonds in alkanes. Molybdenum oxides catalyze the oxidation of methane with high selectivity to produce methanol and formaldehyde.[176a] The reaction mechanism is not clear, but it could be close to the mechanism of hydroxylation of methane at α-oxygen centers formed in the reaction of N_2O with iron-zeolite centers considered in the previous section.

Upon an increase of the temperature, carbon monoxide and dihydrogen are formed in some reactions of methane with the surface oxygen. This is not unexpected for the oxygen atom addition at high temperatures: hydrogen elimination becomes thermodynamically favorable because it corresponds to the considerable increase of entropy (see Section II.2.2). It is of importance that there are examples of *selective* formation of carbon monoxide and dihydrogen according to the stoichiometry:

$$[O] + CH_4 \rightarrow CO + 2H_2 \qquad (96)$$

with little or no water and carbon dioxide. This potentially important industrial process (because it provides the possibility of direct formation of synthesis gas from methane) provides strong indication that there is no hydrogen atom abstraction by the active oxygen atom, and addition is the most probable reaction.

It was reported, for example, that CO and H_2 have been obtained by reaction of CH_4 with preadsorbed oxygen on an Rh/SiO_2 catalyst at 573 to 773 K.[176b]

Another apparently similar reaction was observed in an electrochemical membrane reactor in which oxygen species were pumped to a rhodium anode through a solid electrolite YSZ (yttrium-stabilized zirconia).[176c] Partial oxidation of methane to carbon monoxide and dihydrogen was observed, and the complete oxidation to CO_2 and H_2O was found to be another reaction, independent of the first one. In this case the reaction temperature was 773 K.

Although rhodium apparently is particularly active in this reaction, oxygen activated by other metals can also react with methane in the same manner. For example, Otsuka et al. have obtained CO and H_2 with CeO_2 catalyst at 873 to 1073 K, where methane reacts with oxygen of CeO_2 to produce CO and H_2 reduces cerium dioxide.[176d]

The reaction between methane and dioxygen over platinum and rhodium surfaces in an adiabatic reactor produces mostly H_2 and CO (with greater than 90% selectivity for both) with almost complete conversion of CH_4 and O_2 at short reaction times (about 10^{-3} S).[176e] Rhodium was superior to platinum in producing more H_2 and less H_2O. The temperature in the reactor is high because the reaction is strongly exothermic, but at low O_2-to-CH_4 ratios it may not exceed 1100 K. The reaction mechanism is not completely clear, but it definitely does not include hydrogen atom abstraction by the activated oxygen on the catalyst, otherwise water would have been one of the reaction products.

Finally, catalytic oxidative carbonylation of methane in water solution to selectively produce acetic acid with a homogeneous rhodium catalyst was recently discovered by Lin and Sen.[176f] In all the cases presented above the reaction mechanism has still to be elucidated, but it is tempting to suggest that oxygen is activated in the coordination sphere of metal (e.g., rhodium) to be able to react with methane. At low temperatures the reaction leads to the insertion of the oxygen atom into the C–H bond to produce methanol (or acetic acid by further carbonylation) perhaps via a complex of methane with coordinated O atom, but at high temperatures oxygen displaces H_2 from methane to form formaldehyde, which is decomposed further to carbon monoxide and dihydrogen. A complex formation of methane with CeO_2 was indeed detected at lower temperatures.[176g] Some H–H bonding could be suggested to be formed in five-coordinate carbon intermediate complexes facilitating dihydrogen formation at high temperatures similarly to H_2 formation induced by protonation of alkanes.

The importance of the nature of the metal is manifested by the finding that methane does not react with oxygen species on the anode in the electrochemical membrane reactor mentioned above if rhodium as the material for anode is replaced by gold, but ethane does react and produces acetaldehyde.[177] It is not excluded therefore that metal reacts with methane and directly activates the C–H bond; e.g., the following reaction may be suggested:

$$MO + CH_4 \rightarrow M\langle{}^{CH_3}_{OH} \rightarrow M + CH_3OH \qquad (97)$$

II.5.6 CONCLUSION

The examples considered in the two last sections show that it is not too difficult to involve alkanes in an oxidation reaction coupled with oxidation of a reducing agent, particularly low-valent metal compounds. Alkanes with comparatively low C–H bond dissociation energy, e.g., cyclohexane, are oxidized in the presence of excess reducing agent, mainly via a radical mechanism, often with OH radicals interacting with C–H bonds. The presence of both dioxygen and the electron donor in protic media, e.g., water, gives rise to the chain scheme:

$$R^{\bullet} \xrightarrow{O_2} RO_2^{\bullet} \xrightarrow{e,H^+} RO_2H \xrightarrow{e,H^+} ROH + {}^{\bullet}OH \xrightarrow{RH} R^{\bullet} \qquad (98)$$

In aqueous solutions the reaction may selectively afford an alcohol as practically the only product at the early stages of the reaction. Then naturally the alcohol will also react with OH radicals.

The intermediate formation of hydroperoxide is indicative of such a mechanism, although in some of the systems (e.g., Gif system) some complications may appear that are not always easy to interpret, and that may be due, for example, to active participation of the solvent.

In the systems with metal complexes in the presence of an excess of the oxidation agent (e.g., peroxide) one may observe nonradical oxidation of alkanes, although the probability of radical formation in many cases remains high. Radical participation is particularly probable again in water or in other protic media. In media with small water content, peroxides of the metals present or high-valent metal complexes may be produced, and they can react with alkanes in a nonradical way. The examples are the peroxides of Sn^{IV}, Mo^{VI}, and V^{V}. In these cases nonradical oxidation to alcohols are observed and often ketones are formed in parallel reactions. Alcohols are produced in the process of oxygen atom insertion whereas ketones may be formed as a result of O^+ transfer from the metal complex with participation of an oxidant present, e.g., O_2 abstracting an H atom and producing ketone from an alkyloxy metal derivative. Of course, another source of ketone may be hydroperoxide formed in free radical-chain reaction and dehydrated in the presence of an acid–base catalyst.

So far effective hydroxylation in such reactions are observed only in cases of hydrocarbons with low C–H bond dissociation energy. Although several authors have successfully synthesized dinuclear iron complexes with structure close to that of MMO, their activity towards alkanes, particularly to methane, is not high. Presumably the complexes do not contain free sites to coordinate O_2, water, and methanol and they may not be sufficiently stable in the process of the catalytic cycle or sufficiently strong oxidizing agents.

The best functional model of the MMO known so far evidently is the α-oxygen center produced in reaction of nitrous oxide, N_2O, with iron-zeolite complex. The hydroxylation of methane proceeds with these centers at a fast rate and with 100% selectivity at room and lower temperatures. Perhaps the absence of protons helps to avoid free radical mechanisms. Extremely fast rate and high selectivity is indicative of a nonradical mechanism probably very close to that of the MMO functioning. The reaction proceeds most likely via addition of the oxygen atom, bound to iron, to the C–H bond forming a five-coordinate carbon intermediate with further O atom insertion into the C–H bond. The difference with the MMO is that it is not yet a *catalytic* system and instead of dioxygen, nitrous oxide must be used. Methanol formed is too tightly bound and the reaction stops at one cycle.

However, there are many examples of high-temperature chemistry of methane involving surface catalysts with the selectivity that is difficult to explain by hydrogen abstraction as the first step of the alkane reaction. Oxidation of methane with N_2O as the oxidating agent to form methanol and formaldehyde and reactions of methane with an oxygen atom at the metal surface to form $CO + 2H_2$ without formation of water and carbon dioxide can be better explained by oxygen atom addition (via a five-coordinate carbon intermediate) followed by insertion at comparatively low temperatures or by H_2 elimination at higher temperatures than by an ordinary free-radical mechanism (via a three-coordinate carbon, i.e., methyl radical, intermediate).

Catalytic oxidation or oxidative carbonylation of methane can be also observed in solution of platinum complexes, and in these cases the participation of metal (together with oxygen ligand) in the activation of the C–H bond is likely; e.g., electrophilic substitution may be suggested:

$$MO + RH \rightarrow M\langle{}^{R}_{OH} \rightarrow M + ROH \quad (99)$$

Nonradical O atom addition to methane to produce CH_3OH or $CH_2O + H_2$ or $CO + 2H_2$ seems more promising for creation of new industrial processes than further development of traditional chemistry involving free radical chain mechanisms.

We may also ascertain that the unity of the world is confirmed once again when considering various alkane oxidation reactions. Biological oxidation of alkanes is an example of a much larger field of C–H oxidations via O atom addition, starting with singlet oxygen atom reactions in the gas phase and hydrogen peroxide hydroxylation in superacid media and ending with high temperature reactions at the metal surface. This confirms the view that living organisms do not "invent" new mechanisms. They choose the mechanisms that exist already in nature and effectively use them for their purposes.

So far we tried to find chemical analogues of aerobic oxidation with oxygen as the main active center interacting with the C–H bond. Direct involvement of a metal could be only suspected. In the next section we shall see that metals (both free and participating in metal complexes) can activate the C–H bond directly (also through a five-coordinate carbon intermediate) and this mechanism is likely to be used in biological systems for anaerobic alkane oxidation.

II.6 ALKANE ACTIVATION IN THE COORDINATION SPHERE OF METAL COMPLEXES: MODELS OF BIOLOGICAL ANAEROBIC ALKANE OXIDATION?

II.6.1 INTRODUCTION

In the late 1960s activation of alkanes by metal complexes became a pressing problem both from the viewpoint of basic science (to find the ways for activating a saturated C–H bond) and as an answer to ever-increasing demands of the chemical industry for new selective catalytic transformations of hydrocarbons, particularly methane. Activation of C–H bonds in the coordination spheres of transition metal complexes would be of great interest because all the advantages of metal complex catalysis could be used. Similarly to the search for new reactions in dinitrogen chemistry, we could use a purely chemical approach, in particular, proceeding from the analogy with dihydrogen or aromatics, to search for reactions of C–H activation in alkanes, e.g., by oxidative addition to a metal center. We could also try to follow the example of nature which has solved the problem of selective alkane oxidation, including oxidation of methane, using enzymes present in various living organisms. Soon, however, it became clear that aerobic oxidation of alkanes involves activation of dioxygen rather than alkane molecule by iron complexes: activation of alkane proceeds as the next step in the interaction with coordinated oxygen, probably without formation of metal–carbon bonds. Nevertheless, a purely chemical approach did work. New reactions of methane and other alkanes have been found involving intermediate metal–carbon bond formation, i.e., "true activation" of alkane molecule by a metal complex. The first examples of this kind were found in our work for platinum(II) reactions with alkanes; new reactions with other metal complexes were later discovered. It is still not a very easy task. Usually the reactions involve special ways of preparing this complex, e.g., by photochemical decomposition of more coordinatively saturated complexes or the synthesis of complexes that already contain a metal–carbon bond that can be replaced by another metal–carbon bond from the reacting alkane. Platinum(II) complexes are among the very few kinds of complexes that could react with alkanes under comparatively mild conditions in neutral aqueous solutions containing such ligands as chloride ions and water molecules.

We have seen in Section II.3 that nickel complexes are involved in anaerobic methane formation presumably via a nickel–methyl intermediate. This reaction is partly reversible, i.e., the same nickel complexes appear to activate methane to oxidation. Anaerobic methane oxidation in marine sediments and other anaerobic alkane oxidation by sulfate-reducing bacteria might involve similar mechanisms. Nickel(II) has the same electronic configuration (d^8) as platinum(II). Therefore it is tempting to suggest this kind of configuration as particularly suitable for reaction with alkanes in a "biological" environment.

Our group's search for the complexes able to activate methane was inspired by the existence of biological methane oxidation. Now we see that activation of methane by platinum(II) complexes may indeed have something to do with the biological (this time anaerobic) oxidation of methane, although platinum itself is not known to participate in any enzymes. In this section we briefly describe the results obtained in investigations of activation of alkanes, mainly methane, by transition metal complexes and then try to answer the question of why the d^8 electronic configuration is particularly suitable to activate methane in "biological" conditions. The problems of alkane activation by metal complexes were described in several reviews and monographs.[2,81,178-181]

II.6.2 METATHESIS AND RELATED PROCESSES

Electrophilic reagents containing metal–alkyl fragments can exchange alkyl groups with those of alkanes, including methane, provided there is a free site in the coordination sphere of the complexes. This reaction is known now as metathesis.

Thus, Watson has found that lutetium methyl complex Cp^*_2LuMe and its yttrium analog ($Cp^* = \eta^5 - C_5Me_5$) undergo exchange with methane:[182-184]

$$Cp^*_2\,MMe + {}^{13}CH_4 = \quad Cp^*_2\,M^{13}CH_3 + CH_4 \qquad (100)$$

Both lutetium and yttrium methyl complexes form dimers in solution and in the solid state. In either state (in solution or in the crystalline state) the dimers have unsymmetrical structure in which a methyl group from one metal coordinates to free metal orbitals of a second complex. In cyclohexane about 15% of the compound is in the form of the monomer at room temperature. The monomeric complex reacts with olefins, inducing polymerization of ethylene and oligomerization of propylene. The activity towards ethylene is so great that the reaction can be observed even at –90°C.

Other saturated hydrocarbons besides methane, such as ethane and propane, react more slowly than methane, demonstrating strong steric influence on the reaction rate. Cyclohexane, with a much weaker C–H bond than methane, has no activity and can be used as a solvent.

Tetravalent thorium cycloalkyl complex [$Cp^*Th(CH_2CMe_2CH_2)$] synthesized by Marx et al.[185] was found to be a very reactive compound, in particular when interacting with C–H bonds, including those of methane. The reaction with methane proceeds according to the scheme:

$$\left[Cp_2{}^*Th\langle{}^{CH_2}_{CH_2}\rangle CMe_2\right] + RH \rightarrow \left[Cp_2{}^*Th\langle{}^{R}_{CH_2CMe_3}\right] \qquad (101)$$

The kinetic isotope effect measured with CD_4 as compared with CH_4 is equal to 6, indicating that the C–H bond cleavage is the rate-controlling step. The thorium complex is very air and moisture sensitive and apparently is a strong Lewis acid.

Free positive alkylated ions can be active species in alkane reactions in some of the Ziegler–Natta systems (*vide infra*).

II.6.3 ALKANE ACTIVATION BY TITANIUM AND VANADIUM COMPLEXES IN ZIEGLER–NATTA SYSTEMS

Methane was shown by Grigorian et al. to be activated in Ziegler–Natta systems in hydrocarbon media.[186,187] For example, CD_4 is capable of exchanging D atoms with hydrogen of methyl groups in such systems as $TiCl_4$ + $AlMe_2Cl$; Cp_2TiCl_2 + $AlMe_2Cl$; and VCl_3 + $AlMe_2Cl$. The reactions were carried out in heptane at 20 to 50°C and methane pressure of 0.3 to 4 atm. The products were found to be CHD_3, CH_2D_2, and CH_4. The authors suggested formation of carbene complexes as intermediates:

$$\rangle Ti\langle{}^{CH_3}_{CH_3} = \quad Ti{=}CH_2 + CH_4 \qquad (102)$$

H–D exchange between CH_4 and C_2D_4 was detected in the presence of Cp_2V in benzene solution. The mechanism of oxidative addition with participation of olefin was suggested by the authors to explain the results:

$$Cp_2V \leftarrow \begin{matrix} CH_2 \\ \| \\ CH_2 \end{matrix} + CH_4 = Cp_2V\langle \begin{matrix} CH_3 \\ C_2H_5 \end{matrix} \quad (103)$$

Thus, the reaction proposed involved the reverse process of alkyl groups disproportionation in the metal coordination sphere. The existence of the latter reaction was reported earlier. In some cases an addition of methane to a multiple bond (hydromethylation) was reported. For example, hydromethylation of ethylene was observed in the system ($Ti(OBu)_4$ + $AlEt_3$ in benzene, the product being propane, albeit with rather small yields. Hydromethylation of carbon monoxide (methane carbonylation) was shown to proceed yielding acetaldehyde in the system Cp_2TiCl_2 + $AlMe_2Cl$ in benzene and toluene at 20 to 50°C. The reaction is unfavorable thermodynamically (ΔG°_{298} = +13.2 kcal/mol) and the process was conducted at high pressure (1500 to 3000 atm). In the presence of C_2H_4 the formation of methyl ethyl ketone was observed, apparently formed in the reaction:

$$CH_4 + CO + C_2H_4 \rightarrow CH_3COC_2H_5 \quad (104)$$

The mechanisms of all these reactions are mostly hypothetical and the nature of the active species has not been elucidated; they are probably different in different systems.

In homogeneous Ziegler–Natta systems, such as $Cp_2TiCl_2 + AlR_2Cl$, the reaction between the components is known to produce a complex with alkylated titanium, e.g., for R = CH_3:

$$Cp_2TiCl_2 + AlMe_2Cl \rightarrow Cp_2Ti\langle \begin{matrix} Me \\ Cl \cdots Al(Me)Cl_2 \end{matrix} \quad (105)$$

The complex was shown in an early work[188] to dissociate into free ions even in nonpolar media such as benzene–heptane mixture and even more so in methylene chloride:

$$Cp_2Ti\langle \begin{matrix} Me \\ Cl \cdot Al(Me)Cl_2 \end{matrix} = \left[Cp_2TiMe\right]^+ + AlMeCl_3^- \quad (106)$$

Ethylene and other α-olefins were shown to react with the free ions $[Cp_2TiCH_3]^+$, inserting themselves into titanium–carbon bonds. Formation of free alkylated ions proved later to be a general case for many similar systems catalyzing polymerization and oligomerization of olefins. The same species can be active in reactions with alkanes and dihydrogen. For example, with alkanes the metathesis mechanism may be suggested:

$$\left[Cp_2TiMe\right]^+ + RH = \left[Cp_2TiR\right]^+ + MeH \quad (107)$$

It should be noted that the hypothetical mechanisms suggested for alkanes in Ziegler–Natta systems were later confirmed for other more definite systems.

II.6.4 RHENIUM COMPLEXES

Coordinatively unsaturated low-valent rhenium complexes were found to react with alkanes, including methane. Dehydrogenation of cycloalkanes, found first for cationic iridium(I) complexes, was also demonstrated by Felkin and co-workers[189] for the rhenium complexes L_2ReH_2 (L = PPh_3 or PEt_2Ph). These complexes were shown to dehydrogenate cyclopentane to form $L_2(\eta^5\text{-}C_5H_5)ReH_2$ in the presence of 3,3-dimethyl-1-butene as an acceptor of hydrogen. The reaction in cyclopentane solution proceeds at 80°C according to Scheme II.6a. Linear alkanes could be dehydrogenated in this system to give diene complex, e.g., according to Scheme II.6b.

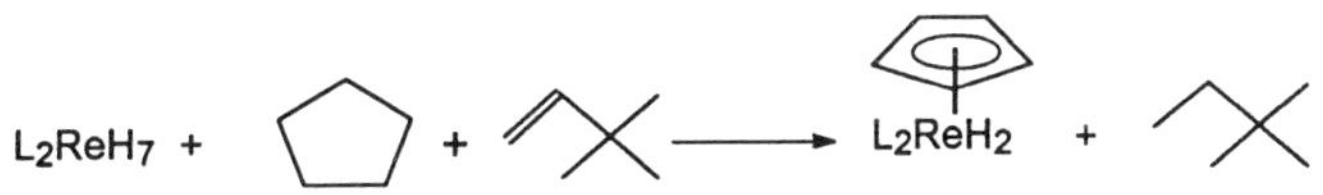

a

SCHEME II.6(a) Dehydrogenation of cycloalkanes in the presence of L_2ReH_7.

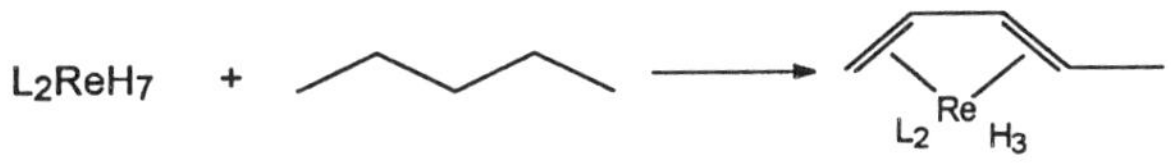

b

SCHEME II.6(b) Dehydrogenation of liner alkanes in the presence of L_2ReH_7.

The diene in the complex can be converted further to monoolefin in the presence of trimethyl phosphite. For example, *n*-pentane can be selectively converted to 1-pentene. In dilute solutions the system can be made catalytic producing several molecules of the olefin per the rhenium complex.[190] The authors suggest that an alkane molecule reacts with an intermediate L_2ReH_3 via oxidative addition:

$$L_2ReH_3 + RH = L_2\overset{\overset{\displaystyle R}{|}}{Re}H_4 \tag{108}$$

Direct observation of oxidative addition of aromatic and aliphatic hydrocarbons including methane, to coordinatively unsaturated rhenium(I) complex was made by Bergman et al.[191] Intermediate rhenium complex was produced photochemically. For example, upon photolysis of $CpRe(PMe_3)_3$ under methane (25 atm) hydridomethyl complex $Cp(PMe_3)_2Re{<}^{H}_{CH_3}$ was produced. Similar reactions proceed with other hydrocarbons containing primary, aromatic, and vinyl C–H bonds, but secondary and tertiary C–H bonds are not reactive (it is not excluded though the weak *sec*-alkyl–metal bonds are formed intermediately [*vide infra*]).

II.6.5 IRON COMPLEXES

Iron atom dimers in an inert gas matrix were shown to react with methane,[192] while separate iron atoms are not reactive. The following reactions were proposed:

$$Fe_2 + CH_4 \xrightarrow{12\ K} HFeFeMe \quad or \quad FeFe\langle^{H}_{Me} \tag{109}$$

$$FeMe \rightarrow \underset{\displaystyle H}{\underset{|}{Fe}}=CH_2 \tag{110}$$

C–H bonds can be activated photochemically in hydrocarbons (arenes, alkenes, and alkanes) with FeH_2L_2 complexes ($L = R_2PCH_2CH_2PR_2$, R = Me or Et).

Oxidative addition of methane to the iron mononuclear complex proceeds upon irradiation of $FeH_2(dmpe)_2$ in liquefied xenone and affords a mixture of *cis*-$FeH(dmpe)_2Me$ and *trans*-$FeH(dmpe)_2Me$.[193]

$$FeH_2(dmpe)_2 + CH_4 \xrightarrow[Xe\ 173\ K]{h\nu} FeH(Me)(dmpe)_2 + H_2 \tag{111}$$

At the early stages of the reaction the *cis*-isomer is the major product, however, over longer time of irradiation the *trans*-isomer is formed almost exclusively. Isomerization to more stable *trans*-isomer was found to be a photochemical process. The low temperature of the reaction shows that C–H bond cleavage in the reaction with coordinatively unsaturated $Fe^0(dmpe)_2$, which is evidently formed under irradiation, requires very low activation energy.

Earlier the hydromethylation of acetylene by methane in the homogeneous system $Fe(acac)_3$ + $AlEt_3$ was reported.[194] Some propylene (up to about 3% per $Fe(acac)_3$) was detected as the product, propylene-d_4 being formed when CD_4 was taken instead of CH_4.

II.6.6 OSMIUM COMPLEX[195]

Pyrolysis of osmium neopentyl hydride complex in cycloalkane solution at 80°C under methane gas at CH_4 pressure of 40 to 60 atm produces a mixture of cyclometalation product and methyl hydride $L_4Os<^{H}_{CH_3}$ with the yields of the latter up to about 16%. Apparently a coordinatively unsaturated Os(O) complex is formed as an intermediate.

II.6.7 COBALT COMPLEXES

Cobalt atoms as well as the atoms of other metals (Fe, Mn, Cu, Ag, Au, and Zn) react with methane when irradiated by UV light ($\lambda < 360$ nm) at 15 K.[196,197] No reaction has been so far reported for low-valent cobalt complexes, but high-valent cobalt(III) trifluoroacetate reacts with methane in CF_3COOH solution at 150 to 180°C and 10 to 40 atm to yield methyl trifluoroacetate.[198] In the presence of dioxygen the reaction becomes catalytic, producing CF_3COOCH_3 corresponding to four cycles with respect to cobalt salt. Apparently the selectivity of the reaction is due to the stability of methyl trifluoroacetate to further oxidation. It is not excluded that a Co–C bond is formed initially in the process of electrophilic substitution

$$Co^{III} + CH_4 \rightarrow Co - CH_3 + H^+ \quad (112)$$

though free methyl radicals could be also suggested as intermediates. Earlier there were some indications (in particular unusual selectivity) that in reactions of cobalt(III) complexes with alkanes, cobalt–carbon bonds might intermediately form (see Reference 2).

II.6.8 RHODIUM AND IRIDIUM COMPLEXES

Dehydrogenation of cycloalkanes in the presence of cationic iridium complexes was discovered by Crabtree et al. in 1979.[199] The complexes $[IrH_2(acetone)_2L_2]BF_4$ and $[IrH_2(H_2O)_2L_2]BF_4$ were studied. Cyclopentane in this system forms a cyclopentadienyl complex $CpIrHL_2$ (30% yield) after 18 h in refluxing 1,2-dichloroethane containing *tert*-butylethylene as a hydrogen acceptor. Under similar conditions cyclooctane gives a cyclooctadiene complex with 70% yield over 4 h. *tert*-Butylethylene is again a necessary hydrogen acceptor.

The mechanism proposed by the authors[200] includes initially hydrogenation of *tert*-butylethylene to form a free coordination site on the iridium atom to which cycloalkane adds oxidatively. With another similar complex $[IrH_2(Me_2CO)_2L_2]SbF_4$ (L = PPh_3 or $P(p\text{-}FC_6H_4)_3$) cyclopentane reacts at 90°C to give $\{IrCpHL_2]SbF_6$ in high yields in 24 h and cyclohexane reacts at 80°C in 24 h to give $[\eta^5\text{-(cyclohexadienyl)}IrH\{(p\text{-}FC_6H_4)_3P\}_2]^+$ (5% yield), $[\text{phenyl } Ir\{(p\text{-}FC_6H_4)_3P\}_2]^+$ (45% yield), and benzene (32% yield). At higher temperature (150°C) the product is mainly benzene.[201]

The first direct observation of oxidative addition of C–H bonds in alkanes to a transition metal complex was made in the works of Janowicz and Bergman[202] and Hoyano and Graham[203] with iridium complexes. One of the systems was based on the dihydridoiridium complex $Cp^*(PMe_3)IrH_2$ ($Cp^* = \eta^5\text{-}C_5Me_5$).[202] Irradiation of the complex with UV light from a mercury lamp (λ_{max} = 275 nm) in hydrocarbon solution produced hydridoalkyl complex according to the mechanism:

$$Cp^*(PMe_3)IrH_2 \xrightarrow{h\nu\ -H_2} [Cp^*(PMe_3)Ir] \xrightarrow{RH} Cp^*(PMe_3)Ir\langle^{R}_{H} \quad (113)$$

Reaction with methane was also observed in cyclooctane at 140 to 150°C in thermal decomposition of hydridocyclohexyl complex:[204]

$$Cp^*(PMe_3)Ir(Cy)H \xrightarrow[\text{cyclo}-C_8H_{16}]{20\text{ atm } CH_4} Cp^*(PMe_3)Ir(CH_3)H \quad (114)$$

In 14 h at 58% yield of hydridomethyl complex was obtained. Obviously the mechanism involves reductive elimination of cyclohexane and oxidative addition of methane to coordinatively unsaturated iridium(I) complex.

Another way to create a coordinatively unsaturated Ir^I complex is irradiation of iridium(I) dicarbonyl:[205]

$$Cp^*Ir(CO)_2 \xrightarrow[+RH]{h\nu-CO} \begin{matrix} Cp^* \\ OC \end{matrix}\rangle Ir \langle\begin{matrix} R \\ H \end{matrix} \qquad (115)$$

In this case methane reaction was also observed in perfluorinated solvent. A solution of $Cp^*Ir(CO)_2$ in perfluorohexane under 8 atm of methane after irradiation for 16 h at room temperature gave the hydridomethyl complex.

The rhodium complex $Cp^*(PMe_3)RhH_2$ on irradiation also undergoes oxidative addition of alkane C–H bonds.[206,207] The reactions are carried out at low temperatures (–30 to –60°C) because the rhodium complexes are less stable than their iridium analogues. There is also some difference in the selectivity in attack on different C–H bonds in alkane molecules. The rhodium complex inserts only into primary C–H bonds; the iridium complex reacts with primary and secondary C–H bonds, the former being more reactive. Neither complex reacts with tertiary C–H bonds.

The energy of the M–C and M–H bonds that are formed seems to be a decisive factor in reactions with alkanes of coordinatively unsaturated species. Perhaps there is an equilibrium with alkyl hydrides formed and secondary alkyl derivatives are produced but undergo further decomposition.

Indeed, such an equilibrium was found for another rhodium complex: tris-(dimethylpyrazoylyl)borato rhodium dicarbonyl $(HBPz^*_3)Rh(CO)_2$ (Pz^* = 3,5-dimethylpyrazolyl). Photochemical decomposition of the dicarbonyl produces an extremely active intermediate able to react with aromatic and saturated hydrocarbons, including methane, at room temperature (Scheme II.7).[208]

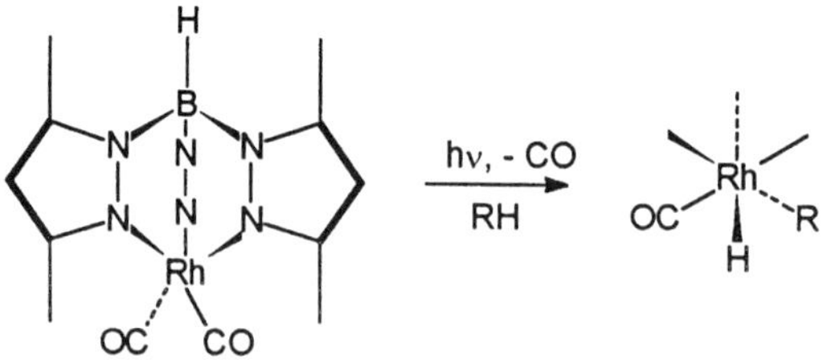

SCHEME II.7 Photochemical alkane reaction with $(HBPz^*_3)Rh(CO)_2$.

Hydridocyclohexyl complex produced under daylight or tungsten illumination reacts with methane reversibly in cyclohexane solution in the dark yielding methylhydride:

$$\mathrm{Rh}\begin{array}{l}/\mathrm{CO}\\-\mathrm{H}\\\backslash \mathrm{C_6H_{11}}\end{array} + \mathrm{CH_4} = \mathrm{Rh}\begin{array}{l}/\mathrm{CO}\\-\mathrm{H}\\\backslash \mathrm{CH_3}\end{array} + \mathrm{C_6H_{12}} \qquad (116)$$

The equilibrium constant for this reaction was estimated to be about 190. Addition of benzene to a cyclohexane solution of hydridocyclohexyl complex results in quantitative conversion to phenylhydride within 10 min.

A quite different reaction of methane as well as toluene was observed with a paramagnetic rhodium(II) porphyrin complex.[209] A dimer of tetra(xylyl)porhyrinato rhodium(II) reacts with methane presumably according to the scheme:

$$[(\mathrm{TXP})\mathrm{Rh}]_2 + \mathrm{CH_4} \rightarrow (\mathrm{TXP})\mathrm{Rh} - \mathrm{CH_3} + (\mathrm{TXP})\mathrm{Rh} - \mathrm{H} \qquad (117)$$

The authors considered the thermodynamics of the reaction to be favorable: D(Rh–H) + D(Rh–C) = 117 kcal/mol; D(Rh–Rh) in the dimer is about 12 kcal/mol. Therefore the reaction is close to thermoneutral. A linear transition state is proposed:

$$\mathrm{Rh}\cdots\mathrm{H_3C}\cdots\mathrm{H}\cdots\mathrm{Rh}$$

A similar reaction is reported for toluene.

Metathesis reaction of alkanes was observed recently with an iridium(III) complex Cp*(PMe_3)Ir(CH_3) (OTf) (OTf = OSO_2CF_3) that reacts with several types of C–H bonds, including those of methane.[210] In CD_2Cl_2 under 2 atm of methane the incorporation of ^{13}C-labeled methane to give Cp*(PMe_3)Ir($^{13}CH_3$)(OTf) had occurred to the extent of 50% after 6 h at 45°C. Neither cyclohexane nor neopentane undergoes the reaction, while with benzene an iridium–phenyl complex is readily formed. (PMe_3)Ir(Me)(Cl) is completely unreactive towards benzene up to 100°C, i.e., the presence of the triflate (OTf) group is important for the efficiency of these reactions. Some ionic dissociation is detected by electroconductivity experiments, therefore it was proposed that free ions Cp*(PMe_3)IrCH_3^+ are formed in the solution and are active in the metathesis reaction.[210]

II.6.9 PHOTOCATALYTIC CARBONYLATION AND DEHYDROGENATION OF ALKANES WITH RHODIUM CARBONYL PHOSPHINE COMPLEXES

In 1986 Tanaka and Sakakura found that terminal methyl group of alkanes could be regioselectively carbonylated in the presence of rhodium carbonyl phosphine complexes, e.g., RhCl(PMe_3)$_2$(CO), to give aldehydes under mild conditions (1 atm CO, room temperature) under irradiation (high-pressure mercury lamp).[211,212] Thus *n*-pentane afforded *n*-hexanal in a yield corresponding to almost 30 turnovers per rhodium after 16.5 h, while the yield of 2-methylpentanal was less than 60% (per Rh) and 2-ethylbutanal was not detected. Cyclohexane was also carbonylated under similar conditions but with smaller yields.

The same catalyst was found to be active in the photocatalytic dehydrogenation of *n*-alkanes to form alkenes and dihydrogen.[213,214] The order of activity of different phosphines was found to be

$$PPh_3 < PEtPh_2 < PEt_2Ph < PEt_3 < PMe_3$$

Relative reactivities of alkanes (per C–H) in dehydrogenation are cyclooctane > cyclohexane > decane ≈ hexane.

The catalytic reaction proceeds effectively without any H_2 acceptor; more than 200 turnovers per hour was observed for cyclooctane at 96°C. In the presence of hydrogen acceptors, such as *t*-butylethylene and styrene, hydrogen can be transferred to the olefins.

The role of the irradiation is the formation of an active 14e intermediate, tri-coordinated $Rh(PMe_3)_2Cl$, with the loss of carbon monoxide, followed by an oxidative addition of a hydrocarbon molecule and insertion of CO or β-elimination of an olefin.[215]

In the system $Rh_2Cl_2(CO)_4$ + PMe_3 the specific rate of the photocatalytic carbonylation of pentane is 20 times higher than in the presence of $RhCl(PMe_3)_2(CO)$.[216] Presumably, several coordinatively unsaturated rhodium complexes may be active in the photocatalytic reactions with different reactivity.

Catalytic activities of other metal phosphine carbonyl complexes (Fe, Ru, Co, Rh, Ir, Ni, and Pt) were found to be much lower than that of $RhCl(PMe_3)_2(CO)$.

II.6.10 PLATINUM, PALLADIUM, AND NICKEL COMPLEXES

II.6.10.1 Platinum(II) Complexes

Activation of C–H bonds in alkanes by transition metal complexes in homogeneous solution was observed for the first time in our work in 1969 when H–D exchange of methane and ethane with solvent D_2O molecules was discovered in the presence of simple platinum(II) chloride complexes.[217] Earlier in the same system H–D exchange was found for aromatic and alkylaromatic compounds but was attributed to π-complex formation of organic molecules with platinum(II) species.[218]

Soon after the H–D exchange the oxidation of alkanes was found by platinum(IV) complexes catalyzed by platinum(II) chlorides.[219] For example, methane reacts according to the equations:

$$H_2PtCl_6 + CH_4 \xrightarrow{PtCl_4^{2-}} H_2PtCl_4 + CH_3Cl + HCl \tag{118}$$

$$H_2PtCl_6 + CH_4 + H_2O \xrightarrow{PtCl_4^{2-}} H_2PtCl_4 + CH_3OH + 2HCl \tag{119}$$

Dehydrogenation proceeds in the case of cyclohexane:

$$3H_2PtCl_6 + C_6H_{12} \xrightarrow{PtCl_4^{2-}} 3H_2PtCl_4 + C_6H_6 + 6HCl \tag{120}$$

H–D exchange and oxidation catalyzed by platinum complexes was then thoroughly investigated by several groups of workers (see References 2 and 177 to 181). The results obtained are summarized below.

H–D exchange with the solvent (D_2O) protons is multiple, i.e., several protons can be exchanged by deuterons in a single contact of an alkane molecule with an active platinum species. Because there is convincing evidence that alkyl platinum intermediate is responsible for the exchange, the results indicate that the intermediate is sufficiently stable to exchange several protons during its lifetime prior to producing the alkane molecule again.

The mass-spectra of deuterated alkanes produced and corresponding deuterium distribution in the alkanes for a given solvent can be used as a "fingerprint" to identify the reaction mechanism. For example, deuterium atom distribution in the molecules produced in H–D exchange indicates the possibility of isomerization forming several alkyl–platinum complexes (*vide infra*).

H–D exchange and the oxidation by H_2PtCl_6 involve the same intermediate alkylated platinum(II) complex, $RPtL_3^{2-}$ (L = Cl or H_2O). The following step results from the competition of $RPtL_3^{2-}$ reactions with D_2O to form a deuterated alkane and with H_2PtCl_6 to produce Pt^{IV}–alkyl complex, which reductively eliminates alkyl chloride or reacts with water to form alcohol (Scheme II.8).

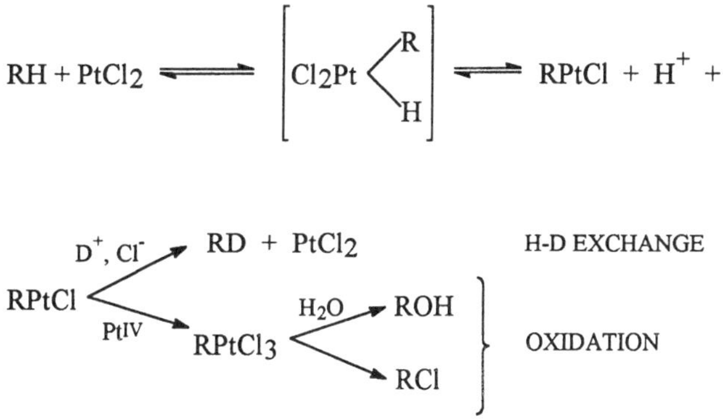

SCHEME II.8 Mechanism of alkane reactions with platinum complexes.

According to Labinger et al.,[220] oxidation of alkyl platinum(II) complex to alkyl platinum(IV) intermediate was found to proceed via electron transfer rather than alkyl group transfer.

Formation of methyl platinum(IV) chloride complex in methane oxidation was confirmed by its NMR spectrum.[221] The rate of methane oxidation was found to be equal to that of the complex decomposition measured separately in concentrations and temperature conditions of CH_4 oxidation. Thus, direct confirmation was obtained for the mechanism of alkane oxidation by platinum(IV) catalyzed by platinum(II) complexes: the reaction is homogeneous and proceeds totally via a mononuclear methyl platinum(IV) complex.

Among the platinum(II) complexes existing in aqueous solutions of $PtCl_4^{2-}$ or specially prepared, positive $PtCl(H_2O)_3^+$ and negative $PtCl_3(H_2O)^-$ are almost equally active, while uncharged $PtCl_2(H_2O)_2$ has the maximum activity. Symmetrical $PtCl_4^{2-}$ and $Pt(H_2O)^{2+}$ are almost completely inactive.

Two main mechanisms can be proposed for the first step of alkane interaction with platinum(II) complexes: (1) oxidative addition followed by proton reversible elimination and (2) electrophilic substitution with simultaneous proton abstraction. Some experimental evidence indicated that oxidative addition is more probable. In particular, the lack of strong rate dependence on polar factors and the acidity of the medium favors oxidative addition rather than electrophilic substitution.

The oxidative addition mechanism was confirmed recently by Zamashchikov et al. in investigation of dimethyl platinum(IV) complex decomposition.[222] Ethane, which is one of the decomposition products in D_2O–DCl, contains several D atoms with similar distribution as in H–D exchange of ethane with D_2O catalyzed by platinum(II) complexes. Moreover, ethyl chloride is formed in the presence of platinum(IV) in the reaction competitive with H–D exchange. Both kinetic and H–D exchange data indicate that the precursor of ethane in dimethyl platinum(IV) decomposition is ethyl hydride platinum(IV) complex and therefore the same complex is the intermediate in the reaction of ethane with platinum(II) complexes.

Two main factors are apparently responsible for differences in different alkanes reactivity towards platinum(II): C–H bond energy and steric factors. The rate increases in the order $CH_4 < C_2H_6 < C_3H_8$. Isolated methyl groups are most active among different groups in alkanes. Methyl and methylene groups situated next to bulky *tert*-butyl groups display very little activity. Tertiary C–H bonds are not active at all. Thus the reactivity is similar to that of iridium(I) and rhodium(I) complexes in the oxidative addition reactions with alkanes described above.

Among different metal complexes that are now found active towards alkanes, platinum(II) still occupies a unique position. The reactions proceed in aqueous solutions and are unaffected by dioxygen. These properties of the interaction of platinum complexes with alkanes continue to attract much interest to these reactions from the viewpoints of their mechanism and synthetic possibilities.

Catalytic oxidation of alkanes can be observed with respect to the platinum if a mediator is used. Thus, dioxygen can be used as an oxidant for methane in the presence of a phosphomolybdic acid ($Na_8HPMo_6V_6O_{40}$) producing methanol and methyl chloride.[223] Electrocatalytic oxidation of *p*-toluenesulfonic acid to the corresponding alcohol, *p*-$HO_3SC_6H_4CH_2OH$, was reported using aqueous $PtCl_4^{2-}$ as the C–H activation catalyst and another phosphomolybdic acid ($H_3PMo_{12}O_{40}$) as a redox mediator.[224]

An interesting possibility of selective alkane oxidation was reported by Sen et al.,[225] who found that whereas unactivated C–H bonds were attacked and oxidized by Pt(II), C–H bonds bound to an oxygen atom in the α position were activated and catalytically oxidized by platinum metal in the presence of dioxygen. For example, platinum(II) was found to oxidize ethane to ethanol and ethylene glycol. In the presence of platinum metal and dioxygen further oxidation of the alcohol occurred to generate carboxylic acid.

II.6.10.2 Platinum(0) Complex[226]

Coordinatively unsaturated platinum(0) complex generated in reductive elimination of neopentane from *cis*-hydridoneopentyl[bis(dicyclohexylphosphino)ethane]platinum(II) reacts with various hydrocarbons, coordinating unsaturated molecules and cleaving C–H bonds in saturated ones, including methane. Thus when the initial Pt(II) complex is heated in cyclohexane solution at 45°C under methane (140 atm) both hydridocyclohexyl and *cis*-hydridomethyl complexes are formed in 10 and 45% yields, respectively.

II.6.10.3 Palladium(II) Complexes

Since 1973 Rudakov and co-workers in a series of paper have reported that alkanes can be oxidized by some metal complexes in concentrated solutions of sulfuric acids.[227,228] Participation of H_2SO_4 is essential since it increases the oxidative ability of the complexes and makes it possible to conduct the reactions even at 25 to 100°C. Solutions of palladium(II), platinum(III), manganese(III), or mercury(II) can be used as oxidants.

Palladium(II) complexes were investigated most thoroughly. They behave very differently from platinum(II) complexes. There is no H–D exchange with D_2O and the reaction rate is very sensitive to acidity. The Hammett equation

$$\log \mathrm{k} = \mathrm{const} - mH_0 \qquad (121)$$

(k and H_0 are the rate constant and acidity function, respectively) holds for different H_2SO_4 concentrations with $m = 1$.

Olefins and carbenium ions are believed to be intermediates in the oxidation and undergo further oxidation.

Evidently, protonated or coordinatively unsaturated palladium complexes are active species in the reaction with alkanes. It follows from the kinetic data that associates involving two or more palladium ions acquire an important role, particularly at high palladium concentrations, palladium ions exerting electrophilic coaction in the associates.

Palladium(II) trifluoroacetate were found to oxidize alkanes in trifluoroacetic acid at 92°C.[229] Palladium is reduced to palladium(0), oxidizing a hydrocarbon. Thus benzene is produced in the oxidation of cyclohexane according to the stoichiometry:

$$3\mathrm{Pd}(\mathrm{CF_3CO_2})_2 + \mathrm{C_6H_{12}} \rightarrow 3\mathrm{Pd}(0) + \mathrm{C_6H_6} + \mathrm{CF_3COOH} \qquad (122)$$

Reaction with methane was reported to produce CF_3COOCH_3 in this system[230a] but apparently was difficult to reproduce.[198]

A mechanism of electrophilic substitution can be suggested for the first step of alkane reactions in the presence of palladium(II) complexes with possible intermediate formation of palladium–alkyl bond. The absence of H–D exchange and strong dependence on the acidity shows that the mechanism is different from that of platinum(II) reaction of alkanes in which oxidative addition is very likely to be the initial step.

However, oxidative addition of various hydrocarbons was reported in their reaction with palladium(II) complexes in nonpolar media by Vedernikov et al.[230b] The complexes of formula $PdX_2(PPh_3)_2$ (X = Cl, Br, or I) were found unexpectedly to react reversibly with saturated and aromatic compounds RH (RH = *p*-xylene, toluene, benzene, *n*-hexane, or cyclohexane), slowly at ambient temperature and rapidly at 70 to 130°C in solutions of corresponding hydrocarbons to produce, reportedly, alkyl and aryl hydrides $Pd(H)(R)X_2(PPh_3)_2$. This is confirmed by the fact that in the presence of bases the products formed eliminate HX to afford organobis(triphenylphosphine)palladium(II) halide $Pd(R)X(PPh_3)_2$. There is no direct evidence that alkyl and aryl hydrides are really produced in the first step, which is found to be reversible. At the same time, benzene was found to react at a much lesser extent than, for example, toluene. Thus the weaker the C–H bond, the higher the yield of the product. It is difficult to explain if aryl hydride is really formed from benzene because the metal–aryl bond is usually very strong. Therefore it may be suggested that the reaction with a hydrocarbon ends with the complex formation with the C–H bond only weakened (*vide infra*) rather than fully cleaved.

It should be noted that in H_2SO_4 the acid is the oxidant of alkanes and formally the reaction is similar to sulfate oxidation of methane and other alkanes in enzymatic anaerobic oxidation in the presence of nickel(II) porphinoid complex.

II.6.10.4 Nickel Complexes

Possible participation of the nickel porphinoid complex in the anaerobic biological oxidation of methane makes it interesting to investigate nickel-containing systems for methane activation. It was found that methane reacted with the product of the interaction of nickelocene, $(\eta^5\text{-}C_5H_5)_2Ni$, with $LiAlH_4$, yielding a methyl–metal complex. Some isotopic H–D exchange was also found in a CH_4–D_2 mixture over a Cp_2Ni–$LiAlH_4$ system.[231]

The product of the reaction of methane with the solution formed in the reaction of $LiAlH_4$ or $LiAlD_4$ with Cp_2Ni in tetrahydrofurane reacts with ethanol (after complete removal of unreacted methane) to yield 0.1 to 0.35 mol methane per mole of initial Cp_2Ni. In an experiment performed with CD_4 instead of CH_4, the alcoholysis of the reaction products by ethanol produces CD_3H and CD_2H_2 as the main products. The treatment of the reaction products in the case of CH_4 by C_2H_5OD yields CH_3D. These results show that methane reacts with the system to produce a complex containing a methyl–metal fragment. The obtained result confirm the conclusion that CH_4 reacts with a nickel complex (probably polynuclear). The structure of the complex active towards methane and its activity have yet to be determined, however, the results confirm that nickel complexes can react with methane in the presence of strong bases. It may be mentioned that polynuclear nickel organic complexes formed in reaction of Cp_2Ni with LiMe are stable even in the presence of water, illustrating high bond energy of the nickel–carbon bond.[232]

II.6.11 MERCURY-CATALYZED OXIDATION OF METHANE

Periana et al. reported[233a] selective catalytic oxidation of methane by sulfuric acid to produce methyl bisulfate at 180°C. The reaction is catalyzed by mercuric ions. Sulfur dioxide is the product of sulfuric acid reduction. At methane conversion of 50%, 85% selectivity to methyl bisulfate is observed. The major side product is carbon dioxide. The mercury turnover efficiency is 10^{-3} s^{-1}. The Hg^{II} ion reacts with methane by the mechanism of electrophilic substitution producing as an intermediate the methylated mercury complex CH_3HgOSO_3H. The complex is formed in appreciable steady-state concentrations and was observed directly by ^{13}C and ^{199}Hg NMR spectrometry. Under the reaction conditions methyl mercuric bisulfate decomposes to produce methyl bisulfate CH_3OSO_3H and the reduced mercurous species Hg_2^{2+}. The catalytic cycle is completed by reoxidation of Hg_2^{2+} with H_2SO_4 to regenerate Hg^{2+} and to form other products, SO_2 and H_2O. In the presence of D_2SO_4 the incorporation of D in methane was observed in the presence of mercuric salts in reaction conditions. Independent reaction of specially prepared methyl mercuris bisulfate confirmed that protolysis occurs at 180°C and this provides the mechanism for isotope exchange. It supports the authors[233a] conclusion that CH_3HgOSO_3H is formed as the intermediate in methane oxidation. Thus mercury reacts with methane in a manner somewhat similar to palladium. Again we may mention that this reaction is in certain aspects similar to a suggested scheme of anaerobic methane oxidation by sulfate-reducing bacteria. Methane is probably activated by nickel(II), which is reduced to nickel(I) in the process of methane oxidation. Reverse reaction produces methane by methanogens. The rather weak oxidant makes the process reversible.

II.6.12 MECHANISM OF C–H ACTIVATION AND CONCLUSIONS

In the past 28 years the majority of transition metals in their suitable ligand surroundings and oxidation states have been found to be able to react with alkanes, including methane, introducing the alkane molecule and its fragments into the metal coordination sphere.

C–H bonds in ligands of many transition metals form so-called agostic bonds with the metal, which is characterized by short contacts of hydrogen atoms in alkyl groups with metals in the complex structure (see References 179 and 233b). Formation of agostic bonds in complexes indicates the possibility of an alkane molecule approach to the metal without serious repulsion and shows the possible direction of this approach. The first step of the alkane interaction with a metal compound is likely to be a complex formation. There is much evidence now of the existence of such complexes[233b] though apparently they are less stable than similar dihydrogen complexes. These complexes are likely to possess five-coordinate carbon atoms, alkane molecules display both donor and acceptor properties, and stability of the complexes is determined by both acceptor and donor properties of the coordinating metals.

The crucial step determining further reactions of the alkane molecule is the cleavage of the C–H bonds (in the general case C–H or C–C bonds). The mechanism of this process can be different kinds of oxidation addition (see Section II.2.2):

$$M + RH = \quad M\langle^{R}_{H}$$

$$M_2 + RH = \quad R-M + M-H$$

or substitution:

$$M-X+RH \rightleftharpoons M-R+HX$$

$$M+RH+B \rightleftharpoons [M-R]^{-}+BH^{+}$$

Formation of a metal–alkyl bond provides the possibility of a variety of further reactions leading to alkane functionalization.

Thus, platinum(II) complexes are not exceptional in their reactions with alkanes: the mechanism of the primary alkane activation and further interactions can be similar to those of the other metal complexes.

At the same time, the reactions of alkanes with platinum(II) complexes have, indeed, some particular features. In most other cases special additional photo- or thermal activation is important to bring about the formation of coordinatively unsaturated species capable of further reactions with alkanes. Being extremely weak bases, alkanes require mainly coordinatively unsaturated metal complexes with which to react. We have seen that it is typical for a number of reactions that the metal–alkyl bond must be already present in a molecule for it to react with alkane either after reductive elimination of another alkane molecule (forming coordinatively unsaturated species) or in metathesis, where one alkyl is replaced by another one.

Meanwhile, the reactions of platinum(II) and its analogues with alkanes do not require special conditions for alkane activation. The reactions with alkanes proceed in neutral water-containing media, often in the presence of dioxygen. The complexes including ordinary chloride and water ligands are most active. It is clearly more probably to organize a catalytic process with those complexes than with the complexes that must be, for example, photoactivated to become active towards alkanes. The reason for this is that platinum(II) complexes with d^8 electronic configuration are very weak Lewis acids (its free $d_{x^2-y^2}$ orbital is closed to external bases by its ligands) and water molecules, being hard bases, do not compete with methane for the active site. The platinum–carbon bond is strong and Pt^{II} complexes behave as strong soft acids.

Formation of the intermediate complexes of Pt^{II} with alkanes is apparently the best way to explain the multiple H–D exchange observed in reactions of alkanes with D_2O catalyzed by platinum(II) complexes. It was found recently that isotope exchange between α-C–H and M–H in an alkyl hydride is faster than reductive elimination of alkane (Scheme II.9).[223c-235]

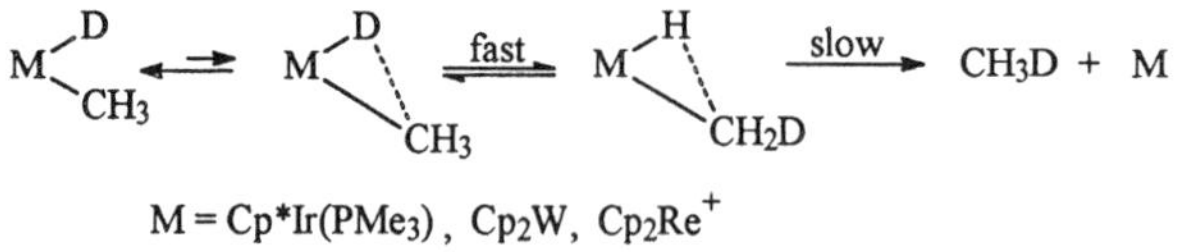

SCHEME II.9 H–D exchange in alkyl hydrides.

An alkane σ-complex is presumably involved as an intermediate in this reaction. The inverse isotope effect observed (k_H/k_D = 0.5 [Ir] or 0.8 [Re]) confirms this proposal. Thus, the alkane complex can be sufficiently stable for the metal to coordinate consecutively different C–H bonds and reversibly undergo oxidative addition. Therefore, if alkyl hydride exchanges its H atom bound to metal with D_2O then several hydrogen atoms of the alkane will be exchanged before the alkane molecule is finally liberated from the complex. The analysis of the distribution of deuteriums in alkane molecules is in quantitative agreement with this mechanism.[222] The shifts 1–2, 1–3, 1–4, etc. proceed in the process of H–D exchange. For example, both methyl groups in ethane are partially exchanged during one contact of ethane with platinum complex, indicating the possibility of isomerization via the alkane complex with platinum center with both carbon atoms entering in turn into contact with platinum.

It might be suggested that hydrogen exchange between alkyl and hydride in alkyl hydride complexes will be accompanied by the loss of chirality for chiral alkanes. A possible mechanism for the exchange is presented in Scheme II.10.

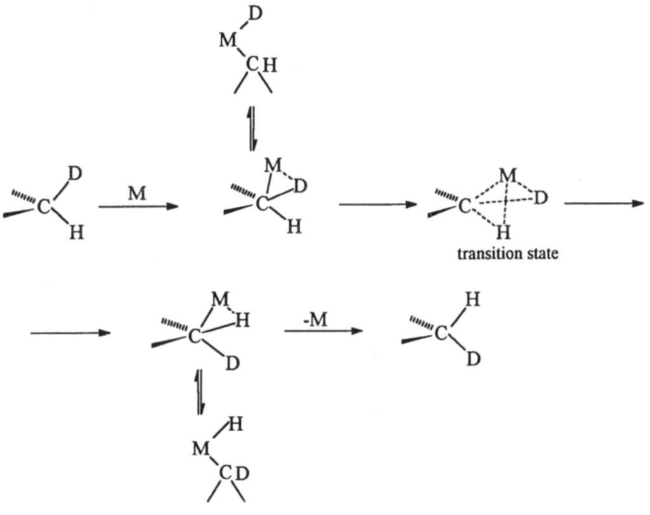

SCHEME II.10 Mechanism of alkane epimerization in the coordination sphere of metal.

A common alkane complex intermediate was proposed recently by Bergman et al. to explain the kinetic results of the interconversion of diastereomeric (hydrido)(dimethylcycloproyl)iridium complexes and deuterium scrambling of their α-deuterio analogues (isomerization with deuterium transfer from alkyl group to the metal to form deuteride).[236] Apparently, the mechanism of the interconversion of the diastereomeric complexes involves a step similar to that for multiple H–D exchange in alkanes catalyzed by platinum(II) complexes: 1–2 shift in the cyclopropyl ring will be sufficient for such an interconversion.

It might be suggested that formation of alkyl hydride complex from the complex with alkane molecules is not necessary and happens only for those metal complexes that are sufficiently strong donors to undergo two-electron oxidation. The H–D exchange with the solvent, electrophilic substitution (abstraction of H^+ by a base), and isomerization of alkanes (e.g., racemization) may be expected already for the alkane molecules bound to metal in the complex.

Other interesting isomerizations can be observed with cyclic compounds. For example, platinum complexes reacting with cyclopropane rings insert platinum into the C–C bond to form platinacyclobutane. Isomerization of the latter complex with a methyl-substituted cycle according to Scheme II.11 has been reported.[237]

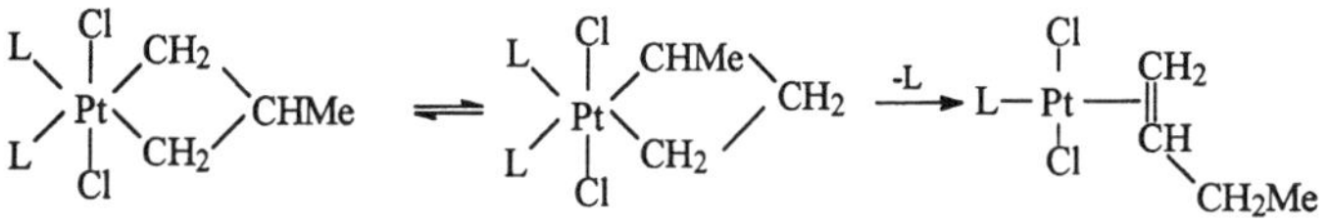

SCHEME II.11 Isomerization of platinacyclobutane into platinum butene complex.

It may be mentioned that a similar reaction in the oxidation of methylcyclopropane rings to form butenols is often regarded as evidence of the intermediate free radical formation (see Section II.2.3). Reaction with platinum complexes is not very likely to involve free radicals. Hence, a careful examination is needed in other cases as well to establish a mechanism leading to the isomerization. Apparently similar non-radical mechanism is not excluded for other species.

In this connection some analogy may be noticed between electron-rich metal atoms and singlet oxygens (oxene) bound to a high-valent metal (iron(IV), manganese(V), etc.). Oxene possesses a free orbital used in not very strong interaction with the metal. This makes oxygen bound to the metal a soft acid reacting with alkane molecules. The similarity to reactions of metal compounds with alkanes when the metal atom itself coordinates an alkane molecule becomes particularly apparent when the possibility of five-coordinate carbon intermediate complex formation in both cases with a hydrocarbon molecule is taken into account (Figure II.18).

It should be noted that "cross-over" events (the loss of chirality, endo–exo H–D isomerization and other isomerizations in the oxygenation reactions) could be explained by the complex formation that is present also in the direct interaction of alkanes with metal in the inner coordination sphere. Insertion of the oxene forming an alcohol is apparently a crucial step in aerobic alkane oxidation performed by monooxygenases and is analogous to oxidative addition of an alkane to metal, which is actually an insertion of the metal into a C–H bond of alkane. The difference is, of course, the greater thermodynamic stability of the alcohol compared with the metal alkyl hydride derivative. This difference leads to the irreversible alcohol formation upon the oxygen atom insertion whereas alkyl hydride can easily produce alkane again. Thus, methane forms complexes (characterized by five-coordinate carbons) both directly with metal centers and with oxygen atoms bound to metal. Apparently there are many more alkane, particularly methane, complexes than so far considered in the literature and their significance as intermediates in a number

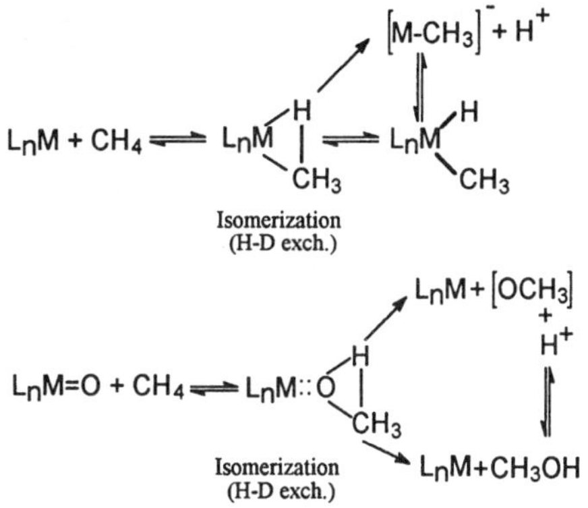

FIGURE II.18 Comparison of alkane interaction with metal and oxygen centers.

of reactions is probably far greater than is presently expected. Perhaps the prediction made by Chatt in 1976 and given as the epigraph of this chapter is close to reality.

The properties of Ni(II) complexes with d^8 electronic configuration could be used by nature to catalyze the anaerobic oxidation of methane and other alkanes via formation of nickel–alkyl bonds. Coordinative unsaturation of nickel complexes might help to react with alkanes in aqueous media and perform a catalytic cycle with a mild oxidating agent. In this case after the initial complex formation of an alkane with the catalyst the next step is probably not oxidative addition but electrophilic substitution to form $Ni^{II}–CH_3$ complex, which undergoes further transformations. Further development of this field using the biomimetic approach might lead to the discovery of new ways of activation and functionalization of alkanes, including methane, and will show whether this suggestion is correct.

REFERENCES

1a. Sheldon, R. A. and Kochi, J. K., *Metal-Catalyzed Oxidations of Organic Compounds*, Academic Press, New York, 1981.
1b. Sheldon, R. A., in *Metalloporphyrins in Catalytic Oxidations*, Sheldon, R. A., Ed., Marcel Dekker, New York, 1994, chap. 1.
2. Shilov, A. E., *Activation of Saturated Hydrocarbons by Transition Metal Complexes*, Reidel, Dordrecht, 1984.
3. Drago, R. S., *Coord. Chem. Rev.*, 117, 185, 1992.
4. Semenov, N. N., *Some Problems in Chemical Kinetics*, Princeton University Press, Princeton, NJ, Vol. 1, 1958; Vol. 2, 1959.
5a. Denisov, E. T., *Mendeleev Commun.*, 1, 1992.
5b. Denisov, E. T., *Kinet. Katal.*, 35, 671, 1994.
6a. Kirmse, W., *Carbene Chemistry*, 2nd ed., Academic Press, New York, 1971.
6b. Nefedov, O. M., Ioffe, A. I., and Menchikov, L. G., *Carbene Chemistry*, Khimiya, Moscow, 1990 (in Russian).
6c. Luntz, A. C., *J. Chem. Phys.*, 73, 1143, 1980.
6d. Arai, H., Kato, S., and Koda, S., *J. Phys. Chem.*, 98, 12, 1994.

6e. Tal'rose, V. L. and Lyubimova, A. K., *Dokl. Akad. Nauk SSSR*, 86, 909, 1952.
7a. Olah, G. A., in *Activation and Functionalization of Alkanes*, Hill, C. L., Ed., Wiley, New York, 1989. chap. 3.
7b. Schreiner, P. R., Kim, S.-J., Shaefer, H. F., and Schleyer, P. V. R., *J. Chem. Phys.*, 99, 3716, 1993; Scuseria, G. E., *Nature*, 366, 512, 1993.
7c. McMurry, J. F. and Lectka, T., *J. Am. Chem. Soc.*, 112, 869, 1990.
7d. Olah, G. A., Yoneda, N., and Parker, D. G., *J. Am. Chem. Soc.*, 99, 483, 1977.
8. Deno, N. C., Jedzeniak, E. J., Messer, L. A., and Tomezsko, E. S., in *Enzyme Action (Bioorganic Chemistry)*, Vol. 1, Van Tamelen, E. E., Ed., Academic Press, New York, 1977, 79.
9a. Shilov, A. E., *Khim. Phys.*, 10, 758, 1991.
9b. Geletii, Yu. V., Zubarev, V. E., Levin, P. P., Lubimova, G. V., and Shafirovich, V. Ya., *Kinet. Katal.*, 31, 802, 1990.
9c. Koldasheva, E. M., Shestakov, A. F., Geletii, Yu. V., and Shilov, A. E., *Izv. Akad. Nauk Ser. Khim.*, 845, 1992.
10a. Shestakov, A. F. and Shilov, A. E., *Khim. Phys.*, 3, 1591, 1984.
10b. Shestakov, A. F. and Shilov, A. E., *Zh. Obshch. Khim.*, 65, 622, 1995.
11. Bouma, W. J., Nobes, R. H., and Radom, L., *J. Am. Chem. Soc.*, 104, 2929, 1982.
12a. Khenkin, A. M. and Shilov, A. E., *New J. Chem.*, 13, 659, 1989.
12b. Hammett, L. P., *Physical Organic Chemistry*, 2nd ed., McGraw-Hill, New York, 1970.
12c. Taft, R. W. and Smith, D. E., *J. Am. Chem. Soc.*, 76, 305, 1954; Kreevoy, M. M. and Taft, R. W., *J. Am. Chem. Soc.*, 79, 4016, 1957.
13. Meunier, B., *Chem. Rev.*, 92, 1411, 1992.
14a. Griller, D. and Ingold, K. U., *Acc. Chem. Res.*, 13, 317, 1980.
14b. Bowry, V. W., Lusztyk, J., and Ingold, K. U., *J. Am. Chem. Soc.*, 111, 1927, 1989.
15. Stuart, R. and Spitzer, U. A., *Can. J. Chem.*, 56, 1273, 1978.
16. Krusic, P. J., Meakin, P., and Jesson, J. P., *J. Phys. Chem.*, 75, 3438, 1971.
17. Melander, L. and Saunders, W. H., Jr., *Reaction Rates of Isotopic Molecules*, Wiley, New York, 1980.
18. Kwart, H., *Acc. Chem. Res.*, 15, 401, 1982.
19. Rudakov, E. S., *Alkanes Reactions with Oxidants, Metal Complexes and Radicals in Solution*, Naukova Dumka, Kiev, 1985.
20. Rudakov, E. S. and Tishchenko, N. A., *Kinet. Katal.*, 32, 274, 1991.
21. Hayaishi, O., in *Molecular Mechanisms of Oxygen Activation*, Hayaishi, O., Ed., Academic Press, New York, 1974, 1.
22. Bach, A. and Chodat, R., *Ber. Dtsch. Chem. Ges.*, 36, 600, 1903.
23. Warburg, O., *Heavy Metal Prostetic Groups and Enzyme Actions*, Oxford University Press, London, 1949.
24. Wieland, H., *On the Mechanism of Oxidation*, Yale University Press, New Haven, CT, 1932.
25a. Groh, S. E. and Nelson, M. J., in *Selective Hydrocarbon Activation*, Davies, J. A., Watson, P. L., Greenberg, A., and Liebman, J. F., Eds., VCH Publishers, New York, 1990.
25b. Miwa, G. T., Walsh, J. S., Kedderis, G. L., and Hollenberg, P. F., *J. Biol. Chem.*, 208, 14445, 1983.
25c. McMurry, T. J. and Groves, J. T., in *Cytochrome P-450, Sturcture, Mechanism and Biochemistry*, Ortiz de Montellano, P. R., Ed., Plenum Press, New York, 1986.
25d. Ortiz de Montellano, P. R., in *Cytochrome P-450, Sturcture, Mechanism and Biochemistry*, Ortiz de Montellano, P. R., Ed., Plenum Press, New York, 1986, 217.
26. Dolphin, D., James, B. R., and Welborn, H. C., *J. Mol. Catal.*, 7, 201, 1980.
27. Bonfils, C., Debey, P., and Maurel, P., *Biochim. Biophys. Res. Commun.*, 88, 1301, 1979.
28. Bangcharoenpaurpong, O., Rizos, A. K., Champion, R. M., Jollie, D., and Sligar, S. G., *J. Biol. Chem.*, 261, 8089, 1986.
29. Chottard, G., Schappacher, M., Ricard, L., and Weiss, R., *Inorg. Chem.*, 23, 4557, 1984.
30. Hjelmeland, L. M., Aronow, L., and Trudell, J. R., *Biochim. Biophys. Res. Commun.*, 76, 541, 1977.
31a. Poulos, T. L., Finzel, B. C., Gusalus, I. C., Wagner, G. C., and Kraut, J., *J. Biol. Chem.*, 260, 16122, 1985.
31b. Poulos, T. L., Finzel, B. C., and Howard, A. J., *J. Biochem.*, 25, 5314, 1986.
32a. Groves, J. T., McClusky, G. A., White, R. E., and Coon, M. J., *Biochim. Biophys. Res. Commun.*, 81, 154, 1978.

32b. White, R. E., Miller, J. P., Favreau, L. V., and Bhattacharyya, A., *J. Am. Chem. Soc.*, 108, 6024, 1986.
33. Groves, J. T. and Subramanian, D. V., *J. Am. Chem. Soc.*, 106, 2177, 1984.
34. Groves, J. T., *J. Chem. Educ.*, 62, 928, 1985.
35. Bowry, V. W., Lusztyk, J., and Ingold, K. U., *J. Am. Chem. Soc.*, 111, 1927, 1989.
36. Ortiz de Montellano, P. R. and Stearns, R. A., *J. Am. Chem. Soc.*, 109, 3415, 1987.
37a. Griller, D. and Ingold, K. U., *Acc. Chem. Res.*, 13, 317, 1980.
37b. Newcomb, M., Le Tadic, M.-H., Putt, D. A., and Hollenberg, P. F., *J. Am. Chem. Soc.*, 117, 3312, 1995.
38a. Gebb, M. H., Heimbrook, D. C., Malkolin, P., and Sligar, S. G., *Biochemistry*, 21, 370, 1982.
38b. Armstrong, W. H., *Metal Clusters in Proteins*, Que, L., Ed., ACS Symposium Series, No. 372, 1988, chap. 1.
39. Hawkins, B. K. and Dawson, J. H., *J. Am. Chem. Soc.*, 114, 3547, 1992.
40a. Groves, J. T. and Myers, R. S., *J. Am. Chem. Soc.*, 105, 5791, 1983.
40b. Mansuy, D., Leclaire, J., Fontecave, M., and Momentau, M., *Biochem. Biophys. Res. Commun.*, 119, 319, 1984.
40c. Collman, J. P., Kodachek, T., and Brauman, J. T., *J. Am. Chem. Soc.*, 108, 2588, 1986.
41. Jorgensen, K. A., *Chem. Rev.*, 83, 431, 1989.
42. Dabrowiak, J. C., *J. Inorg. Biochem.*, 13, 314, 1980.
43. Hecht, S. M., *Acc. Chem. Res.*, 19, 383, 1986.
44. Knorre, D. G., Fedorova, O. S., and Frolova, E. I., *Usp. Khim.*, 62, 70, 1993.
45. Katapodis, A. G., Smith, H. A., and May, S. W., *J. Am. Chem. Soc.*, 110, 897, 1988.
46. Katapodis, A. G., Caspi, E., Piper, J., and Shapiro, S., *J. Chem. Soc. D*, 1981, 76.
47. Fu, H., Newcomb, M., and Wong, C.-H., *J. Am. Chem. Soc.*, 113, 5878, 1991.
48. Van Ravenswaay Claasen, J. C. and van der Linden, A. C., *Antonie van Leewenhoek*, 37, 339, 1971.
49. Katapodis, A. G., Wimalasena, K., Lee, J., and May, S. W., *J. Am. Chem. Soc.*, 106, 7928, 1984.
50. Abbot, M. T. and Udenfriend, S., in *Molecular Mechanisms of Oxygen Activation*, Hayaishi, O., Ed., Academic Press, New York, 1974, 167.
51. Nordland, P., Sjoberg, B.-M., and Eclund, H., *Nature*, 345, 593, 1990.
52. Baldwin, J. E., Adlington, R. M., Pilrouch, N., and Pareira, I. A., *J. Chem. Soc. D*, 105, 1993.
53a. Kurtz, D. M., Shriver, D. F., and Klotz, I. M., *Coord. Chem. Rev.*, 24, 145, 1977.
53b. Wilkins, P. C. and Wilkins, R. G., *Coord. Chem. Rev.*, 79, 195, 1987.
54a. Vincent, J. B., Oliver-Lilley, G. L., and Averill, B. A., *Chem. Rev.*, 90, 1447, 1990.
54b. Holmes, M. A., Trong, I. L., Turley, S., Sieker, L. C., and Stenkamp, R. E., *J. Mol. Biol.*, 218, 583, 1991.
54c. Nordlund, P., Eklund, H., and Sjöberg, B. M., *Nature*, 345, 593, 1990.
54d. Atta, M., Nordlund, P., Aberg, A., and Eklund, H., *J. Biol. Chem.*, 267, 20682, 1992.
55. Fontecave, M., Gerez, C., Atta, M., and Jeunet, A., *Biochem. Biophys. Res. Commun.*, 169, 659, 1990.
56a. Bollinger, J. M., Edmondson, D. E., Huynh, B.-H., Filley, J., Norton, J. R., and Stubbe, J., *Science*, 253, 292, 1991; Bollinger, J. M., Stubbe, J., Huynh, B.-H., and Edmondson, D. E., *J. Am. Chem. Soc.*, 113, 6289, 1991.
56b. Nordlund, P., Dalton, H., and Eclund, H., *FEBS Lett.*, 307, 257, 1992.
57a. Anthony, C., *The Biochemistry of Methylotrophs*, Academic Press, London, 1982.
57b. Dalton, H., in *Methane and Methanol Utilizers*, Murrel, J. C. and Dalton, H., Eds., Plenum Press, New York, 1992.
58. Colby, J., Stirling, D. I., and Dalton, H., *Biochem. J.*, 165, 395, 1977.
59. Tonge, G. M., Harrison, D. E. F., and Higgins, I. J., *Biochem. J.*, 161, 333, 1977.
60. Patel, R. N., Hou, C. T., Laskin, A. I., and Felix, A., *Appl. Biochem.*, 4, 175, 1982.
61. Green, J. and Dalton, H., *J. Biol. Chem.*, 264, 17698, 1989.
62. Woodland, M. P. and Dalton, H., *J. Biol. Chem.*, 259, 53, 1984.
63. Ericson, A., Hedman, B., Hodgson, K. O., Green, J., Dalton, H., Bentsen, J.-G., Beer, R. H., and Lippard, S. J., *J. Am. Chem. Soc.*, 110, 2330, 1988.
64. De Witt, J. G., Bentsen, J. G., Rosenzweig, A. C., Hedman, B., Green, J., Pilkington, S., Papaefthymion, G. C., Dalton, H., Hodgson, K. O., and Lippard, S. J., *J. Am. Chem. Soc.*, 113, 9219, 1991.

65. Rosenzweig, A. C., Frederick, C. A., Lippard, S. J., and Nordlund, P., *Nature*, 366, 531, 1993.
66. Green, J. and Dalton, H., *J. Inorg. Biochem.*, 43, 556, 1991.
67. Gvozdev, R. I., Shushenacheva, E. V., Pylyashenko-Novokhatnii, A. A., and Belova, V. S., *Oxid. Commun.*, 7, 249, 1984.
68a. Woodland, M. R., Patil, D. S. Cammack, R., and Dalton, H., *Biochem. Biophys. Acta*, 873, 273, 1986.
68b. Lee, S.-K., Nesheim, J. C., and Lipscomb, J. D., *J. Biol. Chem.*, 268, 21569, 1993.
68c. Liu, K. E., Wang, D., Huynh, B. H., Edmondson, D. E., Salifoglou, A., and Lippard, S. J., *J. Am. Chem. Soc.*, 116, 7465, 1994.
68d. Liu, K. E., Valentine, A. M., Qiu, D., Edmondson, D. E., Appelman, E. H., and Lippard, S. J., *J. Am. Chem. Soc.*, 117, 4997, 1995.
68e. Shteinman, A. A., *FEBS Lett.*, 362, 5, 1995.
68f. Dalton, H., Wilkins, P. C., and Jiang, Y., *Biochem. Soc. Trans.*, 21, 749, 1993.
69. Shimada, M., Nemoto, S., and Okura, I., *J. Mol. Catal.*, 64, 373, 1991.
70a. Rataj, M. J., Kauth, J. E., and Donnely, M. I., *J. Biol. Chem.*, 266, 18684, 1991.
70b. Nesheim, J. C. and Lipscomb, J. D., *J. Inorg. Biochem.*, 59, 369, 1995.
71. Ruzicka, F., Huang, D.-S., Donnely, M. I., and Frey, P. A., *Biochemistry*, 29, 1696, 1990.
72a. Priestley, N. D., Floss, H. G., Froland, W. A., Lipscomb, J. D., Williams, P. G., and Marimoto, H., *J. Am. Chem. Soc.*, 114, 7561, 1992.
72b. Wilkins, P. C., Dalton, H., Podmore, I. D., Deighton, N., and Symons, M. C. R., *Eur. J. Biochem.*, 210, 67, 1992.
73a. Liu, K. E., Johnson, C. C., Newcomb, M., and Lippard, S. J., *J. Am. Chem. Soc.*, 115, 939, 1993.
73b. Gvozdev, R. I., personal communcation.
74a. Schröder, D., Fiedler, A., Ryan, M. F., and Schwarz, H., *J. Phys. Chem.*, 98, 68, 1994.
74b. Anthony, C., *The Biochemistry of Methylotrophs*, Academic Press, London, 1982.
75a. Reeburgh, W. C., in *The Dynamic Environment of the Ocean Floor*, Fanning, K. and Manheim, F. T., Eds., D. C. Heath, Lexington, 1981, 203.
75b. Rueter, P., Rabus, R., Wilkes, H., Aeckersberg, F., Rainey, F. A., Jannasch, H. W., and Widdel, F., *Nature*, 372, 455, 1994.
76. Zehnder, A. J. and Brock, T. D., *J. Bacteriol.*, 137, 420, 1979.
77. Halcrow, M. A. and Christou, G., *Chem. Rev.*, 94, 2421, 1994.
78. Thauer, R. K., Schwörer, B., and Zirngibl, C., in *Microbial Growth on C_1 Compounds*, Murrell, J. C. and Kelly, D. P., Eds., Intercept, Andover, 1993, 151.
79. Lin, S.-K. and Jaun, B., *Helv. Chim. Acta*, 74, 1725, 1991.
80. Berkessel, A., *Bioorganic Chem.*, 19, 101, 1991.
81. Hill, C. L., Ed., *Activation and Functionalization of Alkanes*, Wiley, New York, 1989.
82. Mansuy, D., Battioni, P., and Battioni, J.-P., *Eur. J. Biochem.*, 184, 267, 1989.
83a. Montanari, F. and Casella, L., Eds., *Metalloporphyrins Catalyzed Oxidations*, Kluwer, Dordrecht, 1994.
83b. Shilov, A. E. and Karasevich, E. I., in *Metalloporphyrin's Catalyzed Oxidations*, Montari, F., and Casella, L., Eds., Kluwer, Dordrecht, 1994, 87.
84. De Poorter, B., Ricci, M., Bortolini, O., and Meunier, B., *J. Mol. Cat.*, 31, 221, 1985.
85. Mansuy, D., *Pure Appl. Chem.*, 62, 741, 1990.
86. Lichtenberger, F., Nastainczyk, W., and Ullrich, V., *Biochem. Biophys. Res. Commun.*, 70, 939, 1976.
87a. Groves, J. T., Nemo, T. E., and Myers, R. S., *J. Am. Chem. Soc.*, 101, 1032, 1979.
87b. Ohtake, H., Higuchi, T., and Hirobe, M., *J. Am. Chem. Soc.*, 114, 10660, 1992.
87c. Groves, J. T. and Roman, J. S., *J. Am. Chem. Soc.*, 117, 5594, 1995.
88. Tabushi, I. and Yazaki, A., *J. Am. Chem. Soc.*, 103, 7371, 1981.
89. Fontecave, M. and Mansuy, D., *Tetrahedron*, 40, 4397, 1984.
90. Lu, W. Y., Bartoli, J. F., Battioni, P., and Mansuy, D., *New J. Chem.*, 16, 1992.
91. Karasevich, E. I., Khenkin, A. M., and Shilov, A. E., *J. Chem. Soc. D*, 731, 1987.
92. Karasevich, E. I., Anisimova, B. L., Rubailo, V. L., and Shilov, A. E., *Kinet. Katal.*, 34, 1993.
93a. Ellis, P. E. and Lyons, J. E., *J. Chem. Soc. D*, 1187, 1989.
93b. Ellis, P. E. and Lyons, J. E., *Catal. Lett.*, 3, 389, 1989.
93c. Ellis, P. E. and Lyons, J. E., *Coord. Chem. Rev.*, 105, 181, 1990.

94a. Ellis, P. E., Lyons, J. E., and Durante, V. A., *Stud. Surf. Sci. Catal.*, 67, 99, 1991.
94b. Ellis, P. E. and Lyons, J. E., *Catal. Lett.*, 8, 45, 1991.
95. Maldotti, A., Bartocci, C., Amadelli, R., Polo, E., Battioni, P., and Mansuy, D., *J. Chem. Soc. D*, 20, 1487, 1991.
96. Cardini, G. and Jurtshuk, P., *J. Biol. Chem.*, 245, 2789, 1970.
97. Karasevich, E. I. and Khenkin, A. M., *Biokhimiya (Moscow)*, 51, 1454, 1986.
98. Ohtake, H., Higuchi, T., and Hirobe, M., *J. Am. Chem. Soc.*, 114, 10660, 1992.
99. Morohashi, K., Sadano, H., Okada, J., and Omura, T., *J. Biochem.*, 93, 413, 1983.
100. Sorokin, A. B., Marakushev, S. A., Khenkin, A. M., Shteinman, A. A., and Shilov, A. E., *Dokl. Akad. Nauk SSSR*, 279, 939, 1984.
101. Groves, J. T. and Neumann, R., *J. Am. Chem. Soc.*, 111, 2900, 1989.
102. Karasevich, E. I., Muradov, N. Z., and Shteinman, A. A., *Izv. Akad. Nauk SSSR Ser. Khim.*, 8, 1805, 1974.
103. Geletii, Yu. V., Karasevich, E. I., Moravsky, A. P., and Shteinman, A. A., *Kinet. Katal.*, 22, 349, 1981.
104. Groves, J. T., Haushalter, R. S., Nakamura, M., Nemo, T. E., and Evans, B. J., *J. Am. Chem. Soc.*, 103, 2884, 1981.
105. Groves, J. T. and Watanabe, J., *J. Am. Chem. Soc.*, 110, 8443, 1988.
106. Khenkin, A. M. and Shteinman, A. A., *Kinet. Katal.*, 23, 219, 1982.
107. Khenkin, A. M. and Shteinman, A. A., *Izv. Akad. Nauk SSSR Ser. Khim.*, 1668, 1982.
108. Khenkin, A. M. and Shteinman, A. A., *J. Chem. Soc. D*, 1219, 1984.
109. Balch, A. L., Chan, Y. W., Cheng, R. J., LaMar, G. N., Latos-Grazynski, L., and Renner, M. W., *J. Am. Chem. Soc.*, 106, 7779, 1984.
110. Balch, A. L., Latos-Grazynski, L., and Renner, M. W., *J. Am. Chem. Soc.*, 106, 2983, 1985.
111. Bortolini, O., Ricci, M., Meunier, B., Friant, P., Ascone, I., and Goulon, J., *Nouv. J. Chim.*, 10, 39, 1986.
112. Bell, S. E., Cook, P. R., Inchley, P., Leanord, D. R., Lindsay Smith, J. R., and Robbins, A., *J. Chem. Soc. Perkin Trans.*, 2, 549, 1991.
113. Sugimoto, H., Tung, H. C., and Sawyer, D. T., *J. Am. Chem. Soc.*, 110, 2465, 1988.
114. Gold, A., Iwey, W., Toney, G. E., and Sangaiah, R., *Inorg. Chem.*, 23, 2932, 1984.
115. Collins, T. J., Powell, R. D., Slebodnick, C., and Uffelman, E. S., *J. Am. Chem. Soc.*, 112, 899, 1990.
116. Meunier, B., *Gaz. Chim. Ital.*, 114, 534, 1988.
117. Traylor, T. G. and Xu, F., *J. Am. Chem. Soc.*, 110, 1953, 1988.
118. Jones, J. P., Rettie, A. E., and Trager, W. F. *J. Med. Chem.*, 33, 1242, 1990.
118a. Groves, J. T., Gross, Z., and Stern, M. K., *Inorg. Chem.*, 33, 5065, 1994.
119. Lindsay Smith, J. R., Balasubramanian, P. N., and Bruice, T. C., *J. Am. Chem. Soc.*, 110, 7411, 1988.
120. Lindsay Smith, J. R. and Lower, R. J., *J. Chem. Soc. Perkin Trans.*, 2, 31, 1991.
121. Gopinath, E. and Bruice, T. C., *J. Am. Chem. Soc.*, 113, 6090, 1991.
122. Traylor, T. G. and Ciccone, J. P., *J. Am. Chem. Soc.*, 111, 8413, 1989.
123. Arasasingham, R. D., Cornman, C. R., and Balch, A. L., *J. Am. Chem. Soc.*, 111, 7800, 1989.
124. Mansuy, D., Bartoli, J. F., and Momentau, M., *Tetrahedron Lett.*, 23, 2781, 1982.
125. Cook, B. R., Reinert, T. J., and Suslick, K. S., *J. Am. Chem. Soc.*, 108, 7281, 1986.
126. Bruice, T. C., *Acc. Chem. Res.*, 24, 243, 1991.
127. Battioni, P., Renaud, J. P., Bartoli, J. F., Reina-Artiles, Fort, M., and Mansuy, D., *J. Am. Chem. Soc.*, 110, 8462, 1988.
128. Panicucci, R. and Bruice, T. C., *J. Am. Chem. Soc.*, 112, 6063, 1990.
129. Traylor, T. G. and Traylor, P. S., *Annu. Rev. Biophys. Bioenerg.*, 11, 103, 1982.
130. Collman, J. P., Gange, R. R., Reed, C. A., Halbert, T. R., Lang, G., and Robinson, W. T., *J. Am. Chem. Soc.*, 97, 1427, 1975.
131. Tabushi, T., Kodera, M., and Yokoyama, M., *J. Am. Chem. Soc.*, 107, 4466, 1985.
132. Beliaev, A. B., Karasevich, E. I., Kuzmin, V. A., Levin, P. P., and Khenkin, A. M., *Izv. Akad. Nauk SSSR Ser. Khim.*, 1877, 1987.
133. Mandon, D., Weiss, R., Franke, M., Bill, E., and Trautwein, A. X., *Angew. Chem.*, 28, 1709, 1989.

134. Grinstaff, M. W., Hill, M. G., Labinger, J. A., and Gray, H. B., *Science,* 264, 1311, 1994.
135. Barloy, L., Lallier, J. P., Battioni, P., and Mansuy, D., *New J. Chem.,* 16, 71, 1992.
136a. Groves, J. T. and Nemo, T. E., *J. Am. Chem. Soc.,* 105, 6243, 1983.
136b. Sorokin, A. B. and Khenkin, A. M., *J. Chem. Soc. D,* 45, 1990; Sorokin, A. B. and Khenkin, A. M., *New J. Chem.,* 14, 63, 1990.
136c. Sorokin, A., Robert, A., and Meunier, B., *J. Am. Chem. Soc.,* 115, 7293, 1993.
137. Traylor, T. G., Hill, K. W., Fann, W.-P., Tsuchiya, S., and Dunlap, B. E., *J. Am. Chem. Soc.,* 114, 1308, 1992.
138. Mansuy, D., Chottard, J. C., Lange, M., and Battioni, J.-P., *J. Mol. Catal.,* 7, 215, 1980.
139. Udenfriend, S., Clark, C. T., Axelrod, J., and Brodie, B. B., *J. Biol. Chem.,* 208, 731, 1954.
140a. Hamilton, G. A., Workman, R. J., and Woo, L., *J. Am. Chem. Soc.,* 86, 3390, 1964.
140b. Hamilton, G. A., in *Molecular Mechanisms of Oxygen Activation,* Hoyaishi, O., Ed., Academic Press, New York, 1974, 405.
141. Ullrich, V. and Staudinger, H., *Z. Naturforsch.,* 24b, 583, 1969.
142. Ullrich, V., *Z. Naturforsch.,* 24b, 699, 1969.
143. Akhrem, A. A., Metelitsa, D. A., and Skurko, M. E., *Usp. Khim.,* 44, 868, 1975.
144. Muradov, N. Z., Shilov, A. E., and Shteinman, A. A., *Kinet. Katal.,* 13, 1357, 1972.
145. Karasevich, E. I., Muradov, N. Z., and Shteinman, A. A., *Izv. Akad. Nauk SSSR Ser. Khim.,* 8, 1805, 1974.
146. Muradov, N. Z. and Shteinman, A. A., *Izv. Akad. Nauk SSSR Ser. Khim.,* 10, 2294, 1975.
147. Denisov, E. T., *Liquid-Phase Reaction Rate Constants,* Plenum Press, New York, 1974, 771.
148. Mimoun, H. and de Roch, S., *Tetrahedron,* 31, 777, 1975.
149. Lavrushko, V. V., Khenkin, A. M., and Shilov, A. E., *Kinet. Katal.,* 21, 276, 1980.
150. Muradov, N. Z., Shilov, A. E., and Shteinman, A. A., *React. Kinet. Catal. Lett.,* 2, 417, 1975.
151. Akhrem, A. A., Metelitsa, D. A., and Skurko, M. E., *Bioorganic Chem.,* 4, 307, 1975.
152. Groves, J. T. and Van der Puy, *J. Am. Chem. Soc.,* 98, 5290, 1976.
153a. Khenkin, A. M., Belova, V. S., and Shilov, A. E., *Catal. Lett.,* 5, 211, 1990.
153b. Fish, R. H., Konings, M. S., Oberhausen, K. J., Fong, R. H., Yu, W. M., Christou, G., Vincent, J. B., Coggin, D. K., and Buchanan, R. M., *Inorg. Chem.,* 30, 3002, 1991.
154a. Goldstein, A. S., Beer, R. H., and Drago, R. S., *J. Am. Chem. Soc.,* 116, 2424, 1994.
154b. Seok, W. K., Dobson, J. C., and Meyer, J. M., *Inorg. Chem.,* 27, 3, 1988.
154c. Parshall, G. W. and Ittel, S. D., *Homogeneous Catalysis,* Wiley, New York, 1992.
154d. Moiseev, I. I., Shishkin, D. I., and Gekhman, A. E., *New J. Chem.,* 23, 683, 1989; Gekhman, A. E., Thesis, Moscow, 1995.
155. Barton, D. H. R., Castiger, M. J., and Motherwell, W. B., *J. Chem. Soc. D,* 41, 1983.
156a. Barton, D. H. R. and Ozbalik, N., in *Activation and Functionalization of Alkanes,* Hill, C. L., Ed., Wiley, New York, 281, 1989; Barton, D. H. R. and Doller, D., *Acc. Chem. Res.,* 25, 504, 1992.
156b. Geletii, Yu. V., Lavrushko, V. V., and Lubimova, G. V., *J. Chem. Soc. D,* 936, 1988; Geletii, Yu. V., Lubimova, G. V., and Shilov, A. E., *Kinet. Katal.,* 26, 1019, 1985.
157. Kulikova, V. S., Khenkin, A. M., and Shilov, A. E., *Kinet. Katal.,* 29, 1278, 1988.
158. Chiou, Y.-M. and Que, L., *J. Am. Chem. Soc.,* 114, 7567, 1992.
159. Wieghardt, K., Pohl, K., and Gebert, W., *Angew. Chem. Int. Ed.,* 22, 727, 1983.
160. Armstrong, W. H. and Lippard, S. J., *J. Am. Chem. Soc.,* 105, 4837, 1983.
161a. Taft, K. L., Masschelein, A., Liu, S., Lippard, S. J., Garfinkel-Shweky, D., and Bino, A., *Inorg. Chim. Acta,* 198, 627, 1992.
161b. Shteinman, A. A., *Mendeleev Commun.,* 155, 1992.
161c. Watton, S. P., Masschelein, A., Rebek, J., and Lippard, S. J., *J. Am. Chem. Soc.,* 116, 5196, 1994.
162. Murch, B. P., Bradley, F. C., and Que, L., *J. Am. Chem. Soc.,* 108, 5027, 1986.
163. Vincent, J. B., Huffman, H. C., Christou, G., Li, Q., Nanny, M. A., Hendrickson, D. N., Fong, R. H., and Fish, R. H., *J. Am. Chem. Soc.,* 110, 6898, 1988.
164. Kitajima, N., Fukui, H., and Moro-oka, Y., *J. Chem. Soc. D,* 485, 1988.
165. Kitajima, N., Ito, M., Fukui, H., and Moro-oka, Y., *J. Chem. Soc. D,* 102, 1991.
166. Belova, V. S., Gimanova, I. M., Stepanova, M. L., Khenkin, A. M., and Shilov, A. E., Belova, V. S., Khenkin, A. M., Postnov, V. N., Prusakov, V. E., Shilov, A. E., and Stepanova, M. L., *Mendeleev Commun.,* 7, 1992.

167a. Kurusu, Y. and Neckers, D. C., *J. Org. Chem.*, 56, 1981, 1991.
167b. Iwamoto, H., Hirata, J., Matsukami, K., and Kagawa, S., *J. Phys. Chem.*, 83, 903, 1983.
168. Khenkin, A. M. and Stepanova, M. L., *Mendeleev Commun.*, 7, 1991.
169. Leising, R. A., Brennan, B. A., Que, L., Fox, B. G., and Münck, E., *J. Am. Chem. Soc.*, 113, 3988, 1991.
170. Dong, Y., Fujii, H., Hendrich, M. P., Leising, R. A., Pan, G., Randall, C. R., Wilkinson, E. C., Zang, Y., Que, L., Fox, B. G., Kauffmann, K., and Münck, E., *J. Am. Chem. Soc.*, 117, 2778, 1995.
171. Leising, R. A., Norman, R. E., and Que, L., *Inorg. Chem.*, 29, 2553, 1990.
172. Nishida, Y., *Chem. Lett.*, 195, 1990.
173. Stassinopoulos, A. and Caradonna, J. P., *J. Am. Chem. Soc.*, 112, 7071, 1990.
174. Shteinman, A. A., *Dokl. Akad. Nauk SSSR*, 321, 775, 1991.
175a. Kharitonov, A. S., Sheveleva, G. A., Panov, G. I., and Paukshtis, E. A., *Appl. Catal. A*, 98, 33, 1993.
175b. Sobolev, V. I., Kharitonov, A. S., Paukshtis, E. A., and Panov, G. I., *J. Mol. Catal.*, 84, 117, 1993.
175c. Panna, O. V., Sobolev, V. I., and Panov, G. I., 2nd Workshop, C_1-C_3 Hydrocarbon Conversion, Krasnoyarsk, 1994.
176a. Pitchai, R. and Klier, K., *Catal. Rev. Sci. Eng.*, 28, 13, 1986.
176b. Nakamura, J., Kubushiro, K., and Uchijima, T., *Stud. Surf. Sci. Catal.*, 77, 373, 1993.
176c. Sato, K., Nakamura, J., Uchijima, T., Hayakawa, T., Hamakawa, S., Tsunoda, T., and Takehira, K., *J. Chem. Soc. Faraday Trans.*, 91, 1995.
176d. Otsuka, K., Uchiyama, T., and Yamanaka, I., *Chem. Lett.*, 1517, 1993.
176e. Hickman, D. A. and Schmidt, L. D., *Science*, 259, 343, 1993.
176f. Lin, M. and Sen, A., *Nature*, 368, 613, 1994.
177. Hayakawa, T., Sato, K., Tsunoda, T., Suzuki, K., Shimizu, M., and Takehira, K., *J. Chem. Soc. D*, 1743, 1994.
178. Shilov, A. E. and Shteinman, A. A., *Coord. Chem. Rev.*, 24, 97, 1977.
179. Crabtree, R., *Chem. Rev.*, 85, 246, 1985.
180. Chipperfield, J. R. and Webster, D. E., *The Chemistry of the Metal-Carbon Bond*, Vol. 4, Hartley, F. R., Ed., Wiley, 1987, 1073.
181. Shilov, A. E. and Shul'pin, G. B., *Russ. Chem. Rev.*, 59, 853, 1990.
182. Watson, P. L., *J. Chem. Soc. D*, 276, 1983.
183. Watson, P. L., *J. Am. Chem. Soc.*, 105, 6491, 1983.
184. Watson, P. L. and Parshall, G. W., *Acc. Chem. Res.*, 55, 18, 1985.
185. Bruno, J. W., Marx, T. J., and Day, V. W., *J. Am. Chem. Soc.*, 104, 7357, 1982.
186. Grigorian, E. A., Dyachkovskii, F. S., and Mullagaliev, I. R., *Dokl. Akad. Nauk SSSR*, 224, 859, 1975.
187. Grigorian, E. A., *Usp. Khim.*, 53, 347, 1984.
188. Dyachkovskii, F. S., Shilova, A. K., and Shilov, A. E., *J. Polymer Sci. Part C*, 16, 2333, 1967.
189. Baudry, D., Ephritikhine, M., and Felkin, H., *J. Chem. Soc. D*, 1243, 1980.
190. Baudry, D., Ephritikhine, M., Felkin, H., and Holmes-Smith, R., *J. Chem. Soc. D*, 788, 1983.
191. Bergman, R. G., Seidler, P. F., and Wenzel, T. T., *J. Am. Chem. Soc.*, 107, 4358, 1985.
192. Barrett, P. H., Pasternak, M., and Pearson, R. G., *J. Am. Chem. Soc.*, 101, 222, 1979.
193. Field, L. D., George, A. V., and Messerle, B. A., *J. Chem. Soc. D*, 1339, 1991.
194. Enikolopyan, N. S., Menchikova, G. N., and Grigoryan, E. A., *Dokl. Akad. Nauk SSSR*, 291, 11, 1986.
195. Harper, T. G. P., Shinomoto, R. S., Deming, M. A., and Flood, Th. C., *J. Am. Chem. Soc.*, 110, 7915, 1988.
196. Billups, W. E., Konarski, M. M., Hauge, R. H., and Margrave, J. L., *J. Am. Chem. Soc.*, 102, 7393, 1980.
197. Ozin, G. A., McIntosh, D. F., Mitchell, S. A., and Garcio-Prieto, J., *J. Am. Chem. Soc.*, 103, 1574, 1981.
198. Vargaftik, M. N., Stolarov, I. P., and Moiseev, I. I., *J. Chem. Soc. D*, 1049, 1990.
199. Crabtree, R. H., Mihelcic, J. M., and Quirk, J. M., *J. Am. Chem. Soc.*, 101, 7738, 1979.
200. Crabtree, R. H., Demou, P. C., Eden, D., Mihelcic, J. M., Parnell, C. A., Quirk, J. M., and Morris, G. E., *J. Am. Chem. Soc.*, 104, 6994, 1982.
201. Burk, M. J., Crabtree, R. H., Parnell, C. A., and Uriarte, R. J., *Organometallics*, 3, 816, 1984.
202. Janowicz, A. H. and Bergman, R. G., *J. Am. Chem. Soc.*, 104, 352, 1982; Janowicz, A. H. and Bergman, R. G., *J. Am. Chem. Soc.*, 105, 3929, 1983.

203. Hoyano, J. K. and Graham, W. A. G., *J. Am. Chem. Soc.*, 104, 3723, 1982.
204. Wax, M. J., Stryker, J. M., Buchanan, J. M., Kovac, C. A., and Bergman, R. G., *J. Am. Chem. Soc.*, 106, 1121, 1984.
205. Hoyano, J. K., McMaster, A. D., and Graham, W. A. G., *J. Am. Chem. Soc.*, 105, 7190, 1983.
206. Jones, W. D. and Feher, F. J., *Organometallics*, 2, 562, 1983.
207. Periana, R. A. and Bergman, R. G., *Organometallics*, 3, 508, 1984.
208. Chosh, C. K. and Graham, W. A. G., *J. Am. Chem. Soc.*, 109, 4726, 1987.
209. Wayland, B. B., Ba, S., and Sherry, A. E., *J. Am. Chem. Soc.*, 112, 1259, 1990; Wayland, B. B., Ba, S., and Sherry, A. E., *J. Am. Chem. Soc.*, 113, 5305, 1991.
210. Burger, P. and Bergman, R. G., *J. Am. Chem. Soc.*, 115, 10462, 1993.
211. Sakakura, T. and Tanaka, M., *J. Chem. Soc. D*, 758, 1987.
212. Sakakura, T., Sodeyama, T., Sasaki, K., Wada, K., and Tanaka, M., *J. Am. Chem. Soc.*, 112, 7221, 1990.
213. Nomura, K. and Saito, Y., *J. Chem. Soc. D*, 161, 1988; Nomura, K. and Saito, Y., *J. Mol. Catal.*, 54, 57, 1989.
214. Sakakura, T., Sodeyama, T., and Tanaka, M., *New J. Chem.*, 13, 737, 1989.
215. Maguire, J. A., Boese, W. T., Goldman, M. E., and Goldman, A. S., *Coord. Chem. Rev.*, 97, 179, 1990.
216. Khannanov, N. K., Menchikova, G. N., and Grigoryan, E. A., *Mendeleev Commun.*, 39, 1994.
217. Gol'dshleger, N. F., Tyabin, M. B., Shilov, A. E., and Shteinman, A. A., *Zh. Fiz. Khim.*, 43, 2174, 1969.
218. Garnett, J. L. and Hodges, R. J., *J. Am. Chem. Soc.*, 89, 4546, 1967.
219. Gol'dshleger, N. F., Es'kova, V. V., Shilov, A. E., and Shteinman, A. A., *Zh. Fiz. Khim.*, 46, 1358, 1972; Es'kova, V. V., Shilov, A. E., and Shteinman, A. A., *Kinet. Katal.*, 13, 534, 1972.
220. Luinstra, G. A., Wang, L., Stahl, S. S., Labinger, J. A., and Bercaw, J. E., *J. Organomet. Chem.*, 504, 75, 1995.
221. Kushch, L. A., Lavrushko, V. V., Misharin, Yu. S., Moravsky, A. P., and Shilov, A. E., *Nouv. J. Chim.*, 7, 729, 1983.
222. Zamashchikov, V. V., Popov, V. G., and Rudakov, E. S., *Kinet. Katal.*, 35, 372, 1994.
223. Geletii, Yu. V. and Shilov, A. E., *Kinet. Katal.*, 24, 486, 1983.
224. Freund, M. S., Labinger, J. A., Lewis, N. S., and Bercaw, J. E., *J. Mol. Catal.*, 87, L 11, 1994.
225. Sen, A., Lin, M., Kao, L.-Ch., and Hutson, A. C., *J. Am. Chem. Soc.*, 114, 6385, 1992.
226. Hackett. M. and Whitesides, G. M., *J. Am. Chem. Soc.*, 110, 1449, 1988.
227. Rudakov, E. S., Zamashchikov, V. V., Belyaeva, N. P., and Rudakova, R. I., *Zh. Fiz. Khim.*, 47, 2732, 1973.
228. Rudakov, E. S., *Alkane Reactions with Oxidants, Metal Complexes and Radicals in Solutions*, Naukova Dumka, Kiev, 1985 (in Russian).
229. Gol'dshleger, N. F., Khidekel, M. L., Shilov, A. E., and Shteinman, A. A., *Kinet. Katal.*, 15, 261, 1974.
230a. Gretz, E., Oliver, T. F., and Sen, A., *J. Am. Chem. Soc.*, 109, 8109, 1987.
230b. Verdernikov, A. N., Kuramshin, A. I., and Solomonov, B. N., *J. Chem. Soc. D*, 121, 1994.
231. Rubtsova, T. B., Kirjakov, N. V., Soloveichik, G. L., and Shilov, A. E., *Mendeleev Commun.*, 89, 1993.
232. Lehmkuhl, H., Pasynkiewicz, S., Benn, R., and Rufinska, A., *J. Organomet. Chem.*, 240, C27, 1982.
233a. Periana, R. A., Taube, D. J., Evitt, E. R., Löffler, D. G., Wentrecek, P. R., Voss, G., and Masuda, T., *Science*, 259, 340, 1993.
233b. Crabtree, R. H., *Angew. Chem. Int. Ed.*, 32, 789, 1993.
233c. Periana, R. A. and Bergman, R. G., *J. Am. Chem. Soc.*, 108, 7332, 1986.
234. Bullock, R. M., Headford, C. E. L., Kegley, S. E., and Norton, J. R., *J. Am. Chem. Soc.*, 107, 727, 1985.
235. Gould, G. L. and Heinekey, D. M., *J. Am. Chem. Soc.*, 111, 5502, 1989.
236. Mobley, T. A., Schade, C., and Bergman, R. G., *J. Am. Chem. Soc.*, 117, 7822, 1995.
237. Puddephatt, R. J., *Coord. Chem. Rev.*, 33, 149, 1980.

Chapter III

PHOTOCATALYTIC DECOMPOSITION OF WATER. CHEMICAL MODELS OF PHOTOSYNTHESIS

Although I believe in the future of atomic energy I think that real breakthrough in solving of the energy problems will happen only when we shall be able to perform the large scale synthesis of molecules similar to chlorophyll or even higher quality. In order to reach this goal one must investigate this type of molecules and the process of photosynthesis.

F. Joliot-Curie
(Statement in the Presidium of the
Russian Academy of Sciences, 1949.)

III.1 INTRODUCTION

At present the search for new energy sources and reevaluation of the old ones is a very pressing problem. This is connected primarily with an inevitable exhaustion of mineral fuel resources in the not too distant future.

Two types of energy with practically unlimited resources attract particular interest. They are thermonuclear and solar energies. The use of the first is dangerous and definitely would lead to the overheating of the planet. Solar energy looks very attractive because we have been accustomed to its use since ancient times. The danger of the heat contamination of the environment in the case of large-scale use of solar energy is much less than that foreseeable for the same scale of the use of nuclear energy. Living nature gives us a remarkable example of the use of solar energy in the photosynthesis that occurs in plants. It originated on Earth spontaneously billions years ago and since then it has supported all life on our planet. It is generally recognized now that the photosynthesis in plants and algae is the main source of dioxygen in the atmosphere, producing annually about 150 Gt (1 Gt = 1 billion ton) of dioxygen as well as organic substances (more than 100 Gt annually). The products of ancient photosynthesis are stored in the form of coal and oil, and those of the photosynthesis now occurring are the main source of food for man and animals. Accumulation of solar energy proceeds in nice green surroundings, in forests, fields, and meadows, and is accompanied by the formation of molecular oxygen, which is used for breathing by all animals, including man. Can we create a purely chemical system that would perform a similar process with the same or even higher efficiency? The experience obtained by scientists shows that it is not at all an easy task and the question could be asked whether it is worthwhile in an artificial system to try to repeat a process that fortunately already exists and satisfies the needs of mankind. Moreover, photoelectronic devices based on semiconductors are being developed successfully and definitely are going to be used (and are already being used) prior to chemical systems based on the principles of photosynthesis.

However, only the future will tell what will be the best way of accumulation and use of solar energy. Perhaps it will be different for different purposes and in different places of Earth. "Hybrid" photovoltaic systems based on semiconductors, involving molecular engineering, and also on the principles of the photosynthesis, have just appeared and evidently will be successfully used for practical purposes. Besides, to understand the process of photosynthesis, which is perhaps the most important among all the processes in living nature, one has to learn how to reproduce it in a simplified way in man-made systems. Also the lessons learned in constructing the artificial systems on the principles of plant photosynthesis may be used in the future for constructing other useful devices. No doubt, therefore, that the problem of artificial photosynthesis will continue to be of great importance and attract many investigators. The emphasis in this chapter, as in the previous ones, is on the aspects connected with the use of transition metal complexes in different reactions modeling the steps of plant photosynthesis. It is generally accepted in the scientific community that to solve this problem is one of the mankind's great goals. Apparently, the goal (to solve the problem of photosynthesis) has very broad meaning. For example, it includes the intensification of the photosynthesis for food and fuel production or the choice of different plants for different places on the earth according to the climate and light intensity. In this section of the book we shall examine only one side of the problem, although not the least important. That is, how to use the knowledge of the mechanism of the photosynthesis to build chemical systems, which, while working on principles of photosynthesis, could perform something similar, i.e., to use solar light to accumulate solar energy in chemical substances synthesized from other substances containing less energy. It means that we wish to perform the reaction against the thermodynamic potential with the help of light. It should be said that we are making only first steps in this direction. Photosynthesis is a process of great complexity. It has been developed during long evolution and in its modern form contains several subsystems that posses different functions, each having its own history, because clearly they were developed separately in different periods of time in several types of living organisms (bacteria) and only later turned out to be united in the whole system of plant photosynthesis. It was natural not to try initially to mimic the total process in its full complexity (which would be hopeless), but to solve at first much simpler problems of creating models for the subsystems with the hope of uniting them eventually into the whole system. The efficiency of solar energy conversion in the course of photosynthesis is known to be at best up to a few percent for the very complicated reaction between CO_2 and H_2O, and we can expect in the future to create more efficient artificial systems of solar energy conversion for the essentially simpler reaction of water decomposition to dihydrogen and dioxygen or for producing electricity. It must be recognized that in the last 20 or 30 years scientists have had much success proceeding along this general strategy.

III.2 NATURAL PHOTOSYNTHESIS

III.2.1 PLANT PHOTOSYNTHESIS[1]

Despite a great amount of work, the real mechanism of the photosynthesis became more or less clear only in the last 40 or 50 years, and some important details are still missing. As in previous cases the work to mimic the natural reactions was conducted without full knowledge of their mechanism and, also similarly to the other cases, the understanding of natural photosynthesis was developed in parallel with chemical analogues of its separate stages, both processes influencing each other.

In this section we shall only briefly mention the principal steps in the understanding of the mechanism of photosynthesis that will be used in the development of chemical systems in further sections. However, let us begin with main events in the history of the photosynthesis investigations, which started long before anything concerning its chemical mechanism became known.

The fact that plants are capable of evolving oxygen gas was established more than 200 years ago. In 1771 Joseph Priestley discovered that plants could "purify" air that was "burned out" by a candle.

In 1779 Jan Ingoenhousz in Holland showed that plants depend on light, and oxygen is evolved only in daytime while at night the plants only "spoil" the air. In 1782 Senebier came to the conclusion that plants need carbon dioxide to produce oxygen and later Ingoenhousz realized that carbon dioxide was the source of organic matter in plants.

In 1845 the German scientist Robert Mayer made an important step forward, having demonstrated that plants store light energy in the form of organic compounds. Thus its immense significance to the living world became clear.

The role of chlorophyll was revealed by the Russian scientist Timiryazev in the 19th century. He put forward a bold (for his time) idea that the chlorophyll participates in photosynthesis not only absorbing the light, but also in some chemical reaction.

However, steps toward a real understanding of the photosynthesis mechanism were made only in the 20th century. In the 1930s Cornelius van Niel (U.S.A.) proposed a scheme of photosynthesis with the primary reaction of the photodecomposition of water.[2] In 1939 Robert Hill (Great Britain) found that photolysis of H_2O and fixation of CO_2 are separate processes.[3] In 1941 it was established with dioxygen isotope $^{18}O_2$ that dioxygen evolves from water molecules.[4]

The structure of chlorophyll was established by Hans Fischer (Germany) in 1940. In 1954 Melvin Calvin proposed the cycle for CO_2 transformation in photosynthesis (Calvin cycle).[5] The existence and separate functioning of two photosystems in plant photosynthesis was proposed in 1961–1966 by British scientists Robert Hill and Fay Bendall[6] and independently by Louis Duysens (Holland). In 1945 A. N. Terenin and A. A. Krasnovsky (Russia) applied spectral investigation to chlorophyll for the first time.[7] Krasnovsky discovered subsequently the reversible photoreduction of chlorophyll (Krasnovsky reaction).[8]

The fact that the photooxidation of chlorophyll is the primary stage of the photosynthesis was established in 1968 by William Parsons (U.S.A.).

In this last quarter of the 20th century the application of new physical methods of investigation has greatly accelerated the progress in the understanding of main chemical reactions of the photosynthesis. The dramatic demonstration of this fact was the establishment in 1984 by Johann Deisenhofer (U.S.A.), Robert Huber, and Hartent Michel (Germany)[9a] of the structure of the primary reaction center of the bacterial photosynthesis by the X-ray method. It is remarkable that by that time both the main features of the reaction center structure and the principles of its functioning were already understood at least in general (e.g., see Reference 9b) and the X-ray structure, a spectacular achievement, at the same time confirmed the results already obtained by other methods.

III.2.1.1 Two Photosystems and Z Scheme

It is generally accepted now that there are two different reaction systems in green plants: photosystem I (PS I) and photosystem II) (PS II). Their presence in green plants and algae followed from the experiments of Emerson who found in 1957 that light rays of two different wavelengths (650 and 700 nm) when they are present together provided higher rates of photosynthesis than the sum of the rates if each light ray with its wavelength was acting separately.[10] As mentioned above, Hill and Bendall and also Duysens formulated the mechanism of the interacting of two photosystems and proposed the so-called Z-scheme (Figure III.1).

According to this scheme an electron from water moves against the thermodynamic potential under the action of two quanta consecutively first in PS II, then in PS I. In 1966 Anderson and Boardman succeeded in the separation of the two photosystems of green plants by the treatment of chloroplasts with a detergent.[11] The light fraction was enriched mainly in PS I and the heavy fraction consisted of PS II. Both the fractions contained various fragments of membranes.

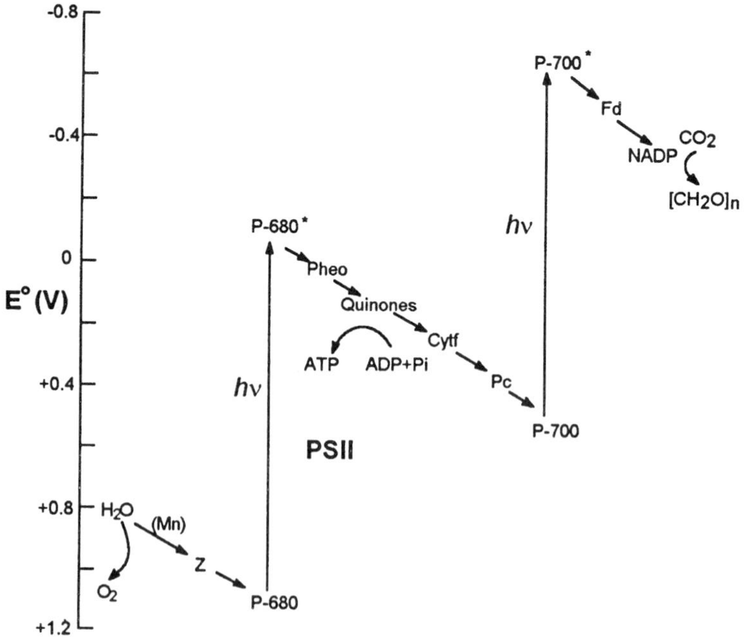

FIGURE III.1 Z scheme of plant photosynthesis.

In PS II water oxidation proceeds to dioxygen with the help of chlorophyll (Chl) as photosensitizer, and formation of reduced electron acceptor A^- takes place:

$$2H_2O + 4A \xrightarrow[\text{Chl}]{h\nu} O_2 + 4H^+ + 4A^- \tag{1}$$

which serves as an electron donor in PS I:

$$CO_2 + 4A^- + 4H^+ \xrightarrow[\text{Chl}]{h\nu} 1/6(C_6H_{12}O_6) + H_2O + 4A \tag{2}$$

Total process of photosynthesis can be written as a catalytic reduction of carbon dioxide by water accompanied by a considerable storage of free energy:

$$H_2O + CO_2 \xrightarrow[\text{Chl}]{h\nu} 1/6(C_6H_{12}O_6) + O_2, \quad \Delta G = 120 \text{ kcal/mol} \tag{3}$$

The thermodynamics and actually the mechanism of the first stage of reaction 3 is close to that of the reaction of water cleavage to dioxygen and dihydrogen:

$$2H_2O \xrightarrow{h\nu} 2H_2 + O_2, \quad \Delta G = 113 \text{ kcal/mol} \tag{4}$$

which can be performed via two photochemical stages:

$$2H_2O + 4A \xrightarrow[\text{S}]{h\nu} O_2 + 4H^+ + 4A^- \tag{5}$$

$$4H^+ + 4A^- \xrightarrow[\text{S}]{h\nu} 2H_2 + 4A \tag{6}$$

where S is a photosensitizer.

PS I provides the redox potential necessary for carbon dioxide fixation, and PS II produces the oxidative equivalents for the water cleavage to dioxygen and protons.

The process of the photosynthesis can be divided into the following steps.[1,13]

1. Light harvesting. Several pigments, mainly chlorophyll, absorb solar light and transfer the energy to a "photosynthetic reaction center" (RC) complex involving chlorophyll *a* molecule. According to the Z scheme water cleavage requires four-electron oxidation of water to produce dioxygen and eight quanta to transfer these four electrons. In the experiments of Emerson and Arnold it was shown that one molecule of CO_2 is reduced (and therefore one O_2 is produced)

per about 2500 chlorophyll molecules absorbing light.[12] Hence there are about 300 pigment molecules per reaction center.

2. Charge separation. In each photosystem an electronically excited chlorophyll molecule ensures the electron transfer from the primary donor to the acceptor with fast further transfer to prevent back-reactions, which would lead to a loss of the accumulated energy transforming it into heat. For this purpose a special arrangement is needed of electron carriers placed at definite distances in phospholipid chloroplast membrane (*vide infra*).
3. Charge transport. In both systems electrons move via several carriers of electron transfer chain located inside the chloroplast membranes to form finally the oxidized manganese cluster at one end of the chain and reduced nicotinamide adenine dinucleotide phosphate (NADPH) at the other.
4. Chemical reactions. Comparatively slow catalytic chemical reactions proceed at both ends of the electron transfer chain: carbohydrate production from CO_2 in PS I and O_2 formation from water in the oxygen evolution center of PS II. Simultaneously ATP is produced in the process of ADP phosphorylation induced by proton transfer through the membrane coupled with electron transfer.

Thus the plant photosynthesis proceeds as a highly organized and very complicated process. However, the Z scheme as it is presented in Figure III.1 may be somewhat misleading. It may seem from the scheme that electrons move one after another through the whole chain of the electron carriers in a very orderly fashion. In fact, electrons can accumulate in the electron pools between the systems, which are not too rigidly bound to each other. Moreover, the ratio of the two photosystems is not exactly one-to-one and can be different in different plants and in different parts of a leaf of one plant.[14] Apparently, the interaction of the two photosystems reflects the origin of the plant photosynthesis, which was the result of the unification of the systems that existed separately.

III.2.2 BACTERIAL PHOTOSYNTHESIS[15,16]

There are a number of bacteria that are known to depend on solar light (they are called phototrophs) and to perform so-called bacterial photosynthesis. The photosynthetic bacteria exist in very diverse forms and apparently arose early in the evolution of phototrophs. Their descendants are, of course, different from the forms that existed billions of years ago, but, as microorganisms, they have been able to survive multiple changes in global conditions and presumably they are closer to ancient organisms than other modern creatures. They can be divided into the following groups.[16]

1. Filamentous photosynthetic bacteria
2. Green sulfur bacteria
3. *Heliobacterium chlorum*
4. Purple bacteria
5. Cyanobacteria
6. Halobacteria

The first four groups have only one type of reaction center and as electron donor use such compounds as H_2S, $S_2O_3^{2-}$, S, H_2, or certain organic substances and do not produce dioxygen. The purple bacteria are the largest group of the anoxygenic photosynthetic bacteria. They can oxidize H_2S and $S_2O_3^{2-}$ and accumulate sulfur, which itself can be further oxidized to sulfate. Many of these bacteria are photoautotrophs, i.e., grow at the expense of carbon dioxide assimilation; the others need some organic substances and are, therefore, photoheterotrophs. A number of photosynthetic bacteria can fix molecular nitrogen.

Cyanobacteria (also called blue-green algae) are the most diverse and ubiquitous among all the photosynthetic prokaryotes. They are true phototrophs and usually have the capacity to use water as an electron donor; hence they evolve dioxygen that requires functioning of two photosystems with two distinct types of reaction centers. Both contain chlorophyll *a* as the main pigment. Several species of cyanobacteria are able to perform anoxygenic photosynthesis using only one reaction center. Cyanobacteria are closely linked to the evolution of chloroplasts in eukaryotes.

The last group, the halobacteria, are archaebacteria, and carry on a photosynthesis based on a nonchlorophyllous protein, bacteriorodopsin. This pigment–protein complex takes part in the synthesis of ATP but does not apparently support the growth of bacteria.

III.2.3 ASSIMILATION OF CARBON DIOXIDE[5]

In 1947, Benson and Calvin reported the results of experiments in which ^{14}C was employed for the first time to follow the fixation of carbon dioxide in plant photosynthesis. This technique, as well as paper chromatography, which was developed shortly before, provided a means of separating and analyzing minute quantities of organic materials formed in primary enzymatic reactions of CO_2 assimilation and the sequence of their following transformations. This work led Calvin to the formulation of the cycle in which carbon dioxide is assimilated in organic compounds. Since then many new details have been discovered, in particular, it was demonstrated that the Calvin cycle is not the only path of carbon dioxide assimilation. Nevertheless, the Calvin cycle is generally accepted to be the main path of carbon transformations in photosynthesis. It is presented in a simplified way in Figure III.2.

Carbon dioxide assimilation proceeds in the dark after the separation of charges and formation of the reduced electron donor NADPH. Before the reduction, carbon dioxide carboxylates diphosphorylated pentose, ribulose diphosphate (RuDP). After addition of CO_2, two molecules of 3-phosphoglyceric acid (PGA) are formed. Electrons from NADPH are transfered to PGA to form glyceraldehyde-3-phosphate. It is of importance that this reduction proceeds as a process coupled with hydrolysis of ATP formed in photophosphorylation of ADP. Further transformations in the pentosophosphate cycle led, on the one hand, to the formation of RuDP and thus to closing the cycle and, on the other hand, to the synthesis of the main products, carbohydrates. Detailed consideration of numerous processes taking part in dark stages of the photosynthesis is out of the scope of this book and apparently primary carbon dioxide fixation does not use transition metals. It could be mentioned that the Calvin cycle functions in so-called C_3-plants. There are other plants, mainly of

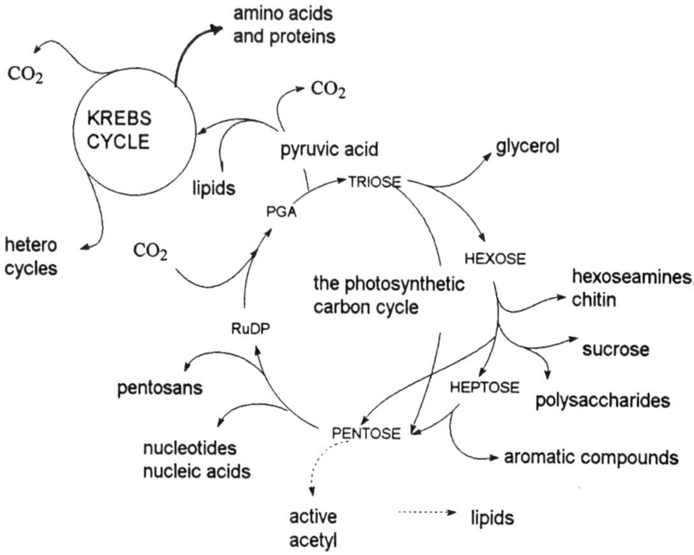

FIGURE III.2 Calvin cycle for carbon dioxide assimilation. (Adapted from Bassham, J. A. and Calvin, M., *The Path of Carbon in Photosynthesis*, Prentice Hall, Englewood Cliffs, NJ, 1957.)

tropical origin, that produce compounds containing four carbons instead of three as in the Calvin cycle.

III.2.4 ABOUT THE ORIGIN AND THE EVOLUTION OF PHOTOSYNTHESIS[16]

There is abundant evidence that present-day plant photosynthesis is the result of the long evolution beginning from the very simple systems. The origin of photosynthesis is probably associated with the origin of life about 3.5 Ga ago (Ga: giga annum = billion years): constant flow of solar energy coming to the surface of the earth created most suitable conditions for the first prebiotic organized molecular systems ("dissipative" systems in Prigogin's terminology[17a]), which ultimately gave rise to the first primitive living creatures. Porphyrin-like pigments may have originated abiologically and the first reaction centers could simply create electron flow in primitive membranes. This could lead to the formation of an ion (e.g., H^+) gradient across the membrane. Later organisms may have coupled the synthesis of ATP from ADP and phosphate with the transport of ions and this provided the means for photoassimilation of certain organic substances. The reaction centers in these photoassimilators may have initiated a cyclic electron transport with a main function of converting light energy into chemical energy in the form of ATP. Indeed, the mechanism of ATP hydrolysis catalyzed by metal ions bears some resemblance to the functioning of enzymes and therefore it might be tentatively suggested that ATP was used already by the primitive organisms in the prebiotic conditions even before RNA and proteins firmly established themselves as the base of all living systems.[17b-d]

The first photosynthetic bacteria originating later, similarly to the extant forms, most probably used such electron donors as hydrogen sulfide or dihydrogen. These substances gradually having been exhausted, the organisms were obliged to use weaker and weaker donors such as sulfur and later probably ferrous ions. The appearance of the two-quantum photosystem, one quantum producing a stronger donor than the initial one, helped photosynthetic bacteria to survive in less and less strong reducing surroundings.

The next major event in the evolution, which happened about 2.5 to 3 Ga ago, was the origination of a photosystem capable of using water as an electron donor (PS II), thus producing dioxygen. This must have led to great advantages for the photosynthetic bacteria because unlimited amounts of the electron donor were available. The first oxygen-evolving organisms may have used natural mineral surfaces to catalyze dioxygen evolution (as well as other minerals as catalysts to produce dihydrogen and fix dinitrogen). Manganese, which is essential for O_2 evolution in photosynthesis, is present in large amounts in the form of MnO_2 in sediments of seas and oceans. Later manganese complexes could be included in the reaction centers of the living cells and this again must have created great advantages for the bacteria, which then could move freely without dependence on the presence of the catalyst in the surrounding. Symbiosis of photosynthetic bacteria that lived in colonies led to the origin of algae and then green plants. Multiple similarities of photosynthetic apparatus of plants and modern bacteria clearly show that they have a common ancestor that was definitely a photosynthetic bacterium. Moreover, some existing eukaryotes contain endosymbiotic cyanobacteria that can be recognized by the still remaining structures of their cell membranes.[18]

This scenario of the development of the photosynthetic organisms, which is not certain in details but apparently true in essentials, has dictated the strategy of the biomimetic approach to the creation of the model photosynthetic systems. It was natural to divide the problem into simpler goals of creating subsystems with the intention of following the main stages of the evolution of natural photosynthesis. First, it was important to understand the principles of the photoseparation and stabilization of charges under the action of solar light to create sufficiently long-living donors and acceptors that would be thermodynamically able to react each in its own compartment with water to form dihydrogen and dioxygen. Then the effective catalytic and photocatalytic systems should be developed to form dihydrogen and dioxygen. Finally, it was essential to find the means to unite these systems in the total system of photocleavage of water. Actually, the work on these problems was conducted not one after another but mostly in parallel in several laboratories. It will be seen that the final problem has not yet been solved in full, but the results obtained so far are encouraging. They show that the approach is promising and the problem will be eventually solved as soon as more effective methods of constructing highly organized molecular systems are created. Moreover, it was mentioned already that molecular engineering based on the principles of the photosynthesis in combination with semiconducting materials, actually playing the role of some parts of the photosynthetic system, helped to create new photovoltaic cells that can successfully compete with conventional semiconductor cells for practical application (*vide infra*).

III.3 PHOTOSEPARATION AND STABILIZATION OF CHARGES

III.3.1 GENERAL CONSIDERATIONS

Similarly to natural photosynthesis, in order to perform endoergonic photocleavage of water we have to use a photosensitizer (S): water does not absorb visible light. When excited by the light quantum to the state S* it can transfer electrons to an acceptor A. The radical cation $S^{\dot{+}}$ formed reacts with water to produce dioxygen, while A^-, also reacting with water, produces dihydrogen. To ensure both processes catalysts are necessary: the average energy of one quantum of solar light (about 40 kcal/mol) is far below the dissociation energy of the H–O bond in H_2O (118 kcal/mol), while the energy of four quanta of the visible light is sufficient to produce one molecule of dioxygen and two molecules of dihydrogen from two molecules of water. Therefore we have to transfer two electrons to a catalyst performing dihydrogen evolution and accept four electrons from the catalyst of dioxygen formation to produce both molecular products without intermediate atoms or radicals. Using such a photosensitizer as natural chlorophyll we can at best have about 32% efficiency of solar energy conversion to the energy of dihydrogen and dioxygen in water cleavage.[19] The main losses are due to different energies of solar light quanta: the quanta with energy lower than 1.229 eV will be lost because they are too small to perform the water cleavage, but the excess of the energy of sufficiently large quanta will be also lost dissipating as heat. But this limiting value of the efficiency is actually very high: the ideal convertor for solar light would have 43% efficiency.[19] Real efficiency of the photosynthesis is much lower and in the best cases (e.g., for chlorella, which is one of the most efficient organisms) reaches only about 6%. Additional losses of energy in natural process are connected with the mechanism of the plant photosynthesis. In particular we have seen that two photosystems are needed in the natural system, i.e., two quanta are used for the transfer of one electron (*vide supra*). Therefore the limiting value of the efficiency of plant photosynthesis must be at least twice less than the value of 32%. However, we are still far from reaching even this "low" efficiency of the natural photosynthesis in our artificial systems. In spite of all its "imperfections", natural photosynthesis is a remarkable achievement of living nature: its *quantum efficiency* (or quantum yield) can approach 100%, i.e., all the absorbed quanta can be used to transfer electrons against the thermodynamic potential and thus lead to the accumulation of solar energy. If we want to reach comparable efficiency in using light quanta we have to understand what are the main reasons for the losses in quantum yields in chemical model systems.

Figure III.3 shows a classic picture of an organic molecule excitation by light quantum.[20a] The energy of the singlet excited molecule ^1S can be lost in fluorescence and nonradiative process of internal conversion. The fluorescent lifetime (τ_0) of the singlet-state excited molecules is very short: about 10 ns. However, it can produce a triplet excited state by a process of intersystem crossing. Rate constants of the latter process can be very high ($k_{ST} = 10^6$ to 10^{10} s^{-1}). Thus in some systems we can have high efficiency of the excited triplet state formation (up to 90%). The lifetime

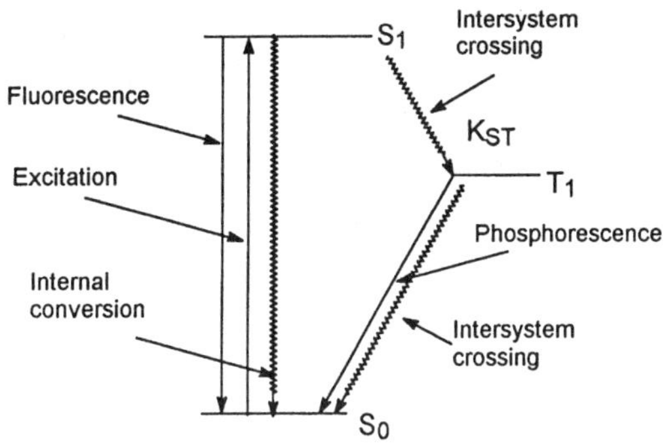

FIGURE III.3 Typical diagram of energy levels of organic molecule. (Adapted from Nadtochenkova, V. A., unpublished results.)

of the triplet state can be sufficiently long to react with an electron acceptor A to produce A^- in the process of the electron transfer

$$^3S + A \rightarrow {}^3[S^+ \cdots A^-] \rightarrow S^+ + A^-$$

The lifetime of the triplet radical ion pair $[S^+ \cdots A^-]$ is long enough for the separation of the radical ions produced because they are formed with parallel spins due to the spin conversation rule. Therefore we can have high yields of separated charges and may expect efficient reactions of S^+ and A^- with respective catalysts even in homogeneous solutions. Thus, the use of the excited sensitizer in triplet state looks more promising for creation of systems converting solar energy into the chemical energy than the use of singlet states in unorganized homogeneous systems. However, further catalytic reactions (in particular formation of H_2 and O_2 from water) cannot be too fast. The first reaction requires the accumulation of two electrons on the catalyst C_1 and subsequent reaction with two protons to form dihydrogen. The second reaction requires another catalyst, C_2, which gives up four electrons, one by one, and then reacts with two water molecules to produce dioxygen. We shall discuss the catalytic mechanisms below, and here it should be mentioned only that even with the best catalysts there are some limiting values of their rate constants because the reactions are not too simple (particularly the one producing dioxygen). Thus, several states of the catalytic particles must coexist in the solution (C_1^-, C_1^{2-}, C_2^+, C_2^{2+}, etc.), which will create a thermodynamically extremely unstable mixture and will react in various cross-processes of the type $C_1^- + C_2^+ \rightarrow C_1 + C_2$. It is easy to show that even for a sensitizer in the triplet state the yield of the final stable products (e.g., H_2 and O_2) will be negligible, i.e., the quantum yields of these reactions will be virtually zero. Very low quantum yields could be expected also in the case of microheterogeneous systems with small particles present in solution if they create a strong reductant and strong oxidant at their surface. However, the possibility of observing only one reaction of the two of the complete photocleavage (i.e., formation of only H_2 or only O_2) still exists in the homogeneous solution. If an electron donor D^s is present in the solution, and can irreversibly react with S^+ to form D_{ox}^s, which does not react with A^-, then the latter will have the possibility of interacting with the catalyst to finally produce dihydrogen. If, on the other hand, an acceptor A^s is added to the solution and can irreversibly destroy the reduced acceptor A^-, then S^+, reacting with H_2O with participation of the catalyst C_2, will produce dioxygen. As the net result we shall have photocatalytic transformations; in the first case the reaction:

$$D^s + H_2O \xrightarrow{h\nu} 2D^s_{ox} + H_2 \qquad (7)$$

and in the second case:

$$4A^s + 2H_2O \xrightarrow{h\nu} 4A^s_{red} + O_2 \qquad (8)$$

Donors of the type D^s and acceptors of the type A^s capable of participating in reactions 7 and 8 are called sacrificial. Amines, EDTA and hydrogen sulfide can be used as sacrificial donors. Irreversible oxidation of a sacrificial donor is ensured by a mechanism in which losing electron the donor particle produces a radical that undergoes a rapid transformation, e.g., is a strong donor itself, and in turn irreversibly reacts with S^+ to produce a terminal stable molecule. For example, for hydrogen sulfide the reactions are

$$SH^- \xrightarrow{-e} {}^{\bullet}SH \xrightarrow{-e} S + H^+ \qquad (9)$$

A similar oxidizing agent, A^s, can be found that does not react with either water or with the catalyst present in the solution but would react with $^*S$ to form S^+. The latter will produce dioxygen catalytically in the presence of the catalyst C_2. We shall see that such compounds as manganese(IV) pyrophosphate or persulfate ion $S_2O_8^{2-}$ belong to this category.

The reactions with sacrificial donors and acceptors accumulate naturally much less light energy (if any) than when full water cleavage takes place producing simultaneously both dihydrogen and dioxygen. However, these reactions are important to test the activity of the catalysts participating in the formation of dihydrogen and dioxygen in conditions of photocatalysis.

III.3.1.1 Charge Separation for Combined Formation of Dihydrogen and Dioxygen

In order to produce simultaneously dihydrogen and dioxygen from water, i.e., effectively perform the water photocleavage, the charges formed initially must be quickly separated spatially from each other to prevent their recombination. This has to lead to their sufficiently long lifetimes to undergo comparatively slow catalytic reactions with water.

One of the possibilities is to use semiconductive materials, which are widely applied now in various photoelectrochemical cells (Figure III.4).[20b] When a semiconductor electrode is in contact with an electrolite solution, this results in formation of a space–charge layer at the semiconductor–solution interface. Within this region the electronic energy bands are bent upward for *n* type semiconductors and downward

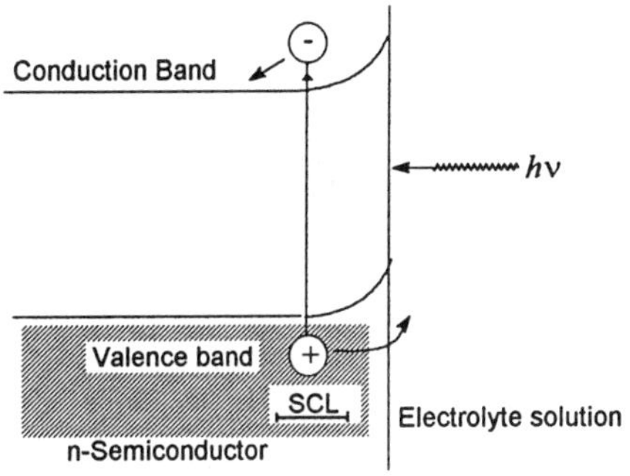

FIGURE III.4 Photoinduced charge separation in the space–charge layer of an *n* type semiconductor in contact with electrolyte solution. (Adapted from Watanabe, T. et al., in *Energy Resources through Photochemistry and Catalysis,* Grätzel, M., Ed., Academic Press, New York, 1983, chap. 11.)

for semiconductors of *p* type. When the electrode receives photons with energy greater than the semiconductor band gap, electron–hole pairs are generated at the interface, and in the case of *n* semiconductors the electric field drives the photogenerated electrons towards the interior of the electrode, while the holes move towards the surface. Electrons and holes therefore become separated, and this phenomenon can be used for construction of the two types of photoelectrochemical cells: photoelectrolytic cells (Figure III.5a) and regenerative photovoltaic cells (Figure III.5b).[20b]

If we have an organic membrane in a solution of electrolite, e.g., in the case of micelles or vesicles (liposomes), the charged surfaces of the organic material in contact with the electrolite solution form a double electric layer similar to that at the semiconductor–electrolite interface. This double layer creates an electric field that stabilizes a photogenerated ion radical pair $D^{+}\cdots A^{-}$ in a similar way as the electron–hole pair is stabilized in the case of semiconductor.

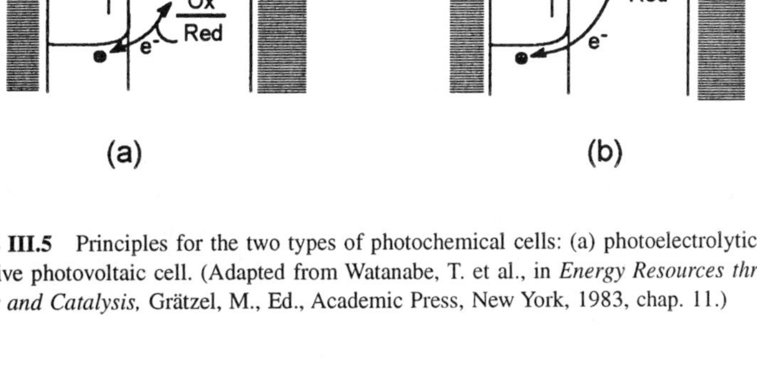

FIGURE III.5 Principles for the two types of photochemical cells: (a) photoelectrolytic cell and (b) regenerative photovoltaic cell. (Adapted from Watanabe, T. et al., in *Energy Resources through Photochemistry and Catalysis,* Grätzel, M., Ed., Academic Press, New York, 1983, chap. 11.)

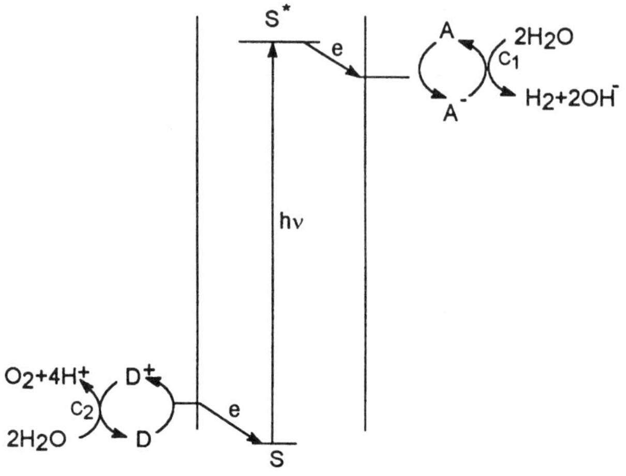

FIGURE III.6 Photocatalytic system of water cleavage with one acceptor.

Another possibility is to use the principles of photosynthesis, which are based on organized molecular systems with spatial separation of donors, photosensitizers and acceptors. The simplest system with one donor and one acceptor is shown in Figure III.6.

Let us suppose that we have a membrane with a photosensitizer placed inside, with donor and acceptor on the boundaries of the membrane and the solution on both sides of the membrane. To increase the efficiency both catalysts, C_1 and C_2, are bound to A and D, respectively. This system, as simple as it is, still requires naturally serious synthetic efforts to combine all the participants, including the catalysts, at appropriate distances from each other. However, even with the best organization the system will again be quite inefficient with respect to the possible quantum yields, at least for the major part of the solar light spectrum. The reason is again the strong possibility of the back-transfer of electrons. Catalytic reactions, even the simplest ones like dihydrogen evolution, are, as already mentioned, comparatively slow (enzymatic processes usually have rate constants on the order 10^3 s^{-1}) and back-transfer of electrons at the distances of the charge separation (e.g., from A^- to S^+ or from *S to D^+) will be much faster. It is easy to show that for reasonable rate constants of the catalytic reactions, back-transfer of electrons will greatly prevail and this will lead to negligible quantum yields.

To significantly increase quantum yields we can use the principle of cascade (e.g., see References 21a and 21b) and place another acceptor (A_2) behind the first one (A_1) with A_2 being capable of accepting electrons of lower energy levels than A_1 (Figure III.7). In this case A_1, accepting the electron, transfers it further to A_2.

If the rate of electron transfer from A_1 to A_2 exceeds that of A_1 to *S or to $S^{\ddagger}$ then the lifetime of A_2^- can be much greater than that of A_1^-. Back-reactions A_2^- +

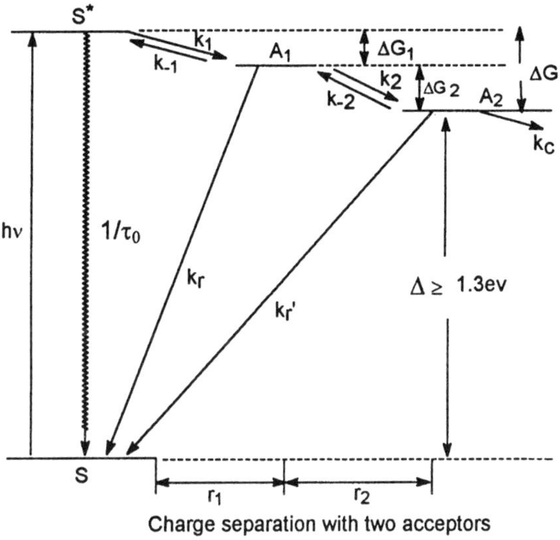

FIGURE III.7 Photochemical charge separation system with two acceptors.

$A_1 \rightarrow A_2 + A_1^-$ and $A_1^- + S^+ \rightarrow A_1 + {}^*S$ will be slow because of energy barriers ΔG_2 and ΔG_1, while direct electron transfer to S^+ will be slowed down because the increased distance between S and A_2 will strongly decrease the electron transfer rate. High quantum yields of A_2^- will be reached if direct electron transfer from A_1 to A_2 is much faster than back-transfer from A_2^- to S^+. If the energy difference ΔG_2 is not very large then the driving force Δ for the back-transfer will be much greater than for the direct transfer to A_2. To prevent the back-transfer in this situation we have to work in the so-called inverted Marcus region (*vide infra*) where the electron transfer rate decreases with the increase in absolute value of the free energy of this process (which corresponds to its driving force). Thus, to prevent or to slow down the back-transfer of electrons through barriers the condition

$$k_c \times \exp\left(\Delta G / k_B T > 1/\tau_0\right. \tag{10}$$

must be fulfilled. If $\tau_0 \approx 10$ ns and $k_c \approx 10^3\ s^{-1}$ then $\Delta G > 0.4$ eV. To estimate quantum yield we have to write a kinetic scheme of the electron transfer:

$$*\dot{S} = [S]\sigma I - [*S](1/\tau_0) - k_1[*S] + k_{-1}[A_1] \tag{11}$$

$$A_1 = k_1[*S] - k_r[A_1] - k_c[A_1] - k_{-1}[A_1] \tag{12}$$

$$A_2 = k_2[A_1] - k_{-2}[A_2] + k_r'[A_2] - k_c[A_2] \tag{13}$$

where σ is the cross section of the light absorption and I is the light intensity. For the stationary state of the process we can estimate the quantum yield of the catalytic reaction (η) if the rate constants of electron transfer processes are known. We can use the classic formula of nonadiabatic electron transfer:

$$k = 2\pi/h \times V^2 FC \tag{14}$$

where V is electronic interaction between the donor and acceptor and FC is the Franc–Condon factor which, according to the Marcus theory,[22] is

$$FC = \left[1/\left(4\pi\lambda_{out}k_B T\right)^{0.5}\right]\exp\left[-\left(\lambda + \Delta G_0\right)^2/4\lambda k_B T\right] \tag{15}$$

Here λ is the reorganization energy ($\lambda = \lambda_{in} + \lambda_{out}$) connected with internal degrees of freedom (λ_{in}) and dielectric properties of surrounding media (λ_{out}) and k_B is the Boltzman constant. It is seen from Eq. 15 that for very exergonic reactions when $-\Delta G_0 > \lambda$ the electron transfer rate constant will decrease with increasing driving force, whereas at smaller absolute values of ΔG_0 ($-\Delta G_0 < \lambda$) there is an increase of the rate upon $-\Delta G_0$ increase. The existence of the region where the electron rate decreases with the increase of $-\Delta G_0$ (inverted Marcus region) was experimentally confirmed for different cases and presumably is the main cause of the much faster direct electron transfer as compared with the reverse process.

V can be estimated as $V^2 = V_0^2 \times \exp[-\beta(r-r_0)]$, where β is the fading factor, approximately equal to 1.2 Å, r_0 is the sum of the donor and acceptor radii, r is the distance between the donor and acceptor, and V_0 is the value of V at the distance r_0.

Estimates of the quantum yields η using the above approximations and taking the plausible values of λ shows that η depends on the value of the light quantum $h\nu$, ΔG_2, and both distances, r_1 and r_2. High quantum efficiencies can be obtained already for visible light.

Figure III.8 shows the dependence of the quantum yield on the distances r_1 and r_2.[23] It is seen that there is a certain combination of both distances that corresponds to high quantum yields, demonstrating that the remarkable quantum efficiency observed in natural photosynthesis finds qualitative theoretical explanation even in the case of only two acceptors.

Actually, the maximum quantum yield should not probably be so sharp as in Figure III.8 due to a certain flexibility of the real triads $S-A_1-A_2$. It should be also taken into account that the theoretical estimate is rather approximate. Nevertheless, if the estimate considered here is close to reality then it is clear that only certain combinations of the distances in a not very broad range correspond to high quantum yields. It may be suggested therefore that modern plants have developed the ability to efficiently use solar light quanta only as the result of long evolution, and initially the quantum yields of photosynthetic organisms were not so high. Therefore the construction of an efficient device on the principle of cascade requires a fine regulation of the nature of the acceptors A_1 and A_2 and their positions relative to the photosensitizer, i.e., is not an easy problem. On the other hand, before plant photosynthesis

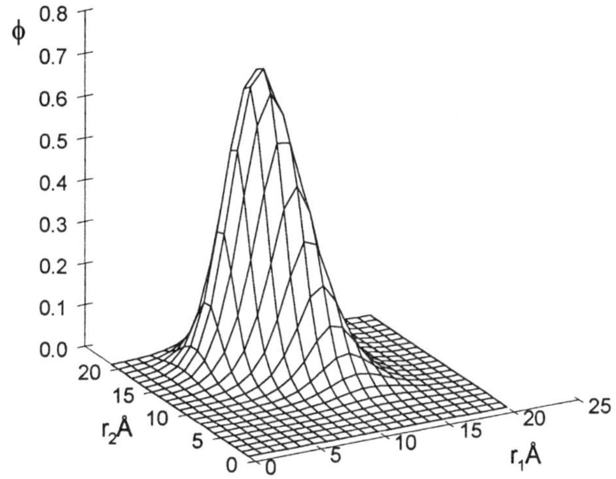

FIGURE III.8 The dependence of the quantum yield of the photochemical system with two acceptors (Figure III.7) on the distances r_1 and r_2.

originated, more UV light reached the earth's surface, because an ozone layer was absent. For larger light quanta the FC factor gives smaller values of electron transfer rate constants, and a similar estimate shows that even one acceptor (A_1) could be sufficient to perform a catalytic reaction, because A_1^- could have enough time to react with a catalyst before back-transfer of electrons. So prebiotic systems could develop and use water cleavage for the accumulation of energy with UV light. Another, more usually accepted, way of involvement of solar light in prebiotic chemical reactions is the use of sacrificial systems. The atmosphere of the earth before plant photosynthesis originated did not contain dioxygen but such electron donors as dihydrogen and hydrogen sulfide, which, as we have seen (Section III.1), are used in modern bacterial photosynthesis. As was suggested in the previous section, these donors were apparently used in photochemical reactions to produce ATP and carbohydrates. As soon as the strong donors became scarce, weaker and weaker donors could be involved, and the reaction centers underwent slow evolution to their modern complex state.

III.3.2 PHOTOSEPARATION OF CHARGES IN NATURAL PHOTOSYNTHESIS[21,24]

The reaction centers from several purple bacteria were isolated in the pure state earlier than from other photosynthetic organisms and investigated most thoroughly. Various physical methods were used to study their structure and kinetics of the electron transfer.

The reaction center in bacterial photosynthesis is situated in the membrane, which is 40 Å thick. In purple bacteria it involves four molecules of bacteriochlorophyll (BChl), two molecules of bacteriopheophytin (Bphe), and two molecules of quinone nature (menaquinone or ubiquinone). These molecules are situated in two subunits (L and M) of a protein that contains 273 and 323 aminoacid residues, respectively.

X-ray diffraction analysis of the reaction center isolated from *Rhodopseudomonas viridus* in the crystalline state was performed in 1984 (Figure III.9).[9a,25] Later the structure of the reaction center from *Rhodobacter sphaeroides* appeared. The structures of both centers bear many similarities. They confirmed very clearly the important conclusions that had been made before, i.e., that electron transfer can effectively proceed between centers separated more than 10 Å from each other and that several electron acceptors need to be placed after the photosensitizer molecule to avoid the back-transfer of electrons and to get quantitative quantum yields. Moreover, as was already mentioned, even the detailed relative arrangement of the electron carriers that had been proposed before using the optical properties of RC[9b] turned out to be essentially correct.

The reaction center of *Rb. sphaeroides* has been studied in particular detail. In this RC, chromophores are arranged in the protein according to the C_2 symmetry. Two molecules of BChl *a* (P_L and P_M) forming a so-called special pair are located close to one another. The special pair is the primary donor P865. The edges of two other BChl (B_L and B_M) molecules in both branches are shifted from the edges of P_L and P_M with minimum distance 3.5 Å, while the distance between centers of P and B is about 13 Å. Only L chromophores are photochemically active and electrons move only via the L branch. Next to B_L and B_M molecules are the molecules of BPhe (H_L and H_M), which, in their turn, are followed by quinones Q_A (in the L subunit) and Q_B (in the M subunit). A strong electron interaction has been established in the special pair of BChl molecules. This interaction and the interaction of the dimer with surrounding protein results in a red-wave shift in the absorption spectrum of the dimer approximately by 1300 cm^{-1}.

The excitation of any of the chromophores in the reaction center results in the energy transfer to the dimer in a time <150 fs and to the electron transfer from the dimer to bacteriopheophytin (H_L) in 2.8 ps with 100% quantum yield.

The molecule of B_L that is next to the special pair is most likely to be another electron carrier between P and H_L[21a] although a superexchange function was also suggested for this molecule. The state of the separate charges $P^+H_L^-$ without the next acceptor, the quinone, exists for about 12 ns and they then recombine to form the triplet state of P. This triplet state is formed as a result of the intersystem crossing in the radical pair $P^+H_L^-$. In the presence of the quinone Q_A, the electron transfer from H_L to Q_A occurs in approximately 200 ps to form the radical pair $P^+Q_A^-$, which (in the absence of Q_B) exists for about 100 ms and then recombines.

Thus, a very fast photoinduced charge separation results in the formation of the intermediate in which the back-transfer of electrons is slower than the following electron transfer to the quinone and this provides a high quantum yield of the charge separation. In the presence of Q_B, the electron transfer from Q_A to Q_B occurs in about 0.1 ms to form stable reduced substances with the energy storage used in the subsequent biochemical reactions. A single iron atom situated between Q_A and Q_B seems to help in the electron transfer between the two quinones without being a real electron carrier in the process.

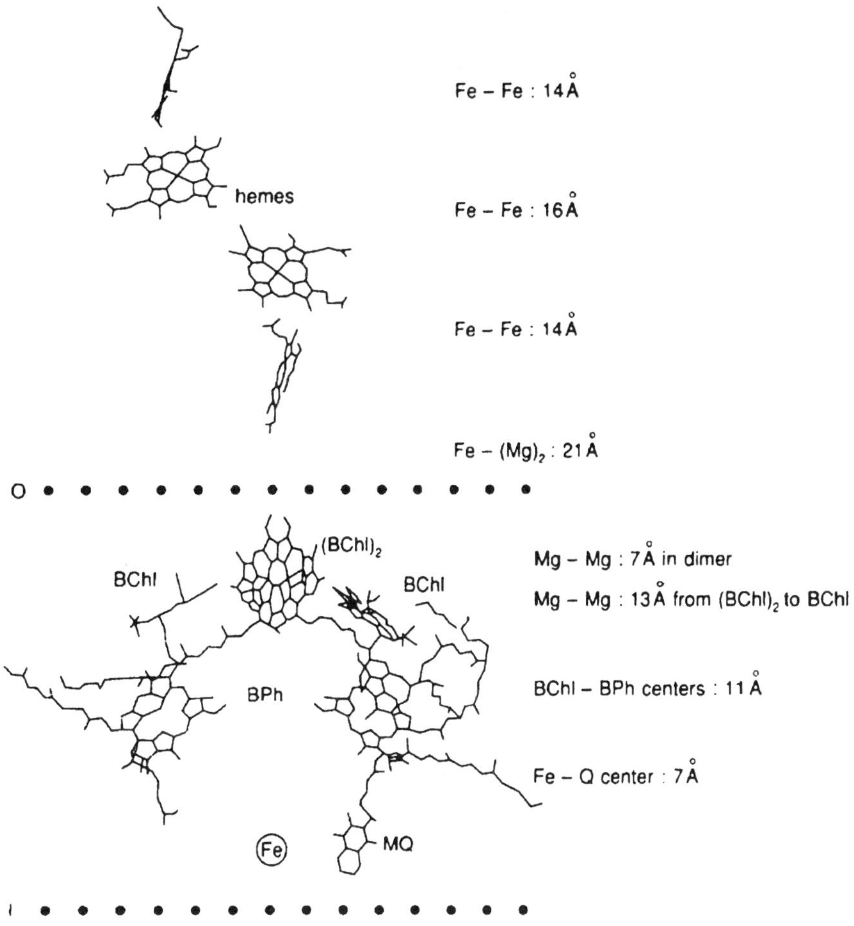

FIGURE III.9 Arrangement of the chromophores in the reaction center of *Rps. viridis.* (From Deisenhofer, J. et al., *Nature,* 318, 618, 1985. With permission.)

Figure III.10 shows the electron transfer rates and redox potentials for RC intermediates in purple bacteria.[24] It is seen that for each step the forward electron transfer is several orders of magnitude faster than the back-transfer to the special pair. Thus, this sequence of electron transfer steps provides the charge separation in the reaction center at a distance of 35 Å to form a long-lived intermediate with a high quantum yield. This is achieved at the expense of some energy. Back-transfer of electrons is slow by both ways: through the same intermediates because of the energy barriers that now exist, and direct electron transfer through space because of the long distance between the special pair and the quinone.

The RCs of PS I in plant photosynthesis include the primary donor P700 (a dimer or more probably a monomer of chlorophyll). The photoexcitation of P700 results in the reduction of the electron acceptors of the chlorophyll A_0 in about 10 ps.

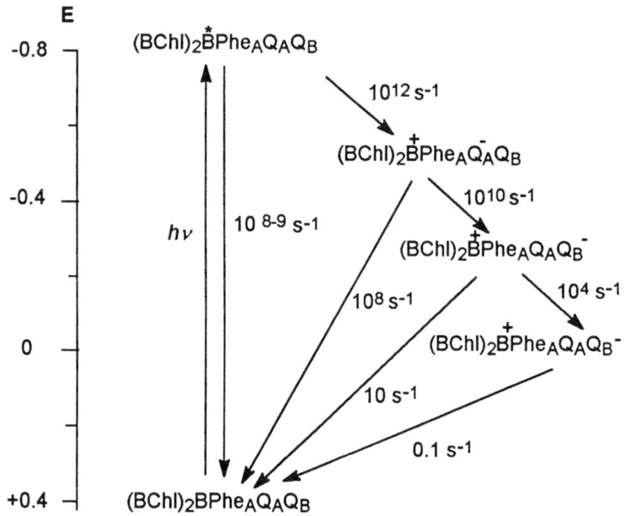

FIGURE III.10 Electron transfer rates and redox potentials for RC intermediates in purple bacteria. (Adapted from Gray, H.B. and Ellis, W.R., Jr. in *Bioinorganic Chemistry,* University Science Books, Mill Valley, CA, 1994, 315.)

Then the electron is transferred from A_0^- to the secondary acceptor A_1, which is likely to be menaquinone, in 40 ps. The lifetime of the radical pair $P700^+A_1^-$ is about 1 ms. Then the electron is transferred to the iron–sulfur clusters.

The RC of PS II was isolated in a sufficiently pure state and characterized. It contains proteins D1 and D2 that demonstrate a homology with L and M proteins of bacterial RC. It was suggested that the complex of D1 and D2 may contain cofactors of the RC of PS II. This was later confirmed and it was found that similarly to bacterial RC there are four chlorophyll molecules, two molecules of pheophytin, and one iron atom. The primary electron donor P680, which seems to be a dimer or monomer of the Chl_a molecule, transfers the electron to pheophytin in 3 ps. In the absence of the quinone, which is the next acceptor, the radical pair $P680^+Ph_a^-$ formed exists for 35 ps and then recombines to form the triplet of P680. The primary quinone acceptor is reduced by Ph_a^- in 500 ps. Thus, the structure and functioning of PS II is analogous to the work of the RC of purple bacteria and this fact confirms their common origin. However, in the content of the proteins there are some differences that apparently reflect a much higher redox potential of $P680^+$ compared with $P865^+$ in bacterial RC. The high redox potential formed is used for the water oxidation to dioxygen in the manganese-containing protein attached to the RC.

III.3.3 PHOTOSEPARATION AND STABILIZATION OF CHARGES IN ORGANIZED MOLECULAR SYSTEMS BASED ON VESICLES

III.3.3.1 Vesicles

The molecular structure of the system of water decomposition may be created by the use of molecules that are capable of "self-organization", i.e., of forming molecular assemblies of a certain structure in solutions. The individual molecules, which are the components of the photocatalytic system, and their complexes, which are synthetic RCs or parts of the electron transport chain, can be incorporated into the organized molecular assemblies.

The simplest organized molecular assemblies are formed in solutions of surfactants, whose molecules have a certain combination of hydrophilic and hydrophobic fragments and charged groups. Photochemical reactions were already performed in micelles, microemulsions, monomolecular layers at the interface, liquid membranes, bilayer lipid membranes at the hole in the teflon diaphragm separating two aqueous volumes (so-called "black" lipid membranes), and vesicles.[26-34]

Vesicles (or liposomes) are of particular interest for the coupling of the processes of charge separation and stabilization with the subsequent dark catalytic reactions. They are spherical or elliptical closed bilayer membranes of 200 to 5000 Å size with the inner water pool. Actually, they are microreactors formed at the molecular level.[35] Lipids are traditional substances for the preparation of vesicles. The ability of lipids to self-organize into bilayer membranes is caused by the existence of the hydrophilic polar or bipolar group ("head") and two long hydrocarbon chains ("tails"). Natural egg lecithin, which is a mixture of phosphatidylcholines, is often used in experimental work. Synthetic lipids, which are individual compounds, are also used, e.g., dipalmitoyl-DL-α-phosphatidylcholine (DPPC), dimiristoyl-DL-α-phosphatidylcholine (DMPC), and others. Lipid vesicles are stable for several days.

Three types of lipid vesicles may be distinguished: multilamellar, large monolamellar (1000 to 5000 Å diameter), and small monolamellar (200 to 400 Å diameter) vesicles. The latter are most frequently used in model systems. In further description we shall use the term *vesicles* to denote small monolamellar vesicles. These vesicles have a spherical form, outer diameter of about 250 Å, and thickness of the lipid bilayer of about 50 Å.

Vesicles may be prepared by the injection of an ethanol solution of the lipid into water or by the ultrasonic treatment of the lipid suspension in water at a temperature higher than the phase transition of the lipid (Huang's method).[36] Components of photosystems may be selectively incorporated into the vesicular wall and inner water pool.

The specificity of reactions with the participation of vesicles is determined by comparatively high local concentrations of the substances in the inner water pool and wall of the vesicle. One molecule in the inner water pool of the DPPC vesicle corresponds to 10^{-3} *M*, and one molecule in the vesicular membrane corresponds to 5×10^{-4} *M*. However, the observed concentration is low, because the inner water pool, for example, for [DPPC] = 3×10^{-3} *M*, equals $^1/_{1000}$ of the total volume of the solution. This makes it possible to perform reactions at concentrations that cannot be realized in a homogeneous solution. For example, one can work with the local concentration of the water-soluble zinc porphyrin of 10^{-2} *M* in the inner water pool of vesicles, while homogeneous solutions with the same concentration are almost unsuitable for the investigation due to the too high optical density of the solution. A large membrane surface, which equal to 1 m^2/ml for 4 mg/ml of the lipid, is another specific feature of vesicles.

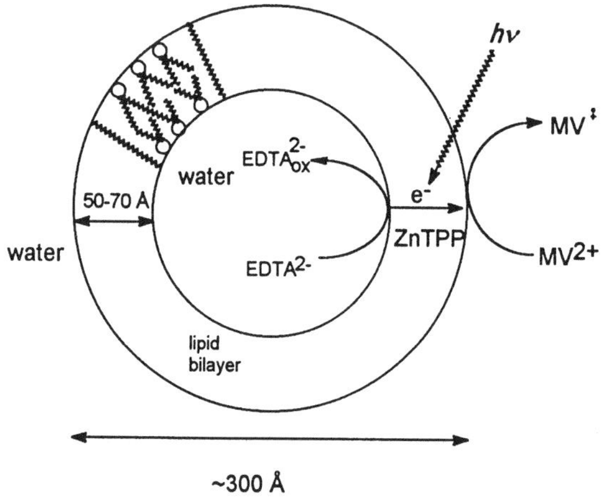

FIGURE III.11 Scheme of electron transfer through the wall of the vesicle with photosensitizer in the wall.

III.3.3.2 Electron Transfer in Vesicular Systems

Vesicles look very attractive for photoseparation of charges: electron transfer through the vesicular wall creates separated and thus long-lived oxidants and reductants that in the presence of suitable catalysts will be able to produce stable molecules, e.g., H_2 and O_2 from water.

Charge photoseparation with the use of vesicles was performed in several works,[37-42a] in which the electron donor (or acceptor) was placed in the inner water pool, the photosensitizer was incorporated into the vesicular walls, and the electron acceptor (donor) was added to the bulk solution (Figure III.11).[42b]

Usually these systems give, however, very low quantum yields (10^{-1} to 10^{-2}%) due to a high probability of back-transfer of electrons resulting in the recombination of charges. The quantum yield can be somewhat increased as the transmembrane electric field is applied, which stabilizes the ion radical pair produced in the reaction of photogenerated sensitizer with the electron transfer agent, e.g., methyl viologen.[42c]

A more efficient system includes the stabilization of separated charges, i.e., the electron transfer from the reduced form of the first electron acceptor to the second electron acceptor, and in these systems direct electron transfer competes more efficiently with the recombination of the primary products of the charge separation. This approach was used for the first time in 1983, when a new type of system for the charge separation and stabilization was reported simultaneously in two papers.[43,44] In these systems, the irreversible electron donor (D), sodium salt of ethylenediamine tetraacetic acid, and photosensitizer (S), $Ru(bpy)_3^{2+}$, were incorporated into the inner water pool of lipid vesicles; the first electron acceptor (A_1), octadecylviologen or cetylviologen, was placed in the vesicular wall, and the second electron acceptor (A_2), various oxidants with $E_{A_2/A_1^-} > E_{A_1/A_1^-}$, was placed in the bulk solution. The oxidant and reductant, S^+ and A_1^-, formed in the photochemical reaction are stabilized due to the oxidation of the reduced acceptor A_1^- by the second electron

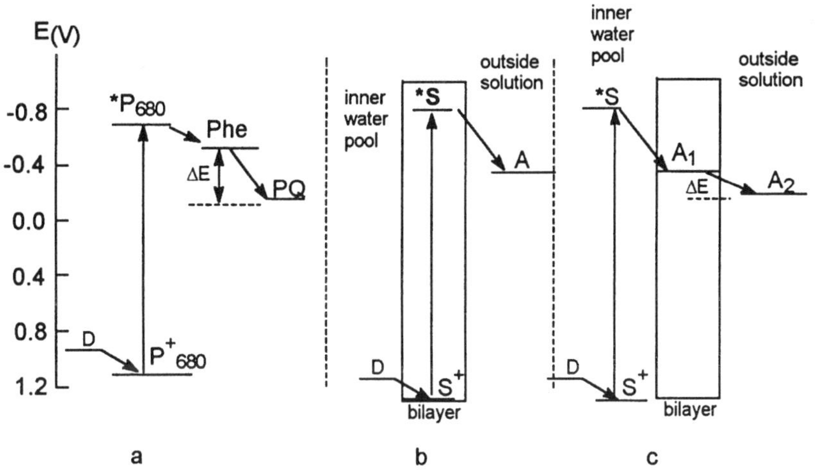

FIGURE III.12 Comparison of electron transfer in PS II of plant photosynthesis (a), vesicle system with one acceptor shown in Figure III.11 (b), and the vesicle system with two acceptors (c).

acceptor A_2. Hence the same principle is used as in natural photosynthesis (Figure III.12).

The existence of a sufficiently high energy barrier for the back-reaction of the electron transfer in the energy profile of the system, similarly to the natural photosystems, allows the reaction of A_1^- with A_2 to compete with the recombination between S^+ and A_1^-. The quantum yield of the electron transfer in this system and in similar systems that were created later[45] is 0.1 to 0.3 under the conditions of steady-state photolysis, thus at least two orders of magnitude higher than in the systems proposed earlier.

The flash-photolysis technique allowed detailed studies of separate stages of the electron transfer in the system to be conducted:[43] quenching of the excited photosensitizer molecule *S (water pool) by the A_1 molecule (vesicular wall), recombination of S^+ (water pool) and A_1^- (vesicular wall), oxidation of A_1^- by the oxidant A_2 (bulk solution), and electron transfer through the vesicular wall (transmembrane electron transfer).[46-48]

The process of the transmembrane electron transfer turned out to be the rate-limiting stage and has been studied in detail.[48] The dependence of the reaction rate of this process on the nature of viologen molecules in the vesicular wall and on the temperature of the solution for a series of various viologens and lipids were obtained.[48] Other electron carriers were also studied.[49] The data obtained allow representation of the transmembrane electron transfer as a complicated process, whose limiting stage was determined as the immersion of the viologen radical cation into the bilayer, which results in a decrease in the distance between the viologen radical and viologen molecule, the latter being located near the outer surface of the vesicle (Figure III.13). As the result of the electron exchange between the viologen radical and viologen molecule, the viologen radical formed is accessible to the external oxidant.

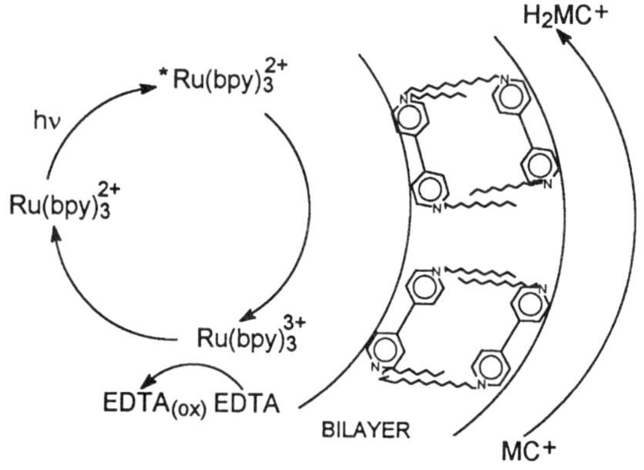

FIGURE III.13 Electron transfer through the wall of the vesicle with two acceptors.

To improve the proposed systems the studies of the regularities of the electron transfer must be continued. It is evident, for example, that vesicular systems for charge photoseparation may be improved by the use of more bulky electron carriers, which "broach" the bilayer. This will permit reducing the time of the electron transfer through the lipid membrane down to values close to the time of the electron exchange between functional groups of the molecule. Another approach to the construction of an effective system for photoseparation of charges is the use of the macromolecules containing not only the electron carrier, but also other components of the electron transfer chain (so-called supramolecules).

III.3.4 ELECTRON TRANSFER IN SUPRAMOLECULES

Much synthetic work was done recently to create various photoactive supramolecules with the hope of increasing the lifetime of the photoseparated charges produced upon radiation by solar light.

Simple covalently linked donor–acceptor systems were used for studying the dependence of the electron transfer rate on the free energy of reaction, distance between the donor and acceptor, their mutual orientation, and the solvent effects.

A large variety of supramolecules, e.g., supramolecules with rigid hydrocarbon bridges and supramolecules based on porphyrins or chlorophylls, and on the other photosensitizers, including ruthenium bipyridine complexes, has been synthesized and studied in a number of works (see Reference 50).

The detailed studies of the electron transfer in the supramolecules with rigid hydrocarbon bridges were performed in nonpolar solvents.[51-55] For example, the following supramolecules were studied: aniline (donor)–substituted ethylene (acceptor),[53] biphenyl (donor)–naphthalene (acceptor) linked by decaline or cyclohexane bridges,[54] and biphenylyl (donor)–benzoquinone or 2-chlorobenzoquinone (acceptor) linked by the steroid bridge.[55] The authors of these works found that the energy of the states with separated charges is sensitive to the change in the energy of the solvent reorganization and to the nature and length of the bridge connecting radical centers of the supramolecule.

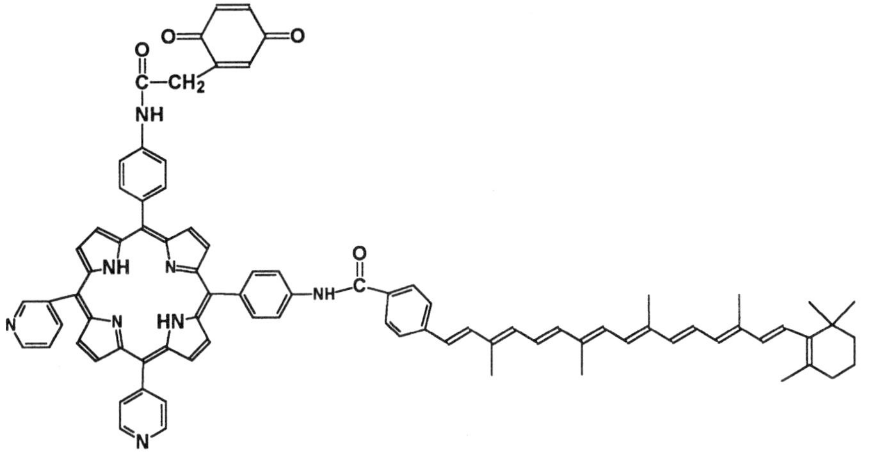

FIGURE III.14 Triad: carotenoid–porphyrin–quinone. (Adapted from Moore, T. A. et al., *Nature,* 307, 630, 1984.)

Development of supramolecules based on porphyrins and chlorophylls may clarify the fundamental problems of the natural photosynthesis and serve as a basis for the creation of model systems designed for photoseparation of charges and their utilization, e.g., for water decomposition under the action of the visible light.

The dimers of porphyrins with a naphthalene bridge,[56] trimers,[57,58] tetramers,[58] and even pentamers[57] of porphyrins linked by phenyl and biphenyl bridges were synthesized. The mixed dimers chlorophyll–porphyrin[59] have also been studied. All these supramolecules were studied by picosecond laser flash photolysis. However, the presence of several porphyrin macrocycles makes the analysis rather complicated.

Quinones and carotenoids have been used by Moore et al.[60] as the electron acceptors in supramolecules based on porphyrins, and such systems attract particular interest as analogues of the reaction centers in photosynthesis. Carotenoids were used as the secondary electron donors (triads carotenoid–porphyrin–quinone) (Figure III.14).

In the tetrad carotenoid–porphyrin–porphyrin–quinone[61] the photoexcitation of the porphyrins results in the formation of the state of the separated charges Car^{+}–Q^{-} with the quantum yield 0.25, which exists for 2.9 μs in anisole. Another very complicated supramolecule is the pentad carotenoid–porphyrin–porphyrin–naphthoquinone–benzoquinone. In chloroform 650-nm excitation yields the state $C^{\bullet +}$–P_{Zn}–P–Q–$Q^{\bullet -}$ with a long lifetime of the separate charges (55 μs) and high quantum yield (0.83).[62a,b] In dichloromethane the quantum yield drops to 0.6 but the lifetime is further increased to 200 ms. The charge-separated states store about half of the energy of the excited singlet state.

Other compounds popular in photochemistry, complexes of bivalent ruthenium and osmium with pyridine derivatives, were also used as photosensitizers in some supramolecular systems. Thus, in dyads and triads investigated in some works[63,64] the photosensitizers (P) were $Ru(ttp)_2^{2+}$ or $Os(ttp)_2^{2+}$, the donors (D) were the amines phenotiazine and di-*p*-anisylamine (DPAA), and the acceptor (A) was methyl

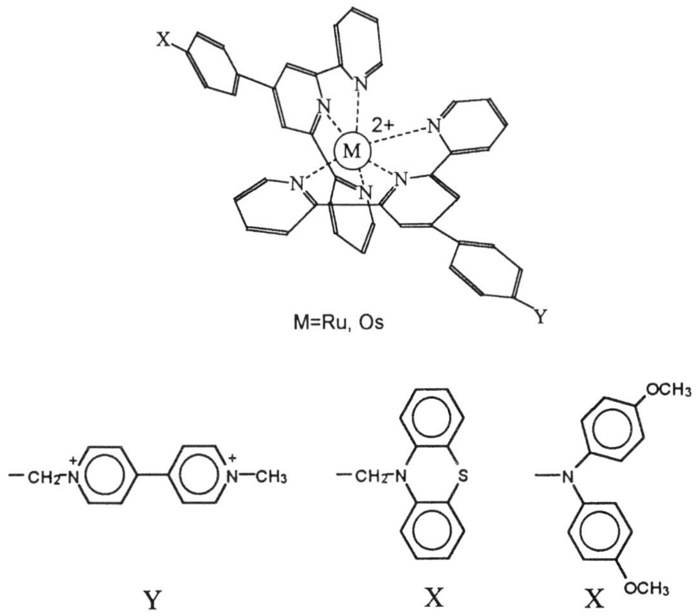

FIGURE III.15 Triad: D–M(ttp)$_2$–MV^{2+}. (Adapted from Collin, J.-P. et al., *Coord. Chem. Rev.*, 111, 291, 1991.)

viologen (ttp = 4′-(*p*-tolyl)-2,2′-6′,2″-terpyridine) (Figure III.15). Transient adsorption spectroscopic studies of DPPAA–Ru(ttp)$_2^{2+}$–MV^{2+} demonstrates that a charge-separated state, D$^+$–P–A$^-$, is formed in the deactivation of the excited state D–P*–A. This state lies at the level of the 1.1 eV and exists for <100 ns.

Using supramolecular systems important information can be gathered concerning the rates of charge separation. The dependence of the charge separation rate on free energy was obtained by measuring electron transfer rate data for a number of molecules of porphyrin–tripticene acceptor. Electron transfer rate constants, k_{cs}, were determined using picosecond transient absorption and emission techniques together with fluorescence quenching measurements.[65,66] In polar liquids the free energy of charge separation $\Delta G'_{cs}$ can be estimated using the one-electron redox potentials of the donor and acceptor and the coulomb stabilization. Thus, the dependence of k_{cs} on $-\Delta G'_{cs}$ can be determined experimentally. Assuming a linear free energy relationship this could be used to obtain the values of ΔG_{cs} in the rigid glass. The ion pair states in rigid glass were found to be 0.80 ± 0.05 eV higher in energy relative to the value found for the polar solvent. This is in very good agreement with prediction based on the dielectric continuum model of the solvent.

In a triad molecule consisting of N,N-dimethylaniline, mesoporphyrin II, and naphthoquinone both free base and zinc-containing complex were studied. Electron transfer rate constants determined by picosecond fluorescence spectrometry were found to be 1.5×10^9 and $>5 \times 10^{10}$ s^{-1}, respectively.[68]

Diad molecules consisting of ZnTPP and quinones with various linkages upon excitation in frozen solutions form triplet states of porphyrins followed by electron transfer to form a triplet radical ion pair as follows from EPR measurements.[69,70] These results show that viscous fluids permit enough solvent motion for solvation of initially formed triplet ion pair on the microsecond time scale of its lifetime.

To follow the example of nature, supramolecular systems based on self-assembling systems were investigated.[71,72] Only preliminary results have so far been obtained. This approach is promising for creation of future systems including catalysts to perform photocleavage of water and other reactions.

Porphyrins can form dimers if they possess charges at their periphery. Self-assembly occurs in media containing a large fraction of water and driven mostly by electrostatics. For example, transient absorption and EPR spectra were measured for a dimer produced from zinc *meso*-tetrakis(4-trimethylaluminium)porphyrin tetrachloride and copper *meso*-tetrakis(p-sulfonatophenyl)porphyrin sodium salt.[73] The dimer exhibits biphasic absorption transient decays with lifetimes of about 150 ns and 30 μs in fluid solution. The long-lived and short-lived transients are attributed to the triplet state of zinc porphyrin and radical ion pair, respectively.

In polar media, the electron transfer rate constant reaches the maximum value, as a rule, at $\Delta G > 1.3$ eV,[74,75] and the rate constant for the back-transfer of electrons becomes greater than the rate constant of the direct electron transfer. Thus, the efficient charge separation in model reaction centers with the participation of singlet states is possible in oil rather than in aqueous media.[76]

An alternative design of model reaction centers, in which efficiency of the charge separation is independent of polarity of the medium, is the use of the spin-forbidden back-transfer of electron reactions. For this purpose, an artificial reaction center is needed with a sufficiently large distance between the photosensitizer and first electron acceptor, so that the intersystem crossing of the excited photosensitizer molecule from the singlet state to the triplet state is possible. In this process the triplet radical ion pair is formed, its recombination is determined by the mechanisms of spin conversion, and the lifetime is considerably longer than that of the corresponding singlet radical pair. In addition, the recombination processes of triplet radical pairs are sensitive to the external magnetic field, and this provides additional opportunities for the control of the recombination processes and, hence, of the efficiency of the charge photoseparation.

This approach was used in a recent work[77] where zinc porphyrin–viologen ($P\text{–}Vi^{2+}$) dyads with the semirigid biphenyl (BPh) or phenylene (Ph) bridges between the photosensitizer (P) and electron acceptor (Vi^{2+}) were synthesized. The photoexcitation of these model reaction centers results in the charge photoseparation with high efficiency, which is almost independent of the medium polarity. The study of the dyad $P\text{–}Ph\text{–}Vi^{2+}$ [78] shows that the intermolecular electron transfer from *P to Vi^{2+} occurs in various assemblies of conformers. The triplet radical ion pair is born in the "closed" conformation, in which intramolecular electron transfer from ^3P to Vi^{2+} becomes favorable due to a short distance between them. Here the exchange interaction (2J) is much greater than the constant of hyperfine coupling (A_{hfc}). The recombination of the triplet radical ion pair may occur via two routes: (1) intersystem crossing to the ground state induced by the spin-orbital coupling; and (2) multistage transition, which involves the conformational transition to the "open" conformation in which $2J \ll A_{hfc}$ and conformational transition to the closed conformation followed

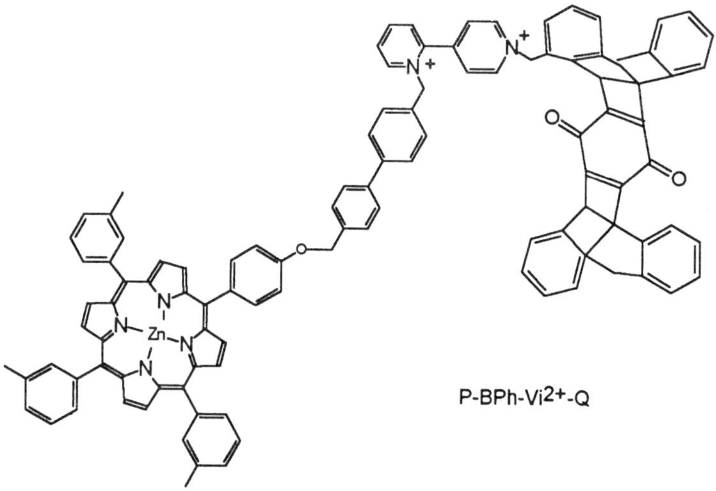

FIGURE III.16 Triad: porphyrin–biphenyl–quinone. (Adapted from Batova, E. E., *New J. Chem.*, 14, 269, 1990.)

by a fast recombination of the singlet radical ion pair. In zero magnetic field the second one is the main pathway of the radical ion pair recombination. Thus, the behavior of the triplet radical ion pair depends on the ratio $2J/A_{hfc}$ in the open conformation. The application of the external magnetic field decreases the recombination rate constant by 3 to 11 times (depending on a solvent),[79] and in a strong magnetic field the first route of radical ion pair recombination dominates. The rate of this transition can be increased by the introduction of a heavy atom, as was demonstrated for micellar solutions of dyad P–Ph–Vi^{2+}.[80]

Thus, recombination of triplet radical ion pairs can be efficiently controlled by the external magnetic field, due to the effect of the heavy atom, and by means of the variation of the solvent viscosity, temperature, and the structure of the bridge molecule.[81,82]

Both triplet and singlet radical ion pairs can be stabilized by the subsequent electron transfer to the second electron acceptor. The photoexcitation of triad P–BPh–Vi^{2+}–Q (where Q is tripticenequinone; Figure III.16) in polar solvents results[83] in the formation of the extremely long-lived triplet radical pair state $P^{+\cdot}$–BPh–Vi^{2+}–$Q^{-\cdot}$ with a lifetime of about 10 ms, which is at least three orders of magnitude greater than the lifetime of the radical ion state of triad carotenoid–porphyrin–quinone and is significantly longer even than the lifetime of the pentad carotenoid–porphyrin–porphyrin–naphthoquinone–benzoquinone considered above. Unfortunately, the quantum yield of this ion radical pair formation does not exceed 0.10 to 0.15.

Supramolecules are of great interest from the viewpoint of their incorporation into lipid membranes and other organized molecular assemblies in order to create the complete cycle of the electron transport. This is still a difficult problem and only preliminary steps have been made so far. Modified carotenoid–porphyrin–quinone triads have been incorporated into black lipid membranes.[84] Hydrophilicity of the porphyrin increased upon the substitution of two phenyl groups of the porphyrin by pyridine groups. Photoinduced transmembrane charge transfer was observed.[84] The intramolecular charge separation was also demonstrated for carotenoid–porphyrin–quinone triads and another porphyrin incorporated into the lipid bilayer.[85] Shafirovich et al.[86,87] incorporated the dyad P–Ph–Vi^{2+} into lipid bilayers of DPPC vesicles. The authors observed the recombination of the triplet radical ion pair in lipid bilayer in zero and external magnetic fields at various temperatures. The temperature dependences of the recombination rate constant and magnetic effect have a typical break in the range of the phase transition of the lipid.

Thus, the development of the supramolecular chemistry provides a possibility to use these substances in various fields of the science and allows us to expect that this will be a basis for the creation of a complete system of water decomposition under visible light in the future.

III.3.5 MOLECULAR ENGINEERING IN PHOTOVOLTAIC SYSTEMS

Molecular constructions for charge separation based entirely on the principles of plant photosynthesis, although definitely feasible, are still too complicated and fragile to immediately use in molecular systems for transformation and accumulation of solar energy, let alone practical applications. Semiconductors using *p–n* transitions are much simpler and are already successfully used for different purposes in various photoconvertors. However, the requirements of very high purity make them at present still very expensive, and this renders a serious obstacle for their wider use on a large scale. Apparently, "hybrid" systems involving semiconductors modified by special synthetic materials, particularly photosensitizers, and arranged in special constructions designed to produce electricity from solar energy are more promising in their use even at the present time. Knowledge of the general principles of plant photosynthesis helps to overcome the immense difficulties that still stand in the way of creation of solar energy convertors based on these principles.

In this connection it is worthwhile to describe the photovoltaic systems based on stable and efficient charge transfer sensitizers that were developed recently by Grätzel and his co-workers[88,89] and are already close to practical use.

The stability of the sensitizers is a particularly crucial condition for their use in the specially constructed cell because, distinct from the living cell, they cannot be easily replaced by new molecules when damaged during their work in the photochemical process. Ruthenium and osmium complexes turned out to be much more stable than natural chlorophyll and its synthetic analogues. After long search the complexes have been synthesized that satisfy the requirements not only of stability, but also of efficient electron transfer to the semiconductor. The complexes have the general formula *cis*-X_2bis(2,2′-bipyridyl-4,4′-dicarboxylate)ruthenium(II), where X = Cl, Br, I, CN, or SCN. As sensitizers the complexes play the same role as chlorophyll in the green leaf: they absorb the incident light and induce the electron transfer. Instead of charge carriers installed in the photosynthetic membrane, a titanium dioxide film is used. The sensitizer is adsorbed on the film and effectively transfers the electrons to the conduction band of the semiconductor. An important problem that had to be overcome was to organize sufficient light absorption, because on a smooth surface only 1% of the incident light is adsorbed by a monolayer of a dye molecule. It must be taken into account that only monolayers of sensitizers can be used because the sensitizer molecule has to be directly attached to the surface of the conducting material. In natural photosynthesis the stacking of thylakoid membranes containing chlorophyll is used to form the grana structures and this helps to effectively absorb solar light. In the artificial device the problem was solved by making a film consisting of colloidal TiO_2 particles with 10 to 30 Å diameter (nanoparticles). Brief sintering of the particles at about 500°C leads to formation of electronic contact between them. In this way a structure with a very high internal surface area is formed. Thus for the 8-μm thick layer the effective surface is 720 times greater than for a smooth film. As a result 98% of the light is adsorbed by the film at the wavelength 536 nm. With *cis*-di(thiocyanato)bis (2,2′-bipyridyl-4,4′-dicarboxylate)ruthenium(II) the light harvesting capacity of these films becomes superior to that of amorphous silicon.

The electrons injected by the sensitizer move through the nanocrystalline film to the conducting glass support serving as current collector. The efficiency of light-induced charge separation at the oxide semiconducting surface is expressed by the quantum yield of charge injection (ϕ_{inj}). This represents the fraction of the absorbed photons converted into the electrons of the conductivity band.

$$\phi_{inj} = k_{inj}/\left(k_{eff} + k_{inj}\right) \tag{16}$$

where k_{eff} is the sum of the rate constants of all nonproductive channels. Typical k_{eff} values are in the range of 10^6 to 10^9 s^{-1}. To achieve a good ϕ_{inj} the value of k_{inj} should be at least two orders of magnitude higher, i.e., higher than 10^{11} s^{-1}. The ruthenium-based sensitizers satisfy these requirements. The functional groups included in dipyridyl rings, e.g., carboxylates, provide an enhanced electron coupling of the sensitizer with the conductivity band of the semiconductor. With the dye molecules mentioned the charge injection is in the picosecond or even femtosecond time region.

To prevent the recapture of the injected electron by the cation of the sensitizer the value of k_{inj} must be much higher than of k_b (the recombination rate constant). Actually, $k_{inj}/k_b > 10^3$. This is due to the fact that molecular orbitals involved in the back-reaction more poorly overlap with the wave function of the conduction band electron than those in the forward reaction. For example, for *cis*-di-thiocyanato)bis(2,2′-bipyridyl-4,4′dicarboxylate)ruthenium(II) bound to a TiO_2 film, the injecting orbital is the π* orbital of the carboxylated bipyridyl ligand because the excited state of the sensitizer has a metal-to-ligand charge transfer character. The carboxylate groups interact directly with the surface titanium(IV) providing good electronic coupling between the π* orbital and the MO formed from the atomic 3*d* orbital of titanium. Therefore, electron injection from the excited sensitizer into the semiconductor membrane is extremely rapid, occurring in less than a picosecond (Figure III.17).

By contrast, the back-reaction of the electrons with the oxidized ruthenium complex involves *d* orbitals localized on ruthenium with small overlap with the TiO_2

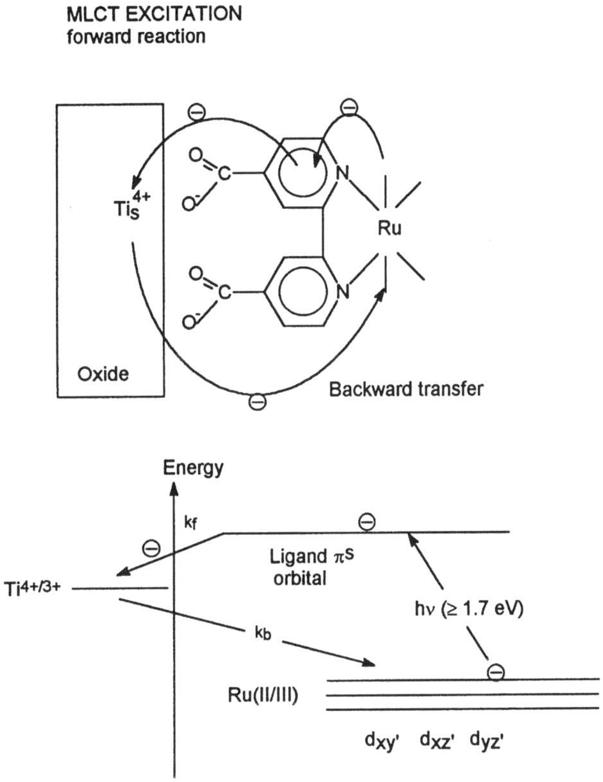

FIGURE III.17 Photoinduced heterogeneous electron transfer cycle with the molecular orbital diagram for ruthenium complexes anchored to the titanium dioxide surface by a carboxylated bipyridyl ligand. (From Grätzel, M., *Platinum Met. Rev.*, 38, 151, 1994. With permission.)

conduction band. Besides, the free energy of the backward electron transfer is large and lies in the inverted Marcus region, therefore backward electron transfer is in the microsecond time region. Thus, similarly to natural photosynthesis, light-induced charge separation is much faster than the back-reaction for essentially the same reasons. Electron movement along the curvature of the flat band potential near the surface of the semiconductor may be expressed in more "chemical" language: titanium(IV) on the surface of TiO_2 is reduced by the excited dye molecule to titanium(III), which is a stronger reductant on the surface than in the bulk of the semiconductor because of the influence of the donor molecules adsorbed. Therefore electrons are moving rapidly into the bulk of TiO_2 particles, away from the sensitizer molecule, and reducing new titanium(IV) species that have lower energy levels than those close to the surface. This results in the rapid electron separation from the sensitizer with some loss of energy. We have seen that a similar explanation for the rapid charge separation can be given for the functioning of the reaction center of the natural photosynthesis but nature had to develop special molecules for electron transfer from the primary donor.

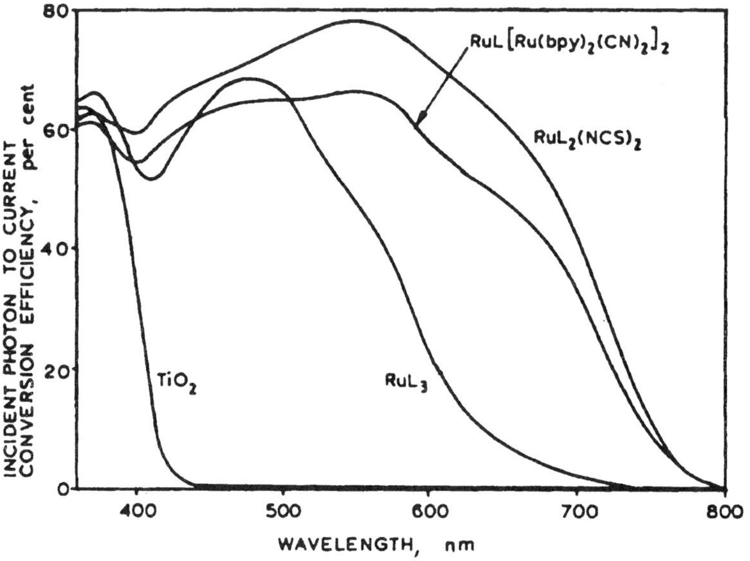

FIGURE III.18 Action spectra of three ruthenium-based sensitizers. (From Grätzel, M., *Platinum Met. Rev.*, 38, 151, 1994. With permission.)

The incident monochromatic photon-to-current conversion efficiency is given by the equation:

$$\eta_i(\lambda) = \mathrm{LHE}(\lambda) \cdot \phi_{inj} \cdot \eta_c \tag{17}$$

where η_c is the charge collection efficiency. With the complexes mentioned that are used as sensitizers adsorbed on nanocrystalline TiO_2 films, solar cells are now available. For these cells all three factors in Eq. 17 are close to unity and therefore $\eta_i(\lambda)$ is also close to 100% within the wavelength range of the absorption band of the sensitizer (similarly to quantum yields of the photosynthesis).

Figure III.18 shows action spectra for three ruthenium complexes illustrating the very high efficiency of current generation obtained with these complexes, exceeding 75%. RuL_3 (L = 2,2′-bipyridyl-4,4′-dicarboxylate) turned out to be efficient and stable as sensitizer in conjunction with the large surface area of a TiO_2 film. It sustained 9 months of intense illumination without loss of performance. However, the visible light absorption of this sensitizer is insufficient for solar energy conversion. A large improvement in the harvesting of light is achieved with the trinuclear ruthenium complex where two peripheral ruthenium moieties were designed as antennas. But *cis*-di(thiocyanato)bis(2,2′-bipyridyl-4,4′-dicarboxylate)ruthenium(II) turned out to be the most successful charge transfer sensitizer investigated so far. It achieves close to quantitative photon-to-electron conversion over the whole visible range. Even at 700 nm the efficiency of current generation is maintained at a 40 to 50% level.

The photovoltage of the cell represents the difference between the Fermi level of TiO_2 under illumination and the redox potential of the electrolyte. Using the triiodide–iodide redox couple in the N-methyloxazolidinone solution, under full sunlight an open-circuit cell voltage of 0.7 to 0.9 V can be measured. The overall efficiency is 10%. The photocurrent measured at 96.4 mW/cm^2 of stimulated solar intensity was 18.3 mA/cm^2, and the open-circuit voltage was 0.72 V.

Figure III.19 presents the layout of the module. The cell consists of two glass plates that are coated with a transparent conducting oxide layer. The nanocrystalline TiO_2 film deposited on one plate functions as a light trap. Visible light is absorbed by a monomolecular layer of the ruthenium complex, which functions as the charge transfer sensitizer. On illumination the electrons pass over the collection electrode into the external circuit where they perform work. Then they are returned to the cell via the counter electrode. The sensitizer film is separated from the counterelectrode by the electrolyte. There is a redox couple in the electrolyte, e.g., I_3^-–I^-, whose role is to transfer electrons to the sensitizer layer, which holds positive charge. The counterelectrode is made of glass covered with a conducting oxide layer that serves as a current collector. A small amount of platinum (5 to 10 μg/cm^2) is deposited onto the transparent oxide layer to catalyze the cathodic reduction of I_3^- to I^-.

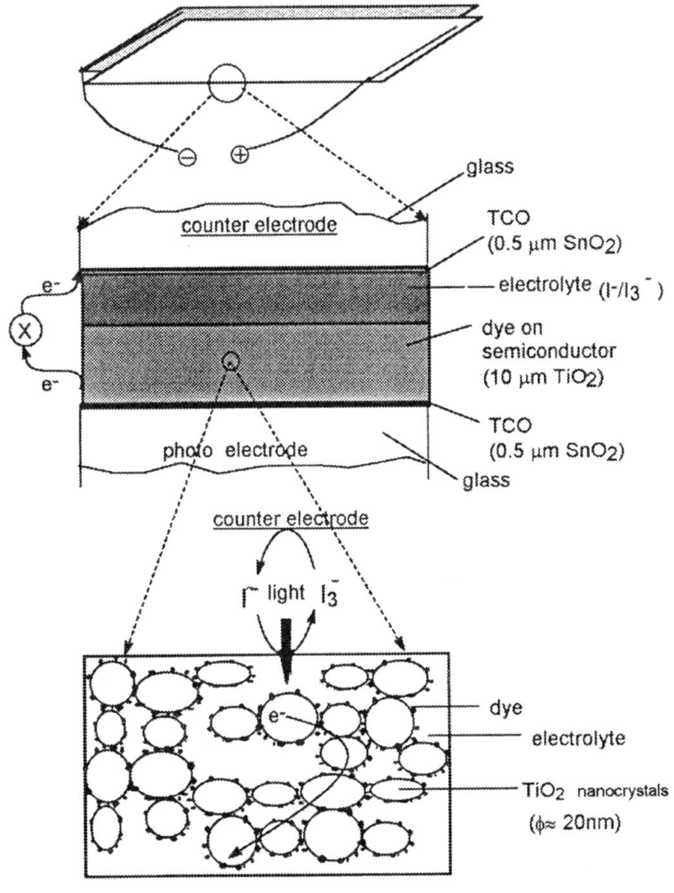

FIGURE III.19 Layout of a nanocrystalline solar cell. (From Grätzel, M., *Platinum Met. Rev.*, 38, 151, 1994. With permission.)

By developing a new way of platinum deposition a very active catalyst was prepared, which attains exchange current densities of more than 0.1 A/cm^2 at platinum loading below 0.1 g/m^2. The electrocatalyst is stable and does not show anodic corrosion for long periods of time.

The ruthenium complex sustained more than 5×10^7 redox cycles without noticeable loss of activity, which corresponds to about 10 years of continuous operation in natural sunlight. By contrast, practically all organic sensitizers tested so far have undergone photobleaching after less than 10^4 redox cycles.

The important point of concern for commercialization of the developed photovoltaic cells could be the cost of ruthenium. However, the required amount of ruthenium is only 10^{-3} mol per square meter of cell module surface and 1 ton of Ru incorporated into the charge transfer sensitizers could provide 1 GW energy under full sunlight. This is more than twice the total photovoltaic capacity that is used at present. It has been reported[88] that industrial companies are already involved in production of such cells and their commercial application will start in the near future. This will be a remarkable example of the use of basic science together with the biomimetic approach for developing new ways of solar energy conversion to electricity for practical purposes. Ruthenium complexes used as photosensitizers turned out to be molecules of "higher quality" than chlorophyll for the purpose of the photovoltaic cells described in this section, and perhaps synthesis of such molecules was in the mind of Professor F. Joliot-Curie when he pronounced his prophecy cited as the epigraph to this chapter.

III.4 PHOTOCATALYTIC DIHYDROGEN EVOLUTION

III.4.1 INTRODUCTION

The reduced acceptor (A^-) produced in the charge separation reaction has to undergo a catalytic reaction with water:

$$2A^- + H_2O \xrightarrow{\text{Cat}} H_2 + 2OH^- \tag{18}$$

This reaction is simpler than reduction of carbon dioxide or dinitrogen. We always observe dihydrogen evolution in parallel with dinitrogen reduction in protic media, which indicates that as a rule it is easier to draw protons of water into this reaction than into other reductions with proton participation when the appropriate potential is reached and transition metal complexes are present. The necessity of a catalyst follows from the two-electron character of this reaction: noncatalytic transformation in two one-electron steps; $H^+ + e \rightarrow H^\bullet$ and $H^\bullet + e + H^+ \rightarrow H_2$, would require a very strong reducing agent for the first step ($E_{H^+/H} = -2.1$ V). We have to stabilize the H atom if we want to use mild reducing agents. The mechanism of this stabilization is not difficult to visualize if we consider the activation of dihydrogen by transition metal complexes, which is known to proceed as oxidative addition via formation of metal–hydrogen bonds:

$$M + H_2 = M\langle^{H}_{H} \tag{19}$$

which may be followed by dissociation in polar media:

$$M\langle{}^{H}_{H} \rightleftharpoons [MH]^{-} + H^{+} \quad (20)$$

and the reaction with an electron acceptor:

$$M - H^{-} + 2A \rightleftharpoons M + 2A^{-} + H^{+} \quad (21)$$

The mechanism of dihydrogen evolution in the presence of an electron donor will involve the same reactions from the opposite site with variations in a sequence of electron and proton additions, e.g.,

$$M \xrightarrow{e} M^{-} \xrightarrow{H^{+}} M - H \xrightarrow{e} M - H^{-} \quad (22)$$

or

$$M \xrightarrow{H^{+}} M - H^{+} \xrightarrow{e} M - H \xrightarrow{H^{+}} MH_2^{+} \xrightarrow{e} MH_2 \quad (23)$$

$$M - H^{-} + H^{+} \rightarrow \left[M\langle{}^{H}_{H}\right] \rightarrow M + H_2 \quad (24)$$

For dihydrogen formation it is natural to choose the same catalysts that are used for activation of dihydrogen, e.g., platinum metals, in particular, platinum itself and its complexes.

Almost 100 years ago its way already known that water is reduced by bivalent vanadium to form molecular hydrogen in the presence of platinum metal.[89a]

Platinum cathode is known to evolve dihydrogen without overvoltage (at zero potential) indicating that a catalytic process involving surface platinum hydrides takes place in the process of electrochemical dihydrogen production. Homogeneous platinum complexes that are active, for example, in catalytic hydrogenations are expected to be catalysts also in dihydrogen evolution in solutions containing protons.

III.4.2 HYDROGENASE AS CATALYST OF DIHYDROGEN EVOLUTION

Hydrogenases are the enzymes of certain bacteria that were found almost 100 years ago to be able to consume or produce dihydrogen. The name *hydrogenase* was proposed in 1931 by Stephenson and Stickland[90] for the enzymes catalyzing the reaction $2H^{+} + 2e \rightleftharpoons H_2$, but the purification of the enzyme was performed only many years later. Since the 1970s many hydrogenase preparations from various microorganisms have been purified. They are very different from the point of view of their catalytic and other properties, but they all seem to contain iron–sulfur clusters. Therefore, initially these clusters were considered active sites where the binding of dihydrogen and the catalysis takes place. However, later there appeared some doubts concerning direct intervention of iron–sulfur clusters in dihydrogen activation and evolution. For example, the inhibitors of the enzyme, such as carbon monooxide, did not influence EPR spectra of the iron–sulfur clusters of certain hydrogenases. The redox potential of iron–sulfur clusters appeared too high for the catalyst site. Therefore another candidate was suspected to play a direct role in H_2 activation and formation.

A possible alternative was found when it was realized that nickel is required for the activity of many hydrogenases. Initially these indications were considered artifacts because nickel had not been included until recently in the list of essential elements for living organisms. In recent years nickel has turned out to be somewhat unexpectedly a participant in the whole range of important enzymes: urease from plants and from bacteria, methyl-Co-M reductase from methanogenic bacteria, carbon monoxide dehydrogenase from anaerobic bacteria, and most hydrogenases so far investigated. Anaerobic alkane oxidation, as we have seen (Section II.2), may also include nickel complex in the enzyme active center to activate C–H bonds.

The first demonstration that nickel is essential for hydrogenases came from the investigation of *Methanobacterium thermoautotrophicum.* In 1979 Thauer et al. found that their growth on H_2 and CO_2 as sole energy and carbon source was dependent on the presence of nickel.[91] Then the authors[92] found that nickel is needed for the formation of active hydrogenase in the growing bacteria and the enzyme isolated contained about 1 Ni per enzyme molecule. Finally, the EPR spectrum of Ni was demonstrated for the enzyme isolated from *M. thermoautotrophicum.* In subsequent years nickel was found to be present in most other hydrogenases, although there are also hydrogenases that do not contain nickel and involve only iron.

It was concluded from the investigation of the EPR spectra of hydrogenase from *Chromatium vinosum* that nickel in the enzyme could exist in three, possibly four, different redox states: 3+, 2+, 1+, and probably zero valent state. However, the EPR signal attributed to nickel(III) was later ascribed to a nickel(II) center (S = 1) that is spin coupled to a thiyl radical to yield an S = ½ signal. The Ni^I coordination is very light-sensitive and the rate of photodissociation in D_2O is nearly six times slower than in H_2O, indicating the existence and breakage of a nickel–hydrogen bond. This observation confirms the view that nickel is involved in the reaction with dihydrogen.[94] Now there is general agreement that the Ni center is the site of reaction with dihydrogen. From the X-ray absorption experiments with the F_{420}-reducing enzyme of *M. thermoautotrophicum* it was concluded that three sulfur atoms are present in the coordination sphere of nickel(III).[95] All Ni hydrogenases contain at least two ferredoxin type Fe_4S_4 clusters per Ni-containing cluster. The structure of the nickel center is unknown and may vary. The nickel is five to six coordinated and bound by at least one and more likely three or four sulfur atoms. Some hydrogenases from methanogens and sulfate-reducing bacteria contain selenium, in addition to nickel and iron–sulfur clusters (see Reference 96a).

Nickel hydrogenases are usually considered "uptake" hydrogenases, i.e., they use dihydrogen as a growth substrate, and in *in vitro* assays they catalyze H_2 oxidation rather than H_2 production.

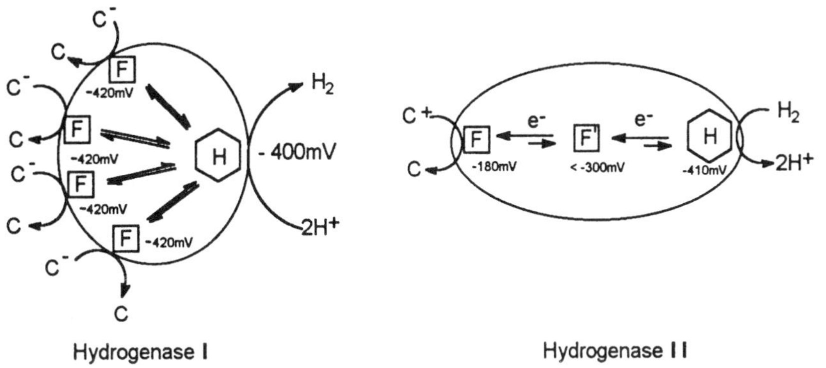

FIGURE III.20 Schemes of action of hydrogenases I and II. (Adapted from Adams, M. W. W. et al., *Proc. Natl. Acad. Sci. U.S.A.*, 86, 4932, 1982.)

When nickel was demonstrated to be important in the majority of hydrogenases there were some doubts that iron-only hydrogenases exist at all. However, later the absence of nickel was unequivocally demonstrated in some hydrogenases from strictly anaerobic bacteria. The nitrogen-fixing bacterium *Clostridium pasteurianum* is unique in that it contains two different hydrogenases, neither containing nickel. Hydrogenase I, first isolated by Nakos and Mortenson,[96b] functions to evolve dihydrogen. Subsequently hydrogenase (II) was isolated by Chen and Blanchard from *C. pasteurianum,*[96c] and this enzyme preferentially catalyzes H_2 oxidation. Both were extensively studied by different physical methods (see Reference 97). Both these enzymes as well as all other such hydrogenases contain two or more [Fe_4S_4] or F clusters and a unique type of iron–sulfur center called an H cluster. The H cluster is similar in all the hydrogenases and is believed to be the site of dihydrogen oxidation and production. It is comprised of at least three iron atoms, and possibly six. In the oxidized state it contains two types iron atoms, has an S = 1/2 spin, and exhibits a novel EPR signal. The reduced cluster is diamagnetic. When being reduced by dihydrogen the H cluster appears to undergo a conformation change with an increase in Fe–Fe distances of about 0.5 Å. The scheme of the hydrogenases I and II action is presented in Figure III.20.[98]

The function of a particular hydrogenase (to evolve or to uptake H_2) is determined by the redox potentials of the F clusters. Presumably F clusters not only transfer electrons to or from H clusters, but also help to perform the redox process of H_2 formation from protons or protons formation from H_2, therefore more than one F cluster is needed to increase the system electron capacity. The situation is somewhat similar to the mechanism of dinitrogen reduction by nitrogenase (see Chapter I). Some kind of *reductive coaction* may be suggested for F clusters, for example, in formation of H_2 by hydrogenase I from iron hydrides formed at the H cluster. Two electrons needed for H_2 formation can be replaced by electrons from the reduced F clusters during or immediately after H_2 is formed.

Hydrogenases, natural catalysts of dihydrogen activation and evolution, in their isolated form can be used *in vitro* in the systems specially designated to produce

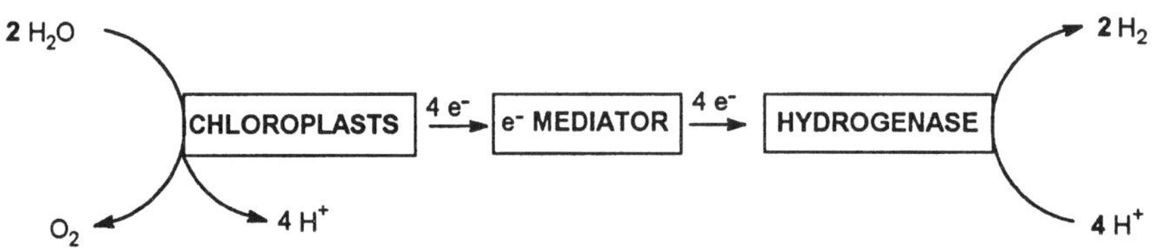

FIGURE III.21 Scheme of a catalytic system for water photolysis. (Adapted from van der Zwaan, Thesis, Laboratory of Biochemistry, B.C.P. Jansen Institute, University of Amsterdam, The Netherlands, 1987.)

dihydrogen. This was demonstrated for the first time in the work of Krasnovsky et al.[99] who found that a mixture of chloroplasts, a suitable electron carrier, and hydrogenase is able to split water into dihydrogen and dioxygen using solar light. Later this approach was developed in a number of works. This field attracted particular interest after the Arab oil embargo in 1973, which greatly stimulated the search for alternative sources of energy. Hydrogenases were considered promising candidates as participants of the catalytic systems for water photolysis. Figure III.21 shows the scheme for such a system.

The systems based on this principle work very well, but there are some limiting factors hampering their practical application at least in the near future, particularly if there are no more oil embargoes: (1) inactivation of the photosynthetic membranes of chloroplasts in the dark and especially in the light; (2) oxidation of the reduced electron carriers by dioxygen produced in the photolysis; and (3) inactivation of hydrogenase by dioxygen.[100] These limiting factors can be made less significant by immobilization of the enzymes taking part in the system on some support, but it is doubtful that they can be easily completely removed.

III.4.3 CATALYSTS OF DIHYDROGEN EVOLUTION FROM WATER

The simple reaction $H_2 = 2H^+ + 2e$ is catalyzed by many metal-containing catalysts. This had been known in chemistry well before this reaction was discovered by biochemists in living nature. The discovery that nickel participates in hydrogenases and is likely to be directly involved in the activation of H_2 and in back-reaction of its formation from protons was not surprising to chemists who have used this metal for many years as an active hydrogenation catalyst (e.g., so-called Raney nickel). This coincidence supports the view that metals for enzymes are chosen by nature for basically the same reasons as they are selected by chemists in their laboratories if in both cases the catalyzed reactions are the same or similar.

In neutral and weakly acidic media, colloidal or finely dispersed noble metals are the most active catalysts of the H_2 evolution from H_2O under the action of one-electron reducing agents (see References 42a and 101 to 103). The catalyst can be chosen by the same criteria used for choosing the cathodic material for hydrogen evolution in electrolysis. Such metals as platinum, iridium, and palladium reveal the smallest or zero overvoltage when they are used as cathodes in the electrolysis of water. Actually, metal colloids when they are used as catalysts of dihydrogen evolution from water may be regarded as *microelectrodes*[104] and a simple electrochemical theory helps to explain and predict the dependence of the rate of dihydrogen evolution on pH and the nature of the mediator and the catalyst.[102]

Colloidal platinum is the most widespread catalyst for dihydrogen evolution from water. A high catalytic activity of the Pt particles makes it possible to evolve H_2 by one-electron reducing agents, whose redox potentials are close to that of the two-electron water reduction, for example, methyl viologen radical cations (E = –0.44 V). The Pt colloids stabilized by various polymeric carriers (polyvinylpyrrolidone, gelatin, polyvinyl alcohol (PVA), and platinum supported on particles of TiO_2) have been studied in details. Among them, Pt colloids stabilized by PVA turned out to be the most active.[105]

Pt colloids with particle size of 10 to 1000 Å can be obtained by the reduction of platinum salts with sodium citrate, alcohol, and $NaBH_4$ and in γ-radiolysis. The dependence of the catalytic activity of the Pt colloids and supported catalysts on the particle sizes shows that the particles of 15 to 30 Å size are most active. The hydrogen evolution rate decreased sharply for the particles of <8 Å diameter. The activity of the supported platinum catalysts decreases in the following sequence of the carriers: $TiO_2 > Al_2O_3 > SiO_2 > Fe_2O_3$.

Other metals of group VIII stabilized by PVA were also used as catalysts of dihydrogen evolution. Metals of the platinum group (Pt, Ir, and Os) demonstrated the maximum activity that corresponded to the lowest overvoltage in the reaction of hydrogen evolution during electrolysis. However, the catalytic activity of other metals cannot be so easily correlated with their overvoltage. They can be arranged in the following sequence in their activity in the dihydrogen evolution:[106]

Ir, Pt, Os > Ru > Rh > Co, Ni, Pd, Ag, Au > Cu, Cd

Chemical possibilities to stabilize the catalysts and to make them highly effective and selective are now enormous. Therefore, artificial catalysts rather than immobilized hydrogenase will likely be used in future practical systems (if and when they are created) for photodecomposition of water.

III.4.4 PHOTOCATALYTIC CHEMICAL SYSTEMS

To observe dihydrogen evolution from water it is sufficient to have a one-electron donor with redox potential more negative than that of normal hydrogen electrode (zero at pH 0 or –0.41 V at pH 7) in water solution in the presence of one of the catalysts described above. For a photocatalytic system producing dihydrogen one has to use additionally a photosensitizer capable of forming a sufficiently strong donor and possibly an electron transfer agent capable of interacting with the catalyst. Until 1976 no such system producing dihydrogen under solar light was known, but there was already enough information to construct such a system. For example, acridine dyes were known to be photosensitizers for the reduction of methyl viologen, MV^{2+}, to the brightly colored radical cation $MV^{\dot{+}}$, which is stable in solution in the absence of O_2, and this reaction was used for the photocolorimetric analysis of the traces of dioxygen. Taking into account that $MV^{\dot{+}}$ is a sufficiently strong reductant to form dihydrogen reacting with water, we added platinum compounds to the above system to produce colloid platinum, and observed dihydrogen evolution in the presence of such sacrificial donors as EDTA, triethanolamine, or hydrogen

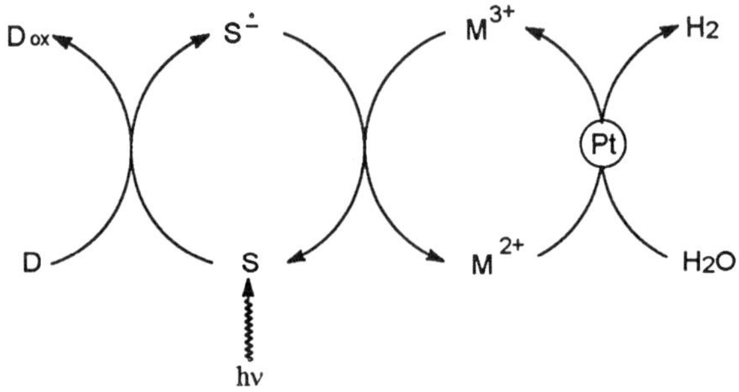

FIGURE III.22 Photocatalytic dihydrogen evolution. D: EDTA, H_2S, cysteine, or triethanolamine; M: MV^{2+}, salicilates of V^{2+}, or Eu^{2+}.

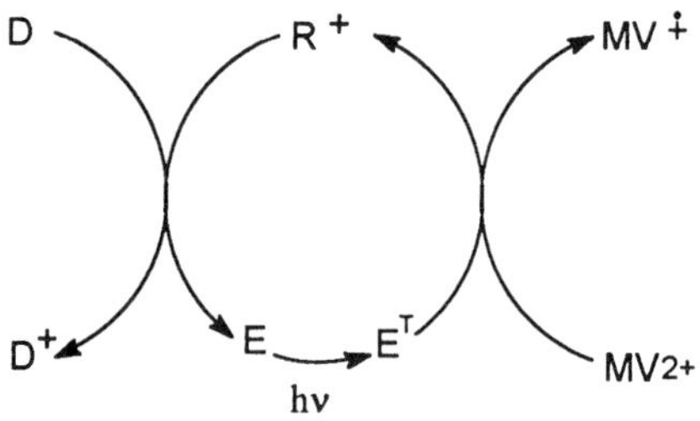

FIGURE III.23 Oxidative mechanism of photochemical formation of reduced cation radical $MV^{\dot{+}}$.

sulfide,[107] the latter being known as one of most common donors for bacterial photosynthesis. Thus, a purely chemical model system for bacterial photosynthesis or PS I of plant photosynthesis was developed in this simple way, although we realized, of course, that only one function was accomplished in these systems, that is, the electron transfer from the sacrificial donor to protons in solution, and many other important functions, such as photophosphorylation were lacking. However, because photocleavage of water was the main goal we could consider this system as an important step towards this goal. Later research has shown, however, that this was only the simplest step forward. Figure III.22 shows the possible scheme of the action of these and similar systems. Metal ions (V^{3+} or Eu^{3+}) could be used instead of methyl viologen because when reduced they form ions (V^{2+} and Eu^{2+}) thermodynamically capable of producing H_2 from H_2O.

The scheme in Figure III.22 demonstrates only one, so-called reductive, mechanism of quenching of excited photosensitizer by a donor molecule to produce a radical anion acting in subsequent redox reactions. Another possible (oxidative) mechanism includes the oxidation of an excited sensitizer by an acceptor (e.g., MV^{2+}) to form a radical cation that returns to its initial state when reacting with a donor. This is shown by the scheme in Figure III.23.

A number of other systems based on different sacrificial donors, photosensitizers, electron transfer agents, and catalysts were developed subsequently by several research groups. For example, proflavine was used as a sensitizer.[108] Tris-dipyridine ruthenium(II) complex, $Ru(bpy)_3^{2+}$, mentioned in the previous section, turned out to be a very convenient photosensitizer[109-111] and was used subsequently in many investigations. $Ru(bpy)_3^{2+}$ is a deeply colored compound (λ_{max} = 452 nm, ε_{452} = 14500 M^{-1} cm^{-1}). The absorption of photon results in its transformation to the excited state (^1d-π), which is characterized by the charge transfer from metal to ligand. The complex is transformed from the ^1d-π state to the lower excited state, whose quantum yield is close to 1 and lifetime equals 0.6 μs,[112] with the rate constant $k_{isc} > 10^{11}$ s^{-1}. According to current knowledge[113,114] electrons in this excited state are almost completely localized at one dipyridyl molecule, thus $^*Ru(bpy)_3^{2+}$ is actually close to $Ru^{3+}(bpy)_2(bpy^{-\bullet})$. This is the excited state that participates in the electron transfer reactions. $^*Ru(bpy)_3^{2+}$ is a strong reducing agent with redox potential $E_{Ru(bpy)_3^{3+}/^*Ru(bpy)_3^{2+}} = -0.84$ V,[115] and $Ru(bpy)_3^{3+}$ is a strong oxidant and may be used for the formation of O_2. Viologens are efficient quenchers of $^*Ru(bpy)_3^{2+}$ via the electron transfer mechanism.[116a] The reduced form of the viologen (for example, methyl viologen $MV^{+\bullet}$) is quite stable in solution in the absence of air oxygen and has a typical blue color (λ_{max} = 605 nm, $\varepsilon_{605} = 1.24 \times 10^4$ M^{-1}cm^{-1}).

The quantum yield for dihydrogen production varies from very small values (0.01) obtained in the first biochemical systems involving chlorophyll as photosensitizer, NADH or NADPH as donors, and hydrogenase as a catalyst, to almost quantitative values in some very efficient purely chemical systems, which were developed as early as in 1983 (see Table III.1).

TABLE III.1
First Photocatalytic Systems Evolving Dihydrogen

Year	Photosensitiser	MV^{2+} electron-transfer agent	Donor	Catalyst	Quantum yield MV^{2+}	Quantum yield H_2	λ, nm	Ref.
1975	Chlorophyll	MV^{2+}	NADH, cysteine	Hydrogenase		0.01	670	99
1975	NADH, NADPH, BNA	MV^{2+}	—	Hydrogenase		0.01	365	99
1976	VCL_3		Alcohols	PtO_2		0.1	280	107
1977	Acridine yellow	V^{III}, Eu^{III} salicylates	EDTA TEA	PtO_2 or Pt colloid	0.6	0.2	480	107
1979	$Ru(bpy)_3^{2+}$	$Rh(bpy)_3^{2+}$	TEA	Pt colloid		0.3	450	116b
1980	$Ru(bpz)_3^{2+}$	MV^{2+}	TEA		0.77	—	450	116c
1981	$ZnTMPP^{4+}$	MV^{2+}	EDTA	Pt/PVA	0.75	0.6	550	116e
1981	$Ru(bpy)_3^{2+}$	MV^{2+}	EDTA	Pt colloid	—	0.1	450	116d
1983	$Ru(bpym)_3^{2+}$	MV^{2+}	TEA		0.85	—	435	116f
1983	$Ru(bpy)^{3+}$ 9-AC	MV^{2+}	EDTA	Pt/PVA		0.85	450	116g
1983	9-AC	MV^{2+}	EDTA	Pt/PVA	0.91	0.93	385	116h
1983	$Ru(bpz)_3^{2+}$	MV^{2+}	TEA		0.98		435	116f

BNA: benzylnicotinamide; bpy: bipyridyl; bpz: bipirazyl; bpym: bipyrimidine; TMPP: tetramethylpyridylporphyrin; TEA: triethanolamine; PVA: polyvinylalcohol; 9-AC: 9-antracene carboxylate; NADH: nicotinamide dinucleotide (reduced); NADPH: nicotinamide dinucletoide phosphate (reduced); EDTA: ethylenediamine tetraacetic acid.

Kinetic laser spectroscopy helps us to understand the causes for the differences in behavior of different sensitizers. According to the theoretical expectations, long-living triplet states of dyes molecules are responsible for electron transfer from the excited state of the sensitizer to the electron transfer agent or from the donor to this excited state. The example with eosin[116i] showed the following main steps of the process (in brackets are the bimolecular rate constants in $M^{-1}s^{-1}$):

$$E^T + MV^{2+} \rightarrow E + MV^{2+} \qquad (3.1 \times 10^9) \qquad (25)$$

$$E^T + MV^{2+} \rightarrow R^+ + MV^{\dot{+}} \qquad (3.5 \times 10^8) \qquad (26)$$

$$MV^{\dot{+}} + R^+ \rightarrow MV^{2+} + E \qquad (6.4 \times 10^9) \qquad (27)$$

$$E^T + D \rightarrow R^- + D^+ \qquad (2 \times 10^6) \qquad (28)$$

$$R^+ + D \rightarrow E + D^+ \qquad (7 \times 10^5) \qquad (29)$$

$$R^- + MV^{2+} \rightarrow E + MV^{\dot{+}} \qquad (1.4 \times 10^9) \qquad (30)$$

Consideration of the rate constants demonstrates that the oxidative quenching of the eosin triplet state (E^T) to form $MV^{\dot{+}}$ (reaction 26) is more effective than reductive quenching by EDTA (reaction 28). Therefore, at the initial stages the process proceeds mainly according to the scheme in Figure III.23. However, $MV^{\dot{+}}$ is accumulating in the system with time, the contribution of the oxidative route diminishes, and the reaction chooses the less effective reductive route. Low quantum yield in the system with eosin (less than 0.03% under stationary irradiation) is due mainly to the low probability of the dissociation of the triplet ion pair formed in the electron transfer in the encounter complex. Presumably, coulombic attraction of the ions (R^+ in the case of eosin has in fact negative charge because eosin is a dianion) helps to keep the ion pair long enough for the intersystem crossing to the singlet pair and then to backward electron transfer to the main part of the encounter complexes. Elaborate investigation of different factors influencing the formation and dissociation of the triplet ion pairs for different dyes confirms the importance of coulombic forces between the participants of the ion pairs.[13] Table III.2 shows the quantum yields for a series of zinc porphyrins with different substituents changing the charge of the porphyrin rings and the Z_AZ_B term from +8 to –8. In qualitative agreement with the expectations, the quantum yield (f ions) changes in this series from 0.75 to less than 0.01.

An effective catalyst for dihydrogen evolution can use the electrons transferred to the acceptor almost quantitatively, ensuring high quantum yields for dihydrogen evolution.

TABLE III.2
Influence of Electrostatic Forces on the Efficiency of Photoreduction of MV^{2+}

Sensitiser	Z_AZ_B	$k_T \times 10^{-7}$ ($M^{-1}s^{-1}$)	f_{ions}
$ZnTMPyP^{4+}$	8+	1.8	0.75
$ZnTMPyP^{+}$	2+	2.1	0.10
$ZnCPP^{-}$	2–	100	0.037
$ZnTCPP^{4-}$	8–	1300	<0.01
$ZnTSPP^{4-}$	8–	1400	<0.01

III.4.5 DIHYDROGEN EVOLUTION WITH THE CATALYST ATTACHED TO MEMBRANE

It was mentioned already that for the photodecomposition of water the catalysts of dihydrogen and dioxygen evolution must be separated from each other to prevent their interaction. One of the possible ways to organize such a system is to encapsulate the catalyst of dihydrogen evolution inside a bilayer vesicle and to place the catalyst for dioxygen formation in the outer solution.

The first attempts to incorporate a catalyst into the inner water pool of a vesicle was made in the works of Fendler et al.[117-121] Platinum dioxide powder was exposed to ultrasonication in the presence of dihexadecylphosphate, which is able to form vesicles in water solution. However, the size of the platinum particles was so large that the system obtained in this way was in fact platinum stabilized by the surfactant rather than the catalyst incorporated inside the vesicle. In another work platinum was included inside the vesicle by preparing it with a solution of K_2PtCl_4. Under irradiation of ultraviolet light platinum was reduced and formed colloidal particles with sizes of about 270 Å. The vesicles were stable and permeable for certain substances.[118] Membrane-bound catalysts of dihydrogen evolution were prepared by photochemical reduction of rhodium salts in suspension of the vesicles prepared from dihexadecylphosphate and contained in the membrane of a semiconducting photocatalyst: colloidal cadmium sulfide. In the process of reduction, particles of rhodium metal are formed and precipitated at CdS. Introduction into the system of an electron donor, thiophenol, and irradiation in the absorption band of CdS results in the oxidation of thiophenol and dihydrogen evolution catalyzed by rhodium.[119-121]

Recently methods were developed by Shafirovich et al.[122-125] to immobilize platinum-containing particles by attaching them to bilayers. Two types of platinum catalysts have been developed, both attached to the walls of lipid vesicles. The catalysts were prepared by ultrasonication of a lipid in the solution of a platinum complex at 60 to 80°C. Under this treatment the platinum complex is reduced and platinum metal particles become attached to the bilayer. In the case of relatively stable $Pt(NH_3)_2Cl_2$ the generation of Pt particles is slow and they are attached mainly to the outside monolayer of the vesicle. The less stable K_2PtCl_4 decomposes faster and therefore a large portion of platinum particles are caged by the vesicles. The size of the particles measured by electron microscopy was found to be 10 to 20 Å.

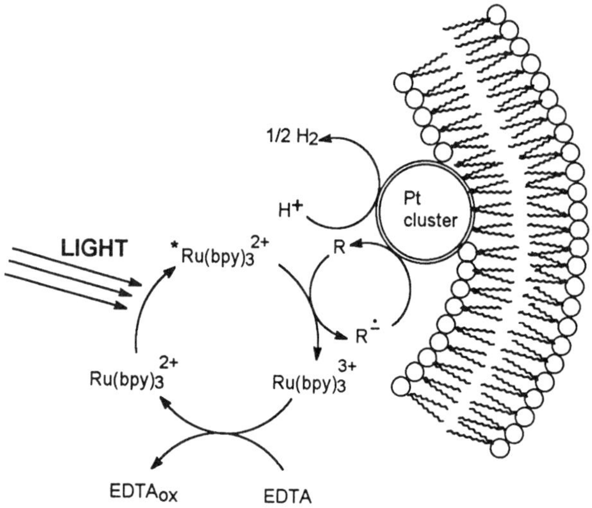

FIGURE III.24 Dihydrogen evolution inside the vesicle.

A photocatalytic system EDTA–$Ru(bpy)_3^{2+}$–electron relay (R) can be encapsulated in the interior water pool of the vesicle in the process of the platinum catalyst preparation. When R is rhodium tris-dipyridine complex, $Rh(bpy)_3^{3+}$, it can be photocatalytically reduced to $Rh(bpy)_3^{2+}$ and the latter evolves dihydrogen interacting with the platinum catalyst. In this system the quantum yield, $\Phi(H_2) = \Phi(Rh(bpy)_3^{2+} =$ 0.02 to 0.04. The mechanism of the system functioning is present in Figure III.24.

The rate of dihydrogen evolution is not affected by addition of an oxidant, $K_3Fe(CN)_6$, into the bulk solution. The redox potential of the bulk solution remains practically the same during the reaction, i.e., *H_2 evolution occurs in the medium with positive redox potential.* In contrast to this, in similar systems without vesicles no H_2 evolution takes place until the potential reaches negative values. Thus the vesicles containing the catalyst can be part of the system of the cleavage of water into dihydrogen and dioxygen.

III.5 CATALYTIC FORMATION OF DIOXYGEN FROM WATER. CHEMICAL MODELS OF PHOTOSYSTEM II

III.5.1 INTRODUCTION

Catalytic oxidation of water to form dioxygen is a four-electron reaction and *a priori* is likely to be a more complicated process than dihydrogen evolution. Redox potentials of one-, two-, three,- and four-electron water oxidations are shown in Figure III.25.

Thermodynamically the most difficult reaction is the first one-electron oxidation to produce OH radical (2.5 V). This is unacceptable for biological systems primarily because OH radical would quickly destroy organic material by actively attacking C–H bonds and secondarily because too strong an oxidating agent is needed to perform the process. Two-electron oxidation via hydrogen peroxide intermediate

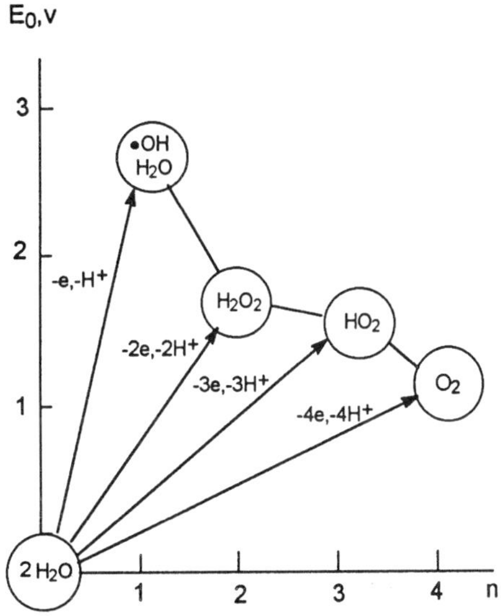

FIGURE III.25 Redox potentials of one-, two-, three-, and four-electron oxidation of water.

formation would be more plausible, but clearly the concerted four-electron oxidation is the most economical from the viewpoint of energy consumption. The problem is somewhat similar to the catalytic dinitrogen reduction and we can try to solve it in a similar manner: an optimal catalyst should be able to give up four electrons, one by one, to four molecules of a one-electron acceptor and then to oxidize two water molecules directly to yield dioxygen. We could expect to find such a catalyst among transition metal clusters uniting several metal atoms, each capable of being oxidized by a one-electron acceptor with redox potential close to that for dioxygen formation from water (+1.229 V at pH 0). Even for the concerted four-electron oxidation the potential must be rather high and require metals capable at least in their highest oxidation state of performing the reaction. Among first-row transition metals in the periodic table only those on the right side of manganese, including manganese itself, can form such catalysts.

Thus, although free hydrogen peroxide might be involved as an intermediate in two two-electron processes, an optimal catalyst should exclude also H_2O_2 and directly produce dioxygen from two water molecules.

Dioxygen evolution can be considered from the viewpoint of the reverse reaction of four-electron reduction of dioxygen to water. This reaction with participation of transition metal complexes is known both in biological and chemical systems. We have seen, in particular, that monooxygenases utilize dioxygen to produce high-valent iron complexes, presumably iron(IV). Iron–porphyrin mononuclear complexes in cytochrome P-450 and dinuclear iron complexes in methane monooxygenase are used to perform this reaction. With more positive redox potentials of the oxidized metal ions in the cluster the same mechanism functioning in the opposite direction would produce dioxygen from coordinated water molecules. Therefore even mononuclear complexes are able in principle to produce dioxygen from water in four-electron oxidation and, thus, to be catalysts for water oxidation. Additional electron donors help to use all four equivalents of dioxygen in its reaction with low-valent metal compounds. Electron acceptors will help to withdraw electrons from water when it is oxidized by high-valent metal complexes:

$$M^{n}(H_2O)_2 \xrightarrow{-2e,\,-2H^+} M^{n+2}\langle{}^{OH}_{OH} \xrightarrow{-2e,\,-2H^+} M^{n+4}(O)_2 \rightarrow M\langle{}^{O}_{O} \rightarrow M^{n} + O_2 \quad (31)$$

However, in O_2 *activation* the purpose was to create a strong two-electron oxidant to attack stable C–H bonds. For dioxygen *evolution* we would like to use the *weakest* oxidant capable, however, thermodynamically of forming dioxygen from water. Therefore, although a mononuclear complex is able in principle to oxidize water, this reaction would require presumably a very strong oxidizing agent. The situation is similar (as a mirror reflection) to N_2 reduction upon protonation: although mononuclear complexes can, as we have seen, reduce dinitrogen, this corresponds to a very low oxidation state of the metal coordinating N_2. For an optimal catalyst we need a cluster containing at least four metal atoms (M) with each M changing its oxidation state by one electron:

$$M_4^{n+}(H_2O)_2 - 4H^+ \rightarrow M_4(O)_2 \rightarrow M_4^{(n-1)^+} + O_2 \quad (32)$$

This mechanism in its general form for M = Mn was proposed for dioxygen evolution in photosynthesis by Semenov et al. as early as 1975.[126] Taking into account that even with a concerted mechanism a rather strong oxidant is needed, we would like to use various approaches to decrease the activation energy of the reaction. If our one-electron oxidant has pH-independent redox potential we can decrease the redox potential of water oxidation to form dioxygen by increasing the pH of the media. According to the Nernst equation for the reaction $2H_2O - 4e = O_2 + 4H^+$ the redox potential E of the pair $O_2/2H_2O$ decreases with pH:

$$E = E_0 - 0.059\ pH \quad (33)$$

Of course, the total energy necessary to cleave water to dihydrogen and dioxygen does not depend on pH because the redox potential of the pair $2H^+/H_2$ will also decrease with pH and a stronger reductant will be needed to produce dihydrogen from water at higher pH. However, because catalytic H_2 formation seems to be achieved more easily than the oxidation of water to dioxygen, the increase of pH could promote dioxygen evolution. Moreover, we have seen in Chapter I that it is the easiest way to combine transition metal ions into polynuclear hydroxides in order to catalyze the two- and four-electron reactions. There is yet another reason to work at high values of pH for dioxygen production. We may need a certain (high) oxidation state of the metal ions participating in the catalyst producing dioxygen. It is thermodynamically easier to oxidize hydrolyzed metal ions with OH^- in the coordination sphere than the same ions with neutral water molecules. Other negative ions in the coordination sphere of the metals, instead of hydroxides (e.g., Cl^-) involved in the catalytic complex, would be also helpful, particularly in the neutral pH region. Coordinative unsaturation is also needed for a complex to be an effective catalyst, and flexible bonding between the metals to adjust for several oxidation states is essential as well.

Finally, the activation energy of the process of dioxygen formation may depend on the number of contacts of oxygen atoms involved in the formation of dioxygen with metal atoms forming the catalyst. In the case of dinitrogen we came to the conclusion that the activation of N_2 to its reduction is more effective when the number of contacts is increased. Here again the situation is similar. O_2 evolution could be facilitated if oxygen atoms come from two rather than from one metal atom in the cluster, and even more metal atoms can be directly involved.

To use this mechanism with the optimal catalyst we have to prepare a special kind of complex with certain distances between metal atoms coordinating water molecules and flexible bonding between the atoms to adjust to several oxidation states and formation and breaking of metal–oxygen bonds. We may hope to use polynuclear metal hydroxocomplexes as catalysts in our artificial systems, but certainly nature had to invent a more sophisticated catalyst for O_2 evolution in photosynthesis.

III.5.2 OXYGEN EVOLUTION CENTER IN NATURAL PHOTOSYNTHESIS

One of the most important functions of plant photosynthesis, dioxygen evolution, remained for a long time the most obscure part of the reaction scheme. In 1957, when the carbon path in carbon dioxide reduction was already in principle established, Calvin presented the mechanism of O_2 formation by the following, not very informative scheme:[5]

$$H_2O \rightarrow ? \rightarrow ? \rightarrow ? \rightarrow O_2 \tag{34}$$

Even now not all the details of the mechanism are clear. Nevertheless, in principle the mechanism is not so enigmatic any more, and we can compare the biological dioxygen evolution with recently found various catalytic chemical reactions of oxidation of water to yield O_2.

After demonstration in 1941 that in plant photosynthesis dioxygen is produced from water and not from carbon dioxide[4] (this was the only information available when Calvin drew his "mechanism"), an important step forward was the formulation of the now famous Kok scheme,[127] preceded by the Joliot experiments with measurements of dioxygen formation in the process of consecutive flashes of light.[128] These experiments demonstrated that O_2 evolved after every four flashes, therefore the oxygen-evolving center (OEC) accumulates holes after each electron transfer induced by P-680 irradiation, increasing the number of holes up to four, and this is followed by a spontaneous evolution of dioxygen returning the center to the initial state. These states were designated by the letter *S* with the corresponding number in the Kok scheme (Scheme III.1).

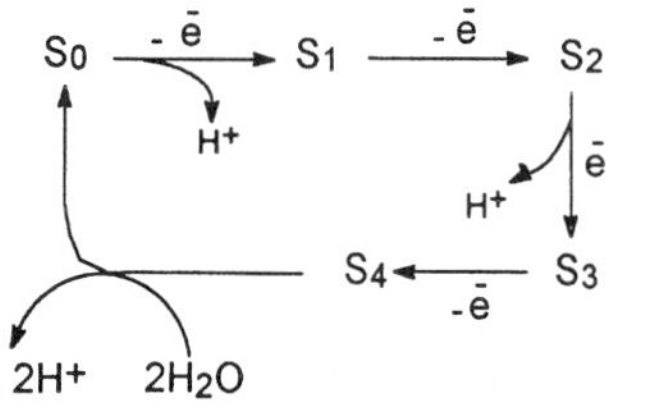

SCHEME III.1 Kok scheme for dioxygen evolution in plant photosynthesis.

The concrete meaning of the "accumulators" of the holes, S_i, became understood, at least in principle, when it was established almost at the same time by Cheniae that manganese is required for dioxygen evolution and there are several atoms of manganese in the dioxygen evolution center.[129] In 1974 Calvin suggested the two-electron oxidation of water via hydrogen peroxide with O_2 as the terminal product[130] and the next year the already mentioned four-electron mechanism was proposed[126] to explain direct dioxygen formation from two water molecules with participation of four Mn ions united in a cluster. At that time we had already good evidence for a four-electron mechanism of dinitrogen reduction by nitrogenase and as soon as manganese became known to be involved in the OEC of plant photosynthesis it was natural to suggest a principally similar reaction for water oxidation. The thermodynamic reasons for such a mechanism (*vide supra*) were clear. Since then the detailed mechanism has continued to be discussed in light of new experimental findings, but there is no doubt that the increasing number of holes corresponds, in fact, at least partially, to the rise of the manganese oxidation state, and there is spontaneous oxidation of two water molecules by the S_4 state to form ultimately O_2 and the initial S_0 state. Manganese ions are contained in the OEC, which has been studied in a number of works.

III.5.2.1 Oxygen Evolution Center[131]

Light-induced photosynthetic water oxidation takes place in a supramolecular thylakoid membrane protein complex (PS II).[132,133] The complex consists of several polypeptides with different molecular masses. The 47- and 43-kDa chlorophyll *a* proteins serve as proximal antennas, the first containing 19 to 26 chlorophyll *a* molecules and the second, 10 to 12 chlorophyll *a* molecules. The low mass (14 to 15 kDa) pigment protein complex is believed to act as an additional antenna complex within PS II. The intrinsic membrane spanning D1 and D2 proteins with molecular mass of 32 kDa contains the photochemically active reaction center of PS II. There are some structural and functional analogies between PS II and the reaction center of purple bacteria located at the electron-accepting side. Figure III.26 shows a current view of the enzyme structure based on the analogy between PS II and the reaction center of purple bacteria;[131a] as was mentioned, it is now not quite clear whether a monomeric chlorophyll molecule or a dimer similar to the "special pair" is actually the primary electron donor.

Three water-soluble membrane proteins with molecular masses of 33, 24, and 18 kDa are essential for full dioxygen evolution activity.[134] The 33-kDa protein is the most important for dioxygen evolution among the three. It protects the manganese

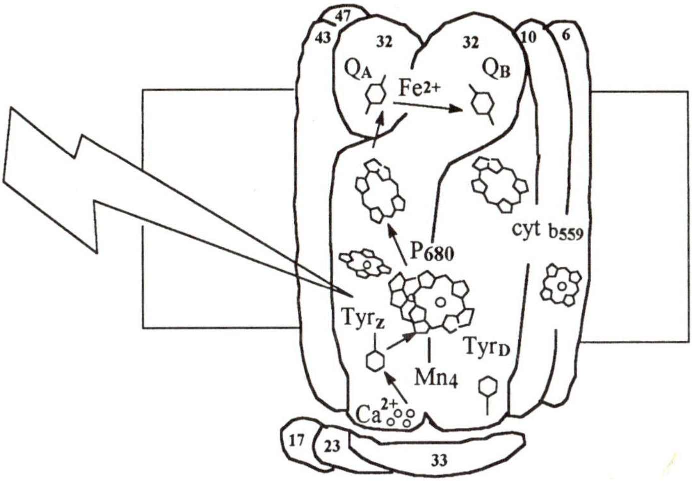

FIGURE III.26 Current view of the OEC in plant photosynthesis. (Adapted from *The Photosystems: Structure, Function and Molecular Biology*, Barber, J., Ed., Elsevier, 1992, 179.)

OEC and maintains it in the configuration optimal for the catalytic efficiency of water oxidation. At the same time, it is not absolutely necessary because oxygen evolution was observed even in the absence of the 33-kDa protein. Ca^{2+} and Cl^- ions are known to be present in the reaction center and essential in the dioxygen evolving activity. The 24-kDa protein functions as a "Ca^{2+} concentrator" restoring dioxygen evolving activity that had been suppressed by low level of Ca^{2+} concentration.[135] The function of the 18-kDa protein is to reduce the Cl^- requirement for dioxygen evolution.[136] The 33, 24, and 18 kDa proteins can rebind stoichiometrically to the OEC depleted of these proteins to restore the dioxygen evolving activity.

III.5.2.2 Coupling Between the Reaction Center and the Oxygen Evolution Center

The reduction of P-680$^+$ proceeds in the nanosecond range of time. For its part, the oxidation of S states in the OEC takes place within the microsecond time range. Therefore, at least one electron carrier has to exist between P-680 and the OEC, which initially was referred to as Z and later as Y_Z. This was revealed by EPR spectroscopy: an EPR signal with g = 2.0046 was detected after excitation of dark adapted chloroplasts with a series of short flashes. The EPR signal was attributed to the oxidized form of Y_Z. The EPR properties of Y_Z^+ are virtually identical to those of the stable PS II free radical, which was shown to be a tyrosine residue.[137]

The EPR studies together with recent UV optical data indicate that PS II contains two redox active tyrosines. One of these reduces the oxidized reaction center P-680$^+$ and transfer the oxidizing equivalents to the OEC.[138] the second is a long-lived radical of unknown function. It was recently established that the two tyrosines are most likely symmetry-related Tyr-161 of D1 polypeptide and Tyr-160 of D2 polypeptide.

Now we have some information concerning the behavior of each of the S states in the Kok scheme. Redox potentials of $S_{i+1/i}$ transitions are estimated as follows:[139a]

Transition	E_0, V
$S_0 \rightarrow S_1$	≤0.24
$S_1 \rightarrow S_2$	0.89
$S_2 \rightarrow S_3$	0.93
$S_3 \rightarrow S_4$	0.87

The average potential per electron oxidation amounts to about 0.73 V, and is therefore close to the redox potential of water oxidation to dioxygen (0.82 at pH 7). However, more recent data suggest higher redox potentials, approaching 1 V for the average value per electron.[139b] It is still lower than the value for P-680$^+$ (1.12 V) and perhaps it could be expected that the average value of the redox potential has to be close to the middle position between the potentials of the pairs P-680$^+$/P-680 and O_2/H_2O.

The highest oxidation state of the OEC corresponds to S_4 and this state is known to convert within about 1 ms to the S_0 state with the evolution of O_2 and the yield of two H^+ in the medium. Both S_0 and S_1 are stable in the dark and present in the resting center in the approximate proportion 25:75. S_2 is unstable and when light is switched off it decays to S_1 by a back-reaction with an electron coming from the electron donors. The lifetime of S_2 depends on the origin of the electron, and usually $t_{1/2}$ is around 30 s. The S_3 state decays via S_2 to S_1 in reactions also involving electron donors and the S_3 lifetime is similar to that of S_2. S_2 and S_3 become longer-lived at low temperatures. The analysis of the electron transfer times from different S states via tyrosine to P-680$^+$ provides some information concerning the states. The kinetics of P-680$^+$ reduction by state S_0 or S_1 is monophasic (20 ns) and by S_2 or S_3, biphasic (50 + 260 ns). Such retardation up to ten times gives clear evidence for the positive surplus charge in the S_2 and S_3 states.[131]

III.5.2.3 Arrangement of Oxygen Evolution Center

Manganese is an essential component of the OEC. Proceeding from the measurements of atomic absorption and EPR of released Mn ions, it has been shown that there are four manganese atoms per active center.[129,140] These four Mn are not equivalent and seem to serve different functions. One pair of Mn ions appears to be more weakly bound than the other. It was also reported that there were two binding sites of Mn^{2+} in Mn-depleted PS II preparations, which differ in their location relative to the position of P-680 and Y_Z.[141]

There is an indication that two of the four Mn ions are replaceable by other bivalent cations, preserving the function of electron transfer to P-680$^+$.[142] It is not clear, however, whether dioxygen evolution continues in the center with two Mn ions instead of four or only the electron flow takes place. Recently Pauly and Witt reexamined the manganese content in the OEC.[143] The experimental results based on the number of OECs, the total number of PS II complexes, the O_2 yield, and the Mn content have led to the conclusion that the ratio of the number of Mn ions to the number of OECs in PS II corresponds to six. It is still unknown, however, whether all Mn ions are located at the same site and participate in O_2 evolution. Probably only some of the manganese is directly involved in the process of water oxidation. Some of the manganese may have a more stabilizing function.

Important information concerning the oxidation state of manganese has been obtained from the EPR spectra of the manganese cluster of the OEC. In 1980 Dismukes and Siderer showed for the first time that the OEC produced an EPR signal attributed to the S_2 state of the cluster[144] (see also References 131 and 142). A broad multiline EPR signal is centered near g = 2 and consists of at least 19 hyperfine lines, each separated by approximately 80 G, apparently superimposed on a broad Gaussian-shaped signal. The amplitude of the signal varies with flash number, being largest in the S_2 state. The lifetime of the signal is also typical for the S_2 state. The similarity of the S_2 EPR signal with that observed for mixed valence di-μ-oxo-$Mn^{III}Mn^{IV}$ dimers led to the proposal that a similar dinuclear core may exist for the manganese cluster in the OEC. Nevertheless, there are some differences between the S_2 signal and the signal of di-μ-oxo-$Mn^{III}Mn^{IV}$ dimers. First, the S_2 signal consists of at least 19 lines instead of 16 for the dimers, and its overall width is over 1800 G compared to 1250 G for di-μ-oxo-$Mn^{III}Mn^{IV}$ dimers. The EPR experiments at different microwave frequencies (9.1 and 35 GHz) show that the hyperfine interactions are the same when the microwave frequency is changed and this ruled out g anisotropy as the origin of additional lines in the S_2 spectrum. The second difference is that the hyperfine lines of the S_2 signal are broader and much less resolved than those of the dimers. So the additional lines may arise from the cluster, which is larger than the dimer. Apart from the multiline signal, another EPR signal at g = 4.1 exists that also belongs to the S_2 state. The g = 4.1 signal has a width of 320 to 360 G and lacks well-resolved hyperfine structure. The g = 2 multiline and g = 4.1 EPR signals are interconvertible. Upon illumination at 140 K the signal with g = 4.1 is observed; the increase of the temperature up to 200 K in the dark results in the loss of this signal and appearance of the multiline one. In the presence of sucrose and some ligands (F^-, NO_3^-, and amines), the g = 4.1 signal is stabilized at 200 K and at room temperature. The g = 4.1 signal disappears in the presence of glycerol, ethylene glycol, and ethanol. There have been several explanations concerning the relationship between both EPR signals of the S_2 state. For example, it was proposed that the signal transition was due to redox equilibrium of two components: a single Mn^{IV} yielding the g = 4.1 signal and $Mn^{III}Mn^{IV}$ dimer yielding the multiline signal. Taking into account the two Mn···Mn distances found in EXAFS studies, 2.7 and 3.3 Å (*vide infra*), this model has lead to a trimer–monomer model. In this model the multiline signal arises from the mixed valence trimer, which can be converted by shifting the redox equilibrium, to the g = 4.1 signal from the Mn^{IV} monomer. However, recent EPR data on the ammonia-treated enzyme at the S_2 signal has demonstrated a hyperfine structure on the g = 4.1 signal, ruling out the monomeric Mn^{IV} as its origin. Apparently the most plausible explanation is that both the g = 4.1 and g = 2 signals come from different conformations of the manganese tetranuclear cluster in which the spin states of the exchange-coupled systems are S = ½ and S = 3/2, giving rise to g = 2 and g = 4.1, respectively. Theoretical studies of the origin of the multiline signal were reported, the tetramer of manganese ions being accepted as the origin of the S_2 signal. For the mechanism of dioxygen evolution it is important to find out the oxidation state of the manganese ions in S_2, which determines the oxidation state of Mn in the S_4 state producing O_2 and returning to S_0. For paramagnetic tetranuclear manganese complex with S = ½ there are two possibilities for the oxidation states: either $Mn_3^{III}Mn^{IV}$ or $Mn^{III}Mn_3^{IV}$. Both of these oxidation states are consistent with other physical data. In the theoretical study of Bonvoisin et al.,[146] $Mn_3^{III}Mn^{IV}$ is considered as the most suitable for the S_2 state, while computer simulations led by Kusunoki to the conclusion that the S_2 corresponds most probably to $Mn^{III}Mn_3^{IV}$.[147]

Modern physical methods, EXAFS, XAS, and XANES, give important structural information concerning the OEC (see Reference 131). EXAFS is known to give bond lengths with a precision of 0.02 Å for the nearest neighbors and about 0.5 Å for more distant atoms. According to EXAFS data there are two different Mn···Mn distances observable: two or three at 2.7 Å and one at 3.3 Å. The 2.7-Å distance strongly indicates a di-μ-oxo-dimanganese unit, which is characterized by this distance. The assignment of the 3.3-Å distance could be attributed to a longer Mn···Mn distance (to another manganese dimer) but also to Mn···Ca distance. EXAFS data can be interpreted in different ways. The simplest interpretation is to present the OEC as a pair of di-μ-oxo-bridged dimers. For example, two kinds of arrangement have been suggested in one paper (Scheme III.2):[148a]

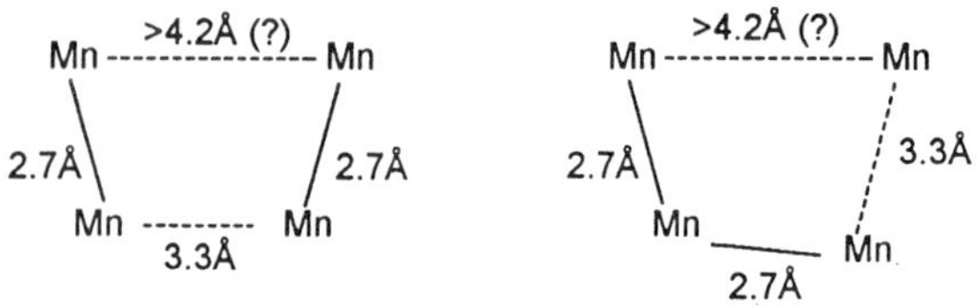

SCHEME III.2 Possible arrangements of manganese atoms in the OEC.

XANES studies of the S_1 state are consistent with an $Mn_2^{III}Mn_2^{IV}$ oxidation state (although an Mn_4^{III} state cannot be ruled out), whereas for the S_2 state the assignment is $Mn^{III}Mn_3^{IV}$, supporting $Mn_2^{III}Mn_2^{IV}$ for S_1. The EXAFS spectra show that there are only small changes accompanying $S_1 \rightarrow S_2 \rightarrow S_3$ transitions, indicating that no serious changes of ligands occur in these steps. In all reported EXAFS studies the environment of manganese ions has been concluded to be very heterogeneous. Analysis of these data indicates that each manganese possesses four to six nitrogen or oxygen atoms at 1.8 to 2.2 Å from manganese.

Chemical modification of amino acids in PS II has indicated that hystidine and carboxylate residues interact with the manganese cluster, possibly being Mn ligands.

A structural model for the teranuclear manganese cluster based on X-ray absorption spectroscopy and all the other known information concerning the OEC was proposed by Yachandra et al. (see Figure III.27).[148b]

The model bears some definite similarities with di-iron centers activating dioxygen (in particular in methane monooxygenase) described in the previous chapter. The similarity may not be accidental: it was mentioned already that a methane monooxygenase active center was developed for the activation and four-electron reduction of dioxygen; the reverse process of four-electron oxidation of water to dioxygen may require a similar structure of the active center (*vide infra*).

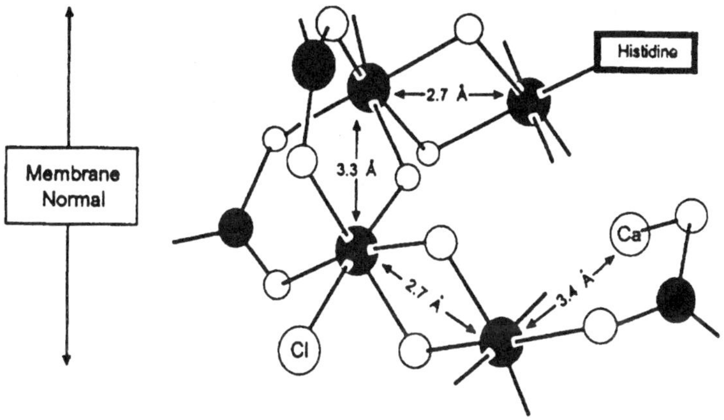

FIGURE III.27 Structural model for tetranuclear manganese cluster in the OEC. (From Yachandra, U. K., *Science* 260, 675, 1993. With permission.)

III.5.2.4 Cofactors of Photosystem II: The Role of Cl^- and Ca^{2+}

Cl^- has been known to play an important role in photosynthetic water oxidation (see References 13 and 149 to 151). In contrast to the absolute specificity for manganese, Cl^- can be replaced by other anions (e.g., Br^-, I^-, or NO_3^-).

There are several types of chloride binding sites in intact thylakoid membranes. The binding sites can be distinguished on the basis of their binding affinities and their influence on the S state transitions. Some of chloride ions are loosely bound and exert an indirect influence on dioxygen evolution, involving nonspecific sites that apparently correlate with changes in the stacking of thylakoids and energy transfer between chlorophyll molecules. Presumably these sites may influence dioxygen evolution rates and increase the efficiency of light utilization. A second pool of chlorides of intermediate binding activity directly influences the kinetics of dioxygen evolution. This pool may be used to neutralize the increasing positive charge on the PS II complex, thus facilitating the cluster oxidation.

The third "tight" binding site is characterized by much stronger Cl^- binding than the "intermediate" one. Photooxidation of PS II samples depleted by Cl^- results in the disappearance of the g = 2 signal. Under these conditions the $S_2 \rightarrow S_3$ transition is blocked and the S_2 state becomes more stable. New evidence of the chloride requirement for dioxygen evolution indicates that Cl^- facilitates the oxidation of the manganese cluster by the Tyr_Z^+ radical.

Ca^{2+} is also an essential cofactor in O_2 evolution (see References 131 and 152 to 155). Calcium is believed to be close to the substrate site of the OEC and participates in regulation of the organization of manganese binding ligands. The replacing of Ca^{2+} by Sr^{2+}, which has larger ionic radius, results in a marked change in the g = 2 multiline EPR signal, with an increase of the number of lines up to 21 to 22 and a reduced hyperfine spacing of 68 G. EPR spectra of the samples modified by strontium are similar to those of Mn clusters with coordinated ammonia. This may indicate that calcium ion is associated with the ligand also bound to manganese.

According to the data of X-ray absorption spectroscopy,[148b] one chlorine atom is directly bound to manganese and one calcium is in the direct vicinity of manganese.

Calcium has been proposed to facilitate the accessibility of the water molecule to the manganese cluster. Ca^{2+} removal may allow carboxylates to bind to the substrate active site instead of water. The Ca^{2+} function was also connected with deprotonation events coupled to the S state-dependent conformational changes or pK_a shifts of carboxylate groups.

A water molecule in the coordination sphere of calcium may be deprotonated to form a $Ca^{2+}OH^-$ moiety in which the OH^- ligand on the calcium ion is a stronger base and a stronger nucleophile than the water molecule. Therefore the transfer of water molecules (actually OH^- ions) to manganese center may be facilitated.[156a]

III.5.2.5 On the Mechanism of Dioxygen Evolution in PS II

In recent years great progress has been achieved in the understanding of the mechanism of O_2 evolution in the OEC of PS II in the multisubunit enzyme water oxidase. Although no X-ray structure is yet available, a certain scheme may be suggested for this reaction, which is of great importance for life, providing almost entirely the dioxygen needed by living organisms on our planet.

According to the evidence described above, presumably four manganese atoms form a cluster consisting of two dimers bound by di-μ-oxo bridges. There are some differences in opinion on the oxidation state of Mn corresponding to each of the S states in the Kok scheme. Most probably the S_2 state corresponds to $Mn^{III}Mn_3^{IV}$. Then the Kok scheme can be represented by Scheme III.3.

SCHEME III.3 Possible oxidation states of manganese in the Kok scheme.

Another (less probable from our point of view) alternative with S_2 corresponding to $Mn_3^{III}Mn^{IV}$ points to the states whose valencies are two units lower.

There are some indications that in the S_3 state a ligand is oxidized on one of the manganese ions in the manganese cluster instead of manganese itself. This ligand is presumably histidine and it forms a free radical in the vicinity of the manganese cluster (see Reference 131). If this is the case then the real oxidation state of manganese in the cluster is one unit less than in the above scheme. It may be also that O–O bonds can be formed not only in the S_4 state, but also earlier (*vide infra*). The formation of the O–O bond will lower the oxidation state of manganese two units more. Nevertheless, the oxidation state of manganese shown in the above scheme indicates that the manganese in the S_4 state able to oxidize water must be in a higher oxidation state than Mn^{IV} or contain additional oxidant in close proximity.

Besides, it is not excluded that several oxidation states of manganese are in equilibrium in the S_4 state, including also the state with Mn^V. We shall see that the situation is similar with the manganese-containing chemical systems oxidizing water with dioxygen evolution.

Manganese-containing catalase with two Mn atoms in the active center may be an example of a possible structure of the dimer in the OEC that ensures the final stage of dioxygen formation when the O–O bond is already established and at the same time ensures the cleavage of this bond from the second H_2O_2 molecule. The X-ray structure of T catalase from extremely thermophilic bacterium *Thermus thermophilus* reveals[156b] that certain similarity exists between this enzyme and such proteins as apoferrytin and hemerithrin. Two Mn atoms are situated between four α helices similarly to what is observed for two iron atoms in hemerithrin. The distance between the Mn atoms was estimated as 3.6 Å, which is close to the ~3.3 Å found for the distance between two Fe atoms for different oxidation states of hemerythrin (and, in fact, both distances could be even closer taking into account that the resolution in the case of T catalase was 3 Å). Therefore it may be suggested that manganese atoms are bound by carboxylate bridges similar to hemerithrin, and both hydrogen peroxide molecules form peroxo bridges between two manganese atoms. The two O_2^{2-} ligands can then disproportionate in the coordination sphere of manganese, $2O_2^{2-} \rightarrow O_2 + 2O^{2-}$, to produce dioxygen and ultimately two water molecules.

Formation of an O–O bond from two separate oxygen ligands on manganese atom(s) is apparently the crucial event in dioxygen production in the OEC. To perform the step of O–O bond formation we have to overcome coulombic repulsion of the negative oxygens and to bring two oxygen atoms sufficiently close to each other to ensure their recombination. Another approach considers combination of two O^{2-} ligands, which leads to the occupation of strongly antibonding π_g^* and $2\sigma_u^*$ orbitals.[148a] Oxidation of manganese diminishes the charges to bring down the repulsion between the oxygen atoms (or helps to remove the electrons from the antibonding orbitals). If oxygen atoms are bound to different manganese atoms (which is likely to be the case) then ligands uniting manganese atoms must be sufficiently flexible to ensure bringing of oxygen atoms to a distance necessary to unite them. Coordination sites for two water molecules should be available and when dioxygen is removed the sites should be again occupied by water. Actually, it might be thought that water replaces dioxygen in the coordination sphere of manganese, compensating the energy lost in dioxygen release. Finally, it should be noted that dioxygen is formed in its ground triplet state, whereas water and presumably coordinated peroxide are singlets.

Recent theoretical analysis with a semiempirical one-electron molecular orbital procedure (extended Hückel method) shows that coordination of an intermediate peroxide ligand (formed from two O^{2-}) with three or four manganese atoms lowers the energy barrier for water oxidation as compared with dinuclear complexes.[148a] However, even for tetranuclear complexes, which are preferred, barriers of up to 4.25–4.38 eV are found. These huge energy barriers are calculated when starting from the equilibrium structures for the manganese clusters known in the literature. Therefore the authors[148a] suggested that the activation of water to form dioxygen in the OEC may be due to distorted coordination of water ligands far from the equilibrium structure for synthetic manganese complexes, and this lowers the activation energy.

It was mentioned above that a number of facile reactions *reverse* to the dioxygen evolution from two water molecules in the process of their oxidation are known. Certain oxidases, such as laccase, ascorbate oxidase, or ceruloplasmine, provide the examples of dioxygen reduction to two water molecules (e.g., see Reference 21b). Copper is involved in the reaction centers of these enzymes, combining four to eight copper atoms. In the process of functioning, substrates reduce copper(II) to copper(I), and then dioxygen reoxidizes the latter to copper(II), producing two water molecules without free hydrogen peroxide as an intermediate. In this process copper atoms reveal different functions, two of them presumably forming a complex with dioxygen, while others providing electrons to ensure the O–O bond cleavage. Other examples of dioxygen activation on monooxygenases were considered in Chapter II. Methane monooxygenase is particularly relevant because it activates dioxygen in a dinulcear complex. The analogy of the OEC with both types of enzymes is profound: two electrons are added from an electron donor (distant Cu^{I} in oxidases or NADH in monooxygenases) to ensure the full use of all four equivalents of activated dioxygen to form finally two O^{2-} ligands. Two of four manganese atoms may transform two O^{2-} into an O–O ligand (possibly bridging) while the other two may accept two electrons to ensure full use of all four electrons and to form dioxygen. Thermodynamically the oxidases and monooxygenases are suitable for dioxygen reduction to water, while the OEC oxidizes water to dioxygen. But activation energies for dioxygen reduction are definitely low, because the reaction rates in all the cases are high. Therefore given that the redox potential of oxidases or monooxygenases is increased (e.g., by placing them on the surface of anode in an electrochemical cell), these enzymes should be able to evolve dioxygen *with low activation energy.* It was already mentioned that the analysis based mainly on X-ray absorption spectroscopy (apparently independently of any analogy with MMO) leads to the structure of the OEC,[148b] which is in many aspects similar to that of MMO (Figure III.27). It might be thought that manganese is chosen by nature for PS II just because it can acquire the necessary redox potential in a four-nuclear cluster for water oxidation. Therefore the ligand surrounding of the metal centers of the dioxygen-reducing enzymes, e.g., of the di-iron complex activating dioxygen in MMO (see Chapter II), are suitable for forming a dimanganese complex and producing dioxygen from water (provided of course that organic ligands are not oxidized instead of water) if electron acceptors, i.e., in this case two other high-valent manganese atoms, are situated nearby. The example of T catalase may be remembered, with two manganese atoms cleaving the O–O bond to produce water.

There is, of course, a great number of various possible combinations leading to dioxygen formation from two oxygen atoms bound to one, two, three, or four manganese atoms, but in view of considerations presented above, it is tempting to choose two manganese atoms bound by two carboxylate bridges with a distance of 3.3 Å between them as the ones directly involved in dioxygen formation. The following scheme presented in Figure III.28 might be tentatively suggested for the mechanism of this reaction.

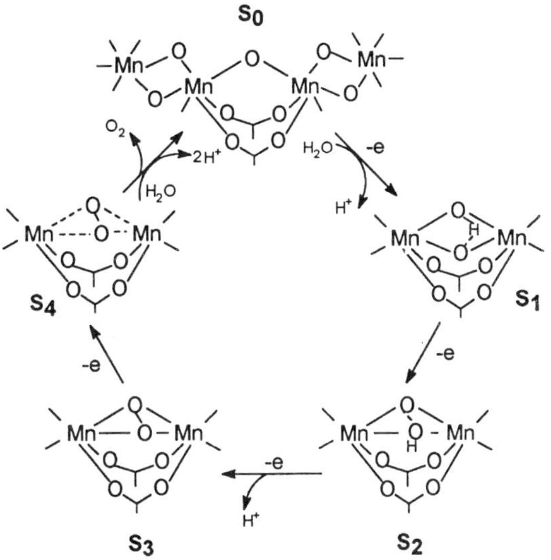

FIGURE III.28 Possible scheme for the mechanism of dioxygen evolution in the OEC.

Carboxylate bridges look attractive not only because they are apparently sufficiently flexible to keep different oxidation states of manganese and to form Mn–O–O–Mn bridges during the catalytic cycle, but also because they are stable to oxidation and therefore presumably will sustain high oxidation states of manganese atoms.

More definite conclusions about the detailed mechanism will be possible after the precise structure of the OEC has been established.

Other elements in the composition and the structure of the catalytic complex in the OEC may apparently be used for increasing its activity. Cl^- and Ca^{2+} ions may facilitate oxidation of the manganese cluster and the addition of water molecules, respectively (Cl^- facilitating the oxidation of one manganese to Mn^V and Ca^{2+} increasing the nucleophilic properties of water by deprotonation of H_2O in the coordination sphere of calcium).

Formation of an O–O bond can be expected from two oxygen ligands after reaching the S_4 state and dioxygen evolution will proceed immediately after (thus the whole reaction will be a two-step process), or an O_2 molecule is produced in a concerted one-step reaction. There is, however, some indication that an O–O bond is formed before reaching the S_4 state. Hydrogen peroxide was detected by Klimov et al. after addition of acid (H_2SO_4) to membrane fragments-containing PS II, and part of the H_2O_2 is definitely formed at or near the manganese cluster at the level of S_2 formation.[156c] Although it is not excluded that hydrogen peroxide is generated as a side reaction if the manganese cluster is disturbed and accessible to bulk water, it could be also suggested that the acid destroys an intermediate containing an Mn–O–O–Mn fragment (perhaps being in equilibrium with two M=O) similarly to the action of acid on functioning nitrogenase, which produces hydrazine. Another possibility is that the acid induces the formation of peroxo bridge from two oxo ligands on manganese.

Thus, some analogy of the proposed mechanism may be noticed if we compare four-electron oxidation of water to dioxygen with six-electron dinitrogen reduction to ammonia. Freshly prepared hydroxides were described in Chapter I to be active in N_2 reduction, with metal ions united by hydroxo and oxo bridges. A certain flexibility is apparently necessary to activate dinitrogen and to perform several steps of the reduction and protonation to form the final product. Therefore, for example, two di-μ-hydroxodimers of vanadium(II) were suggested to activate and reduce dinitrogen, both united by flexible O–Mg–O bridges. Now, manganese hydroxo complexes may be suggested as the catalysts of dioxygen evolution from water, the more so as they contain the necessary hydroxo or oxo ligands to produce O_2.

It could be suggested that simple inorganic manganese complexes present in nature, e.g., polynuclear hydroxocomplexes, were sufficiently active catalysts of dioxygen evolution for primitive life to start with and to develop these catalysts in the process of evolution to the modern state of the PS II. Carboxylate bridges between two di-μ-oxo dimers look natural from this point of view since formation of carboxylates could be envisaged in the oxidation of organic material catalyzed by manganese complexes.

If this suggestion is correct then chemical model catalysts based on simple manganese hydroxo complexes will be capable of performing water oxidation to produce dioxygen, and investigation of this catalysis may help us to understand why manganese complexes were chosen by nature to perform the catalytic action in dioxygen evolution in photosynthesis.

III.5.3 CATALYTIC AND PHOTOCATALYTIC OXIDATION OF WATER IN THE PRESENCE OF TRANSITION METAL COMPLEXES

Catalytic dioxygen formation in the oxidation of water by strong oxidants has been known for a very long time. Electrolysis of water to dihydrogen and dioxygen depends on the material of electrodes: it has been known for a long time that an anode covered, for example, by ruthenium dioxide requires much less overvoltage than other materials; therefore RuO_2 reveals a catalytic effect on the surface of the electrode (e.g., see Reference 157). MnO_2 is another well-known catalyst for dioxygen evolution. Its catalytic action in dioxygen evolution, e.g., in decomposition of potassium perchlorate, is a classic example of the catalysis phenomenon described in chemistry textbooks.

It was already mentioned that transition metal hydroxocomplexes are promising candidates for catalysts of dioxygen evolution from water: they are polynuclear and contain oxygen in the form of H_2O, OH^-, or O^{2-} as their ligands. It is almost obvious that at the increase of redox potential these ligands will produce dioxygen at some point. Therefore metals on the right side of manganese in the periodic table, including manganese itself (metals on the left side of Mn are too weak oxidants even in their highest oxidation states), could be tried as catalysts in the form of their hydroxo complexes. However, *a priori* it is not clear whether they are already close to *optimal* catalysts and are able to effectively catalyze dioxygen evolution from water at pH close to 7 under the action of comparatively mild oxidants such as P-680$^+$ (E_0 = 1.12 V). Another important question to solve (provided they are really good catalysts for dioxygen evolution) is whether being able to oxidize water they will not oxidize organic substances at the same time to form carbon dioxide instead of dioxygen. We have seen that mono- or dinuclear high-valent metal complexes containing O^{2-} ligands (including manganese complexes) are able to effectively insert oxygen even in C–H bonds of alkanes. This activity would greatly limit the search for possible ligands for such catalysts if we want to prepare individual complexes as catalysts and would put in doubt the idea that in living nature the hydroxo complexes could be used initially before developing on their base more elaborate complexes suitable to be involved as cofactors in enzymes. The next question concerns manganese. Is it really particularly suitable in the form of simple hydroxo complexes to be a catalyst and photocatalyst for O_2 formation? Could we explain why it was chosen by nature to form the reaction center in the PS II for dioxygen evolution? What is the reason for lower activity of other metals?

These questions are dealt with below.

III.5.3.1 Nonmanganese Catalysts for Water Oxidation

One of the most studied reactions of catalytic dioxygen evolution is water oxidation by ceric perchlorate in the presence of many metal oxides, such as oxides of Mn, Co, Pt, Pd, Ru, and Ir.[158] In this series RuO_2 is the most active catalyst for acid media oxidation.[159,160] The stoichiometry of the reaction can be expressed by the following equation:

$$\mathrm{Ce(IV)} + 2\mathrm{H_2O} \xrightarrow{\mathrm{RuO_2}} 4\mathrm{Ce(III)} + 4\mathrm{H^+} + \mathrm{O_2} \tag{35}$$

In extensive research of this catalysis (see Reference 161) it has been established that the usual hydrated form of $RuO_2(RuO_2{\cdot}xH_2O)$, which contains more than 24% H_2O, undergoes a "corrosion" and is partially oxidized to RuO_4, i.e., together with reaction 35, another reaction proceeds simultaneously:

$$4\mathrm{Ce(IV)} + \mathrm{RuO_2} \cdot x\mathrm{H_2O} + \mathrm{H_2O} \rightarrow 4\mathrm{Ce(III)} + \mathrm{RuO_4} + x\mathrm{H_2O} + 4\mathrm{H^+} \tag{36}$$

More than half of the oxidant is spent in reaction 36 of ruthenium(IV) oxidation. Some annealing of the catalyst is necessary to prevent or to slow down the corrosion.

In the presence of organic substances their total oxidation to CO_2 is also effectively catalyzed by RuO_2. Thus, when $Ru(bpy)_3^{3+}$ is used as an oxidant and RuO_2 as a catalyst, the oxidation of bipyridyl ligands takes place in parallel with dioxygen evolution from water to produce carbon dioxide according to the formal scheme:[162]

$$\mathrm{C_{10}H_4N_2} + 20\mathrm{H_2O} - 48e \rightarrow 10\mathrm{CO_2} + \mathrm{N_2} + 48\mathrm{H^+} \tag{37}$$

RuO_4 was also one of the products in this system.

In order to prepare a catalyst more selective for water oxidation the samples of RuO_2 have been attached to such supports as TiO_2, Al_2O_3, zeolites, or specially prepared colloids. Active catalyst, stable to corrosion under the action of strong oxidants, is thermally activated hydrate ($RuO_2 \cdot yH_2O^*$). The optimal form of thermally activated catalyst is RuO_2 hydrate baked out at about 140°C for 5 h in air. This material is considered to be the most active stable heterogeneous catalyst developed. In its presence dioxygen is evolved quantitatively according to Eq. 34. Such oxidants as MnO_4^- and IO_4^-, similarly to Ce^{IV}, evolve dioxygen rapidly and quantitatively in the presence of this catalyst. $RuO_2 \cdot yH_2O^*$ is much less hydrated than the initial $RuO_2^{\cdot}$ xH_2O: water content in $RuO_2 \cdot yH_2O^*$ is less than 12%. At higher annealing temperatures than the optimum 140°C, the catalyst remains stable to the corrosion but its catalytic activity decreases. It is very low after baking at 700°C and virtually disappears after baking at 900°C. Anhydrous, highly crystalline RuO_2 is inactive as a catalyst.[161]

The use of the individual ruthenium complexes gives the possibility of investigating more closely the detailed mechanism of water oxidation. Some of the mono-, bi-, and trinuclear ruthenium complexes were found to be catalysts for water oxidation. For example, mononuclear complexes $Ru(bpy)_2(H_2O)_2^{2+}$ and $Ru(NH_3)_5Cl^{2+}$ catalyze the dioxygen evolution from water (the first with PbO_2, the second with Ce^{IV}.[163-165] The mechanism of dioxygen formation apparently involves the oxidation of ruthenium and recombination of deprotonated oxygen ligands.

$$\left[(bpy)_2Ru^{II}(H_2O)_2\right]^{2+} - 4e - 4H^+ \rightarrow \left[(bpy)_2Ru^{VI}(O)_2\right]^{2+}$$

$$\left[(bpy)_2Ru^{VI}(O)_2\right]^{2+} \rightarrow \left[(bpy)_2Ru^{IV}\langle\begin{smallmatrix}O\\|\\O\end{smallmatrix}\right]^{2+} \xrightarrow{H_2O} \left[(bpy)_2Ru^{II}(H_2O)_2\right]^{2+} + O_2 \tag{38a}$$

Dimeric ruthenium complexes $[(bpy)_2(H_2O)Ru]_2O^{4+}$ and $[(bpy)_2(NO_2)Ru]_2O$ in $HClO_4$ solutions of Ce^{IV} were also shown to be able to catalyze the oxidation of water to dioxygen.[166-168] Ru^V complexes formed in the oxidation of initial ruthenium complexes were found to be responsible for dioxygen formation:

$$\left[(bpy)_2(H_2O)Ru^{III}ORu^{III}(H_2O)(bpy)_2\right]^{4+} \xrightarrow[-4H^+]{Ox} \left[(bpy)_2\overset{\overset{O}{\|}}{R}u^{V}O\overset{\overset{O}{\|}}{R}u^{V}(bpy)_2\right]^{4+} \tag{38b}$$

$$\xrightarrow{2H_2O} \left[(bpy)_2(H_2O)Ru^{III}ORu^{III}(H_2O)(bpy)_2\right]^{4+} + O_2$$

The dimers undergo partial decomposition during the catalysis. The catalytic turnover number is not very high and the total yield of dioxygen in these systems remains low.

The trinuclear complexes $\{[(NH_3)_5RuO]_2Ru(NH_3)_4\}^{n+}$, where n = 6 or 7, are far better catalysts for water oxidation than the dimeric complexes, both in homogeneous solutions and supported on Kaolin.[168-170] The cyclovoltagram of $\{[(NH_3)_5RuO]_2Ru(NH_3)_4\}^{8+}$ corresponds to five one-electron oxidation steps (numbers above arrows show oxidation potentials in volts):

$$\text{III,IV,III} \overset{0.72}{=} \text{IV,III,IV} \overset{0.84}{=} \text{IV,IV,IV} \overset{0.95}{=} \text{IV,V,IV} \overset{1.08}{=} \text{V,IV,V} \overset{1.16}{=} \text{VVV} \tag{39}$$

The following scheme is proposed for the catalysts of water oxidation in the presence of these complexes:

$$Ru^{III}-O-Ru^{IV}-O-Ru^{III} \xrightarrow{Ox} Ru^{V}-O-Ru^{V}-O-Ru^{V} \rightarrow Ru^{III}\!\!\begin{array}{c}\diagup O \diagdown\\ \diagdown O-O \diagup\end{array}\!\! Ru^{III}-ORu^{V} \rightarrow Ru^{III}-O-Ru^{IV}-O-Ru^{III}+O_2 \tag{40}$$

Spectral evidence tentatively suggests that the reactive catalyst may be actually a hydrolyzed Ru^{IV} polymeric species.[170]

At high molar ratio of Ce^{IV} to Ru the catalyst was found to undergo a decomposition to a monomeric ruthenium. The trimeric ruthenium complexes, when they are absorbed into a Nafion membrane, are highly stabilized against the decomposition. The enhanced O_2 evolution for $\{[(NH_3)_5RuO]_2Ru(NH_3)_4\}^{6+}$ embedded in a Nafion film was observed when Ce^{IV} was electrochemically generated.[164]

Thus, polynuclear complexes of ruthenium are good models for detailed understanding of the catalytic oxidation of water. Comparatively little difference between the redox potentials of five consecutive one-electron steps makes them very good examples of the catalysis based on a transition to a four-electron mechanism in the catalysis of water oxidation by a one-electron oxidant. It is not clear whether O_2 formation is a concerted one-step, four-electron reaction or a peroxide O–O bond is formed primarily producing O_2 in the next two-electron step. In the latter case the peroxide fragment is apparently stabilized by the interaction with the metal atoms and no free hydrogen peroxide is formed as an intermediate.

In strongly alkaline solutions dioxygen evolution has been established in the case of oxygen-containing oxidants, e.g., permanganate, perruthenate, hypochlorite, and persulfate.[171-174] These reactions proceed with remarkably high rates in the presence of traces of Co^{II} and Ni^{II} and the isotopic composition of dioxygen evolved is practically identical to that of water; therefore these reactions can be considered as catalytic water oxidations.

To perform water oxidation in chemical model systems functioning in neutral media, similarly in this aspect with PS II of plant photosynthesis, it is convenient to use as oxidants tris-bipyridine and tris-phenantroline complexes of Fe^{III}, Ru^{III}, and Os^{III}. These complexes are one-electron oxidants and have suitable redox potentials, 1.26, 1.06, and 0.84 V, respectively,[175] i.e., close to that of the pair $P\text{-}680^+/P\text{-}680$ (1.12 V).[176] In aqueous solutions the complexes are reduced spontaneously with almost quantitative formation of the corresponding tris complexes of Fe^{II}, Ru^{II}, and Os^{II}. This reaction was observed almost 100 years ago by Blau, who reported an ozone odor after transformation of blue solutions in Fe^{III} complexes to red solutions of Fe^{II} ones induced by additions of alkaline solutions.[177] The reaction was later reinvestigated by Dwyer and Gyarfas who detected hydrogen peroxide formation after reduction of Fe^{III} and Os^{III} complexes.[178] However, further studies did not confirm formation of ozone or hydrogen peroxide. Instead, in more recent studies dioxygen evolution was found in the case of $Fe(phen)_3^{3+}$. Then Anbar and Pecht[174] reported that the isotopic composition of dioxygen evolved is very close to that of water. The mechanism of water oxidation developed in this work as well as in further publications[179-181] included addition of OH^- to bipyridine or phenantroline rings followed by reduction of the central metal ion and subsequent redox transformation of the complex organic radical produced to form dioxygen. It was difficult to accept the suggested mechanism primarily because it postulated reversible formation of C–O bonds with formation at the end of O_2 (and not CO_2). Detailed studies of dioxygen evolution by tris-bipyridine and tris-phenantroline complexes by Shafirovich et al. demonstrated that the reactions were in fact catalytic and O_2 was produced only in the presence of hydroxo complexes of Fe, Co, Ni, and Cu (even traces of these complexes were sufficient to facilitate O_2 evolution).[182-184] The necessity of the catalysts for the O_2 evolution was subsequently confirmed by Harriman et al.[185] and by Sutin and Creutz.[175]

Without the catalyst the only reaction is the oxidation of an organic ligand. For example, $Ru(bpy)_3^{3+}$ was found to be reduced to $Ru(bpy)_3^{2+}$ almost quantitatively over a broad range of pH (0 to 14) without O_2 evolution. Instead, the oxidation of the small part of the coordinated ligand occurs to form finally CO_2. In neutral media the following expression may be used to describe the reaction:[182]

$$(1-x)\mathrm{Ru(bpy)}_3^{3+} + 3x\mathrm{bpy\ (coord.)} + 0.2\mathrm{H_2O} \rightarrow (1-x)\mathrm{Ru(bpy)}_3^{2+} + 0.1\mathrm{CO_2} + 0.48\mathrm{H^+} + \text{tarlike products} \quad (41)$$

where $x = 0.01$ to 0.05 represents the fraction of the complex decomposed. The lifetimes of $Ru(bpy)_3^{3+}$ in aqueous solutions change over a large time scale, from tens of minutes in acid solutions to a few milliseconds in alkaline ones.

Upon addition of hydroxocomplexes of Fe^{II}, Fe^{III}, Co^{II}, Ni^{II}, and Cu^{II} the reduction of $Ru(bpy)_3^{3+}$ to $Ru(bpy)_3^{2+}$ accelerates and is accompanied by dioxygen evolution. At pH 8, the activity of iron and cobalt hydroxocomplexes is so high that noticeable rates of O_2 evolution are obtained at hydroxo complexes concentrations as low as 10^{-6} *M*. Therefore, the concentration of iron ions in ordinary distilled water is sufficient to facilitate O_2 evolution under the action of $Ru(bpy)_3^{3+}$. The upper limit of the activation energy can be estimated for iron hydroxo complexes as 5 to

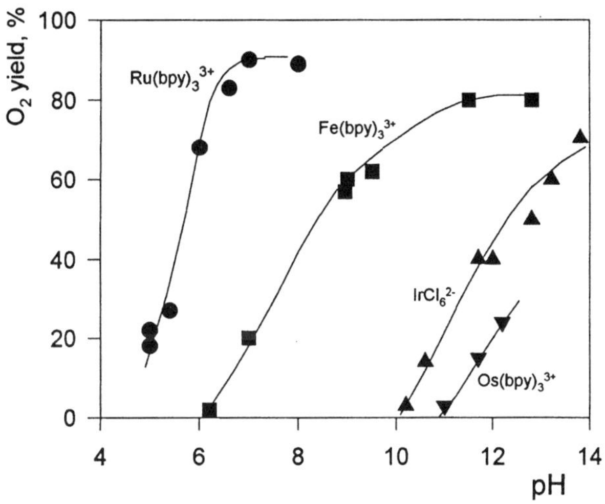

FIGURE III.29 Dependence of dioxygen yield on pH in water oxidation catalyzed by cobalt hydroxo complexes.

10 kcal/mol. Hence the rate constant of dioxygen formation by a high-valent complex is comparable to that by the reaction center in the PS II. Figure III.29 shows the dependence of the O_2 yield on pH in dioxygen evolution in the presence of cobalt hydroxo complexes under the action of several one-electron oxidants.[182-184]

It can be seen that the weaker the oxidant, the higher the pH for the start of dioxygen evolution. It should be remembered that higher pH corresponds to lower redox potentials of the reaction of water oxidation. Therefore if the redox potential of the system oxidant–cobalt hydroxo complex is lower for weaker oxidants then the thermodynamics of the O_2 evolution becomes more favorable at higher pH. On the other hand, for the water oxidation a definite oxidation state of cobalt may be essential (e.g., Co^{IV}). Apparently, a weaker oxidant is needed to oxidize more deprotonated hydroxo complexes (having more negative charge) to this oxidation state. Probably the latter cause is particularly important.

It may seem surprising that hydroxo complexes of so many metals are able to catalyze O_2 evolution under the action of $Ru(bpy)_3^{3+}$. According to the mechanism of catalysis, the redox potential in the highest oxidation state for a four-electron mechanism has to be higher than that of the O_2/H_2O pair and at the same time this redox potential must not be too high, otherwise the external oxidant will not be able to oxidize the catalyst to the active state. These two conditions should seemingly limit the number of the catalysts. Presumably, by increasing the number of the metal atoms in a hydroxo complex in a given oxidation state we decrease simultaneously the redox potential of the first oxidation step of the complex, thus increasing the probability of the reaction with the external one-electron oxidant. With a sufficiently large number of metal atoms we shall have a large number of energy levels for molecular orbitals occupied by electrons. In all cases, by reacting with the catalyst the one-electron oxidant will create a potential close to its own, but for multielectron oxidation. Therefore the probability is high for a cluster consisting of a large number of transition metal atoms to be a catalyst for O_2 evolution if three conditions are fulfilled: (1) the redox potential of the highest oxidation state of a metal cluster should be higher than that of the pair O_2/H_2O; (2) water molecules must have free access to the complex to replace evolving dioxygen; and (3) two oxygen ligands should occupy a suitable position to form dioxygen without high activation energy. All three conditions are met in the case of hydroxo complexes of the metals located on the right side of manganese in the periodic table. The experiment shows that the energy barriers can be low. The situation is somewhat similar with dihydrogen evolution catalyzed by polynuclear complexes. In all the systems with low redox potentials considered in the previous sections dihydrogen evolved nonspecifically.

III.5.3.2 Manganese Compounds as Catalysts for Dioxygen Evolution

Manganese is among a very few elements (and probably the only one among those used in enzymes) that are able of changing their oxidation states by five units (from Mn^{II} to Mn^{VII}), forming species sufficiently stable in water solutions. In this aspect manganese is similar to ruthenium, which, as we have seen, is readily oxidized to the highest oxidation state in RuO_4 and at the same time forms active homogeneous and heterogeneous catalysts for dioxygen evolution from water.

Permanganate ion, being a strong oxidant ($E_0 = +1.51$ V), is able thermodynamically to oxidize water to dioxygen. However, at ambient temperatures aqueous solutions of MnO_4^- are known to be quite stable. High concentrations of strong mineral acids promote dioxygen evolution in aqueous solutions of MnO_4^-. It has been shown that permanganate in 4 to 16 *M* solutions of sulfuric acid evolves dioxygen efficiently at room temperature.[186] At these concentrations of H_2SO_4 and at appropriate concentrations of permanganate the reaction is homogeneous and no precipitates ($MnO_2 \cdot nH_2O$ or $Mn_2O_3 \cdot nH_2O$) have been found after complete decomposition of MnO_4^-. Kinetic measurements have shown that at high acid concentrations dioxygen formation proceeds already in a unimolecular reaction, presumably protonated acid $H_2MnO_4^+$ being the active species.

Kinetics of O_2 evolution shows that the reaction is autocatalytic.[186,187] In the first step Mn(III) is formed alongside O_2:

$$H_2MnO_4^+ + 2H^+ \rightarrow Mn^{3+} + O_2 + 2H_2O \qquad (42)$$

Then Mn^{IV} produced in the reaction acts as a catalyst in the further O_2 evolution. Mn^{IV} is the product of the interaction of MnO_4^- with Mn^{III}:

$$MnO_4^- + 3Mn^{III} + 8H^+ \rightarrow 4Mn^{IV} + 4H_2O \qquad (43)$$

It is of interest that Mn^{III} is an established intermediate in the dioxygen evolution, and Mn^{IV} is a sufficiently stable product. Hence, the only pathway for Mn^{III} formation is four-electron reduction of Mn^{VII} compound. Experiments with $H_2{}^{18}O$ have shown that the isotopic composition of dioxygen is close to that of water due to the fast isotope exchange:

$$MnO_4^- + 4H_2{}^{18}O \rightarrow Mn^{18}O_4^- + 4H_2O \quad (44)$$

Rapid coordination of water is, as it was mentioned, an important condition of its catalytic oxidation.

At elevated temperatures (70 to 80°C) Mn^{IV} in 4 to 16 *M* H_2SO_4 is known to produce O_2 from water:

$$4Mn^{IV} + 2H_2O \rightarrow 4Mn^{III} + 4H^+ + O_2 \quad (45)$$

It has been shown[188] that this reaction involves the intermediate formation of MnO_4^- due to disproportionation of Mn^{IV}:

$$4Mn^{IV} \rightarrow 3Mn^{III} + MnO_4^- + 8H^+ \quad (46)$$

and the main route of O_2 evolution in this system is the reaction:

$$MnO_4^- + Mn^{IV} \rightarrow Mn^{III} + Mn^{IV} + O_2 \quad (47)$$

In this case the isotopic exchange with permanganate is relatively slow; nevertheless, the isotopic composition of dioxygen evolved is again close to that of $H_2{}^{18}O$ used. This fact was explained by the mechanism involving the $Mn^{IV}Mn^{VII}$ complex capable of fast oxygen exchange in the coordination sphere of Mn^{VII} with water's oxygen due to reversible coordination of an additional H_2O molecule to the Mn^{VII} complex.

Hence the complex $Mn^{IV}Mn^{VII}$ reveals the ability to easily evolve dioxygen as well as to exchange oxygen ligands with water. This shows that there is a certain flexibility of oxygen ligands necessary to form O_2 in the process of their recombination.

Thus, dinuclear complexes containing Mn^{VII} and Mn^{IV} evolve dioxygen seemingly producing $Mn^{IV}Mn^{III}$ complex. The complex $Mn^{IV}Mn^{VII}$ is equivalent to $Mn^{V}Mn^{VI}$ and the latter may be compared with the S_4 state of the PS II, particularly if the S_4 state corresponds to $Mn_3^{IV}Mn^{V}$. Two units of the oxidation state of Mn^{VI} in the dinuclear complex are compensated by two more Mn^{IV} atoms in the tetranuclear one. The activation energy of the dioxygen evolution in the case of the dinuclear complex (17 kcal/mol) is higher than in the OEC of the photosynthesis, but is sufficiently low to facilitate efficient O_2 evolution.

Manganese dioxide has been shown to be a catalyst for water oxidation to dioxygen. For example, MnO_2 catalyzes dioxygen evolution in $HClO_4$ solution.[189] The increase of $HClO_4$ concentration in the range of 0.05 to 0.5 *M* results in the increase of the rate of O_2 evolution. Mn^{IV} formed by the oxidation of Mn^{2+} salts by PbO_2 in acid solutions also becomes the catalyst of dioxygen evolution.[190] It has been shown that O_2 is produced when MnO_4^- is in contact with Mn^{IV}. The catalytic oxidation of water can be also performed in neutral and slightly acidic media with sodium periodate as a two-electron oxidant.[191] The system was shown to produce both dioxygen and permanganate. At the same time, the observed rate is higher than that in the reaction of MnO_4^- with MnO_2. Thus dioxygen formation occurs in neutral media apparently via the formation of high-valent manganese complex bound to the surface of MnO_2 without necessary formation of free MnO_4^- escaping to the solution.

$Ru(bpy)_3^{3+}$ can oxidize water in the presence of MnO_2 in neutral media.[191-193] However, the use of MnO_2 as the catalyst results in smaller than quantitative yields of O_2 with respect to the oxidant (10 to 40%). Analysis of the gaseous products revealed that carbon dioxide was produced together with dioxygen, showing that MnO_2 is also a catalyst of the deep oxidation of organic molecules. Bipyridyl molecules were shown to be oxidized in the reaction according to Eq. 40. Acetate ions, which are often used to prepare buffer solutions, can be also oxidized to produce carbon dioxide. For example, if the oxidation is conducted in 0.5 *M* acetate buffer, only CO_2 is formed instead of O_2. Thus the catalysts of water oxidation (RuO_2, MnO_2, and hydroxo complexes of Fe, Co, Ni, and Cu) can oxidize organic substances present in the solution, instead of producing O_2, while the manganese cluster in the OEC of PS II does not oxidize surrounding proteins. It is important to find out whether the selective dioxygen evolution can be achieved if the catalyst is supported by an organic material, e.g., similar to the thylakoid membranes in PS II.

To check this possibility a manganese catalyst was prepared by sonication of dipalmitoyl-DL-α-phosphatidylcholine to form vesicles in buffer solution in the presence of Mn^{2+} salts.[156,194,195] During the sonication an irreversible binding of manganese ions to the membrane occurs due to the oxidation by air. The oxidation state of Mn in the catalyst was found to be III: the electron spectrum of manganese bound to vesicles formed is typical for Mn^{III} complex. The oxidation of Mn^{III} bound to vesicles is accompanied by the appearance of a broad asymmetrical band at 300 nm. Changes in UV spectra are similar to those observed by Dekker et al. in the OEC of PS II.[196,197]

Dioxygen is evolved in the presence of this catalyst under the action of $Ru(bpy)_3^{3+}$. Mn^{IV} bound to vesicles cannot by itself oxidize water. No spontaneous dioxygen evolution was detected in the absence of $Ru(bpy)_3^{3+}$ when the oxidized catalyst solution was acidified or heated at 80°C. The redox potential of the couple Mn^{IV}/Mn^{III} bound to vesicles depends strongly on pH and was found to be equal to 0.5 V at pH 7. It is not sufficient for four-electron water oxidation. It may be suggested, therefore, that to oxidize water manganese should be oxidized to a higher oxidation state than Mn^{IV}, presumably Mn^{V}, or, alternatively, the oxidant ($Ru(bpy)_3^{3+}$) is directly involved in the dioxygen-evolving complex. When a weaker oxidant, $Fe(bpy)_3^{3+}$, is used O_2 evolution starts at higher pH value. It is of importance that whereas the catalytic activity with respect to dioxygen evolution is increased with Mn hydroxo complex incorporated into the vesicle as compared with ordinary MnO_2, the activity towards oxidation of organic material is lost: *the catalyst provokes the selective water oxidation to dioxygen.* The situation with RuO_2 is completely different: when it is incorporated into the vesicles the oxidation of organic molecules takes place in the presence of $Ru(bpy)_3^{3+}$ instead of O_2 evolution and only CO_2 but not O_2 is formed.

Phospholipid vesicles are not unique matrices for stabilization of manganese hydroxo complexes as effective catalysts for dioxygen evolution from water. The

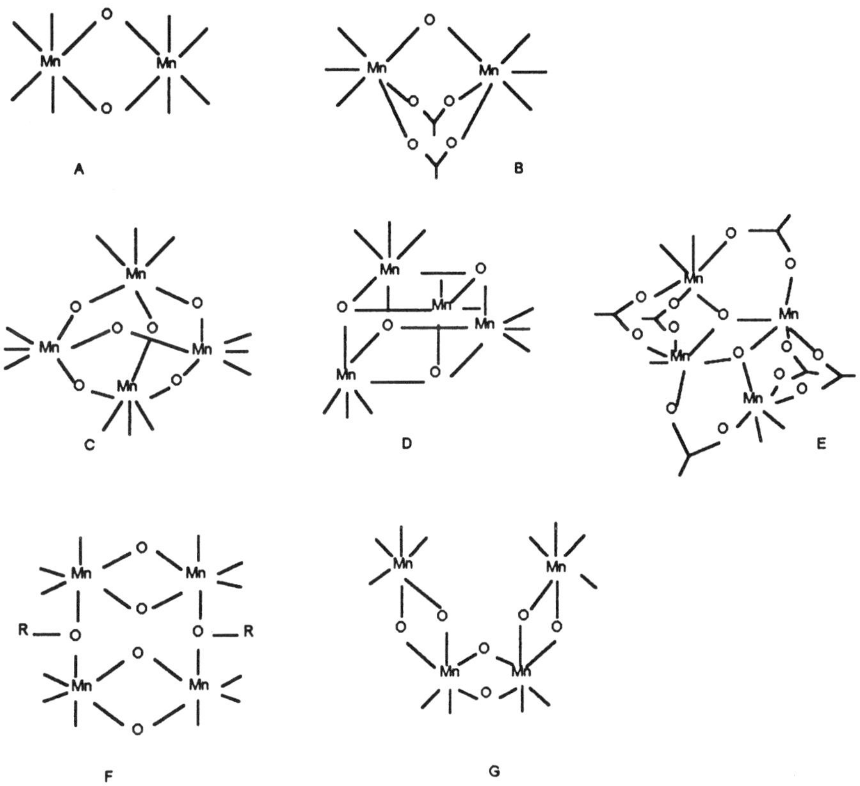

FIGURE III.30 Some polynuclear manganese complexes: possible structural models of the OEC. (Adapted from Wieghardt, K., *Angew. Chem. Int. Ed. Engl.*, 33, 725, 1994.)

manganese catalyst stabilized by cyclodextrins also shows high catalytic activity and selectivity.[198] The spectral properties of Mn hydroxo complexes stabilized on cyclodextrins are similar to those of the complexes bound to vesicles. Mn^{IV} on cyclodextrins should be oxidized to a higher oxidation state (probably Mn^{V}) to form the active state for water oxidation.

It should be stressed that the sonication is not crucial for preparation of an active catalyst of water oxidation supported by an organic material. It is just a convenient method of preparation. Other methods, e.g., precipitation by the change of the solvent without sonication, lead to active catalysts as well.

III.5.3.3 Polynuclear Manganese Complexes: Structural Models of the OEC

Many bi- and polynuclear manganese complexes were synthesized recently in the works of Armstrong, Wieghardt, Crabtree, Christou, and others (see References 131b and 198a, b, and c) in the hope of finding structural models for the oxygen evolution center of photosynthesis. The most interesting structures are shown in Figure III.30.[131b]

Some of them immediately attracted much interest as possible models for the structures of the active center of the OEC, particularly, such symmetrical structures as adamantane, cubane, and butterfly frameworks (structures **C**, **D**, and **E**). However, later it became clear that though very beautiful they were lacking some important features that would make them active catalysts for dioxygen evolution. They are probably too rigid to be the candidates for catalysts of multistep processes of O_2 evolution. The EXAFS data provide conclusive evidence against symmetric cubane and butterfly structures.[148b] Complexes **F** and **G** are both "dimers of dimers" and may bear more similarities with the active center of the OEC. Complex **F** was shown to be able to exist in four different oxidation states. The state $Mn^{III}Mn^{IV}Mn^{III}Mn^{IV}$ corresponds to the S_1 state and its EPR spectrum is similar to that of the OEC S_1 level.[199] This result supports the existence of a low-lying S = 1 state of the OEC S_1 level, and a similar oxidation state ($Mn^{III}Mn^{IV}Mn^{III}Mn^{IV}$) is likely to correspond to the S_1 state of the Kok scheme.

It would be desirable, of course, to find a catalyst of dioxygen evolution among the complexes synthesized as structural models of the OEC. So far no definite activity has been observed for such complexes. Calvin reported[130] that di-μ-oxo-bridged dinuclear complex $Mn_2O_2(bpy)_4^{3+}$ was able to oxidize water photochemically, but initial claims of the success with photolytic dioxygen evolution had to be later withdrawn.[200] Kaneko et al. found that di-μ-oxo-bridged dinuclear complexes $Mn_2O_2(bpy)_4^{3+}$ and $Mn_2O_2(phen)_4^{3+}$ oxidize water to form dioxygen when they are suspended in water as heterogeneous catalysts supported on Kaolin clay in the presence of Ce^{IV} in solutions of 0.1 *M* HNO_3 and 0.1 *M* NaOH, although the same complexes are not active in homogeneous solutions.[201,202] In the process of water oxidation the complexes undergo a decomposition to form MnO_4^-. The complexes are stable in aqueous solutions only in the presence of excess of a ligand. Probably the complexes could be hydrolyzed to MnO_2 or another manganese hydroxo complex and it is a hydrolyzed or partially hydrolyzed form that is actually the catalyst. This could explain the poor activity of the complexes in homogeneous solutions.

Several polynuclear Mn complexes were tested as catalysts of dioxygen evolution in water oxidation by $Ru(bpy)_3^{3+}$. However, dioxygen evolution was observed only in the cases of complexes that are partially hydrolyzed to form the products, so the partially hydrolyzed forms are apparently the real catalysts of the process.[203a]

Structure **B** is probably close to the catalytic site of the OEC, because, as was mentioned above, carboxylate bridges are likely to unite two manganese atoms directly involved in O_2 evolution. They have Mn···Mn distance equal to about 3.2 Å, i.e., close to the distance found by EXAFS, which is likely to correspond to the closest distance between two manganese dimers in the OEC. Several complexes with $\{Mn_2(O_2CR)\}^{2+,3+}$ and $\{Mn_2O(O_2CR)_2\}^{2+,3+}$ cores were synthesized and new complexes of this kind continue to appear.[203b] However, to become catalysts of dioxygen evolution they have to have one more manganese atom added to each manganese atom present in the complex via di-μ-oxo bridges and have other ligands easily replaceable by water to form dioxygen upon oxidation.

TABLE III.3
The Quantum Yield (f) of $Ru(bpy)_3^{3+}$ Generation under Irradiation (λ = 436 nm) of $Ru(bpy)_3^{2+}$ Solution with Various Electron Acceptors

Acceptor	E_0 vs. NHE	ϕ	Medium (pH)	Ref.
$Co(NH_3)Cl^{2+}$	0.52	0.44	0.3	211
		0.41–0.43	1.2	
		0.34–0.39	2.2	
$S_2O_8^{2-}$	?	2.0	0.3	212
Tl^{3+}	?	0.67	0.3	211
		0.26	1.0	
		0.17	1.7	
Mn^{IV}-pyrophosphate	1.07	0.23–0.26	6-8	208

III.5.3.4 Photoinduced Water Oxidation

Similarly to chlorophyll in plant photosynthesis, metalloporphyrins are expected to be photosensitizers in chemical model systems. However, at present there is no unambiguous examples of using metalloporphyrins as photosensitizers in catalytic water oxidation. Efficient intersystem crossing to form long-lived triplet states that can participate in electron transfer reactions takes place in excited porphyrins. Photoexcitation of the metalloporphyrins in dilute solutions containing an appropriate electron acceptor, such as $S_2O_8^{2-}$, Fe^{III}, or $Co(NH_3)_5^{2+}$, can result in the formation of metalloporphyrin π-radical cations with high yields. The π-radical cations are powerful oxidants (in some cases redox potentials exceed +1 V). However, the lifetimes of π-radical cations are too short for efficient reactions with catalysts due to the secondary reactions resulting in the formation of π-dications and isoporphyrins.[204]

Photoexcitation of $Ru(bpy)_3^{2+}$ leads to the formation of the relatively long-lived (620 ns) excited state, $*Ru(bpy)_3^{2+}$. The quenching of the latter by an appropriate electron acceptor (A) can be successfully used for generation of $Ru(bpy)_3^{3+}$:

$$*Ru(bpy)_3^{2+} + A \rightarrow Ru(bpy)_3^{3+} + A^- \quad (48)$$

For steady-state photogeneration of $Ru(bpy)_3^{3+}$ a sacrificial electron acceptor has to be used, e.g., $Co(NH_3)_5Cl^{2+}$,[162,185,205] $S_2O_8^{2-}$,[206] Tl^{3+},[206] or Mn^{IV} pyrophosphate complex.[207-209] The quantum yield of $Ru(bpy)_3^{3+}$ generation (ϕ) in the presence of these electron acceptors are summarized in Table III.3. In the case of $Co(NH_3)_5Cl^{2+}$ the value of ϕ is rather high due to the decomposition of the unstable Co^{II} amine complex according to the scheme:[210]

$$*Ru(bpy)_3^{2+} + Co(NH_3)_5Cl^{2+} + 5H^+ \rightarrow Ru(bpy)_3^{3+} + Co_{aq}^{2+} + Cl^- + 5NH_4^+ \quad (49)$$

The rapid hydrolysis of $Co(NH_3)_5Cl^+$ initially formed prevents the back-reaction of $Co(NH_3)_5Cl^+$ with $Ru(bpy)_3^{3+}$ and the product of the hydrolysis, Co^{2+}_{aq}, cannot be oxidized by $Ru(bpy)_3^{3+}$ ($E_0(Co^{3+}_{aq}/Co^{2+}_{aq})$ is equal to 1.81 V, i.e., it is higher than $E_0(Ru(bpy)_3^{3+}/Ru(bpy)_3^{2+})$ which equals 1.26 V). Co^{2+} ions formed are themselves effective catalysts for water oxidation, as mentioned above.

A very effective sacrificial acceptor for $Ru(bpy)_3^{3+}$ formation is the persulfate anion.[211] The high quantum yield arises from the additional oxidation of $Ru(bpy)_3^{2+}$ by a sulfate radical:

$$*Ru(bpy)_3^{2+} + S_2O_8^{2-} \rightarrow Ru(bpy)_3^{3+} + SO_4^{2-} + SO_4^{\bullet -} \quad (50)$$

$$Ru(bpy)_3^{2+} + SO_4^{\bullet -} \rightarrow Ru(bpy)_3^{3+} + SO_4^{2-} \quad (51)$$

The $SO_4^{\bullet -}$ radical is itself a powerful oxidant and can react not only with the ruthenium complex, but also with organic substances present in the solution. Besides, there exists a possibility that dioxygen evolved originates from persulfate ions, because it has been established that transition metal ions can catalyze the decomposition of $S_2O_8^{2-}$ in aqueous solutions.

For the system $S_2O_8^{2-}$–$Ru(bpy)_3^{2+}$–Mn (bound to vesicles) it was demonstrated with $H_2{}^{18}O$ that water molecules are the source of O_2, but the quantum yield is low due to the oxidation of lipid molecules instead of water, carbon dioxide being produced in the process of the photolysis.

Talium(III) can be used only at pH < 3; at higher pH it undergoes hydrolysis to $Tl(OH)_3$. Tl^{3+} is a poor electron acceptor for $*Ru(bpy)_3^{2+}$ and the quantum yield for O_2 formation even with very active colloid RuO_2 catalyst is less than 1%, compared with 20% when $Co(NH_3)_5Cl^{2+}$ was used.

Mn^{IV} pyrophosphate complex is stable at neutral pH and starts to hydrolyze at pH > 8. The couple Mn^{IV}/Mn^{III} pyrophosphate undergoes a reversible oxidation with $E_0 \approx 1.1$ V at pH 7, thus close to that of P-680$^+$ ($E_0 = 1.12$ V). This electron acceptor was used with Mn hydroxo complex as the catalyst.[208,209] For the hydroxo complex of manganese bound to vesicles the quantum yield of dioxygen evolution (0.17) is close to that of $Ru(bpy)_3^{3+}$ formation (0.26). This is the evidence for the high activity of the catalyst. Figure III.31 illustrates the mechanism of dioxygen evolution in this photochemical system. At present this system remains probably the best functional model of the OEC of PS II of plant photosynthesis.

III.5.4 COMPARISON OF PS II AND ITS FUNCTIONAL MODELS

As the result of the recent development we have now a situation similar to that which existed in the field of dinitrogen reduction in protic media after the discovery of first hydroxo complexes capable of N_2 activation but before individual polynuclear complexes active in the reduction were found, such as vanadium(II) catechol, or alkoxymolybdenum(III) complexes.

Now we can easily observe dioxygen evolution catalyzed by hydroxo complexes of Mn, Fe, Co, Ni, and Cu, yet, in spite of many efforts, no individual complexes

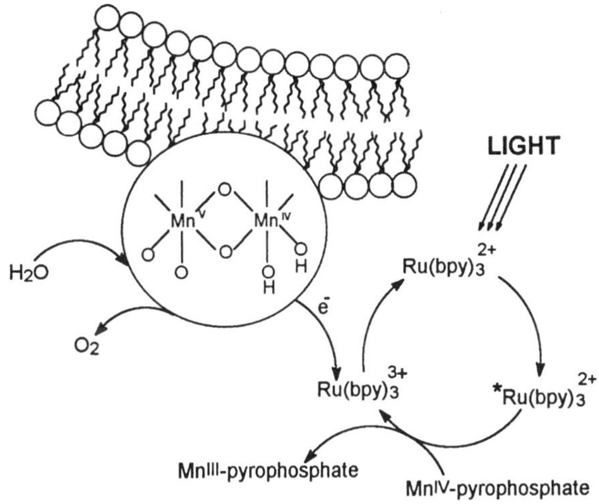

FIGURE III.31 Dioxygen evolution catalyzed by manganese hydroxo complexes supported by vesicle.

have been found that would catalyze *efficient* O_2 evolution from water under the action of mild oxidative agents. Individual ruthenium and manganese complexes evolve dioxygen in homogeneous solutions only in strongly acid solutions.

In PS II of plant photosynthesis the dioxygen evolution starts at the redox potential of the active center close to that for four-electron O_2 formation from water. There is no or very little activation energy. This seemingly remarkable achievement of living nature is easily matched in the functioning of simple hydroxo complexes: all of them, beginning with Mn hydroxo complexes, are capable of evolving O_2 when their redox potential reaches the value of $E^0_{O_2/H_2O}$. The energy barriers are apparently also very low, therefore even very small concentrations are sufficient to facilitate efficient O_2 evolution when the redox potential of an external one-electron oxidant is close to that of P-680$^+$. The activity of hydroxo complexes is due primarily to the ease of the achievement of the necessary redox potential.

It was mentioned already in the beginning of this chapter that in a metal cluster there must be always a broad range of energy levels occupied by electrons and there is a high probability that a suitable level for the catalysis will be also present.

For manganese (and ruthenium) hydroxo complexes, which are easily oxidizable, the necessary redox potential is reached already at pH 7. This is apparently connected with the small positive charges of nuclei of the atoms of the elements among those capable of oxidizing water in their highest oxidation states. For other metals the redox potentials of high oxidation states have larger positive values and differences ΔE^0 between different states are also larger. By increasing pH we decrease both the corresponding E^0 and ΔE^0. So each of these metals in its polynuclear hydroxo complex becomes a catalyst at some (high) pH value.

Hydrated MnO_2 and RuO_2 are in some aspects very similar as catalysts of dioxygen evolution from water. Both can undergo a "corrosion", i.e., transformation to Mn^{VII} (in the form of MnO_4^-) and Ru^{VIII} (in the form of RuO_4) and escape from the catalyst surface to the solution. Hence, unusual oxidation states (Mn^V and Ru^V are apparently easily achieved as the intermediates in the process of oxidation and they are further oxidized to the highest oxidation states in samples containing much water with comparatively weak bondings of Mn atoms and their neighbors. There must be a certain degree of freedom for the metal atoms and their oxygen surrounding to be oxidized and to leave the surface of the catalyst. When annealed or bound to a support they undergo less corrosion but can be stili good catalysts for dioxygen evolution. Apparently, a more rigid structure prevents the corrosion but is still able to unite two oxygen atoms to form O_2. When the structure is too rigid, e.g., after annealing the sample at higher temperature, the activity for the catalysis is lost because not only metal but also oxygen atoms are too tightly bound and cannot be replaced by water to produce dioxygen.

When comparing manganese and ruthenium it could be mentioned that both are able to oxidize organic material in parallel with water oxidation, but RuO_2 produces more active catalyst for C–H oxidation and therefore is unsuitable for O_2 evolution in the presence of organic substances. We have seen in the previous chapter that the activity of oxidized ruthenium in C–H oxidation can be due to less steric hindrances and also to electronic effects resulting in more effective activation of C–H bonds.

Catalytic action of hydrated MnO_2 must involve either participation of an external oxidating agent in the catalytic complex or the oxidation of one or several Mn atoms to Mn^V because MnO_2 itself was found to be unable to evolve O_2 from water at neutral pH. Involvement of Mn^V seems to be more likely in view of the results with the corrosion: intermediate formation of Mn^V stabilized in the cluster is apparently difficult to avoid in this reaction.

Thus, hydroxo complexes of manganese are very promising as candidates for best catalysts of oxygen evolution from water. Could they have been used in the first attempts of photosynthetic organisms to use water as electron donor and to produce dioxygen?

There is now a certain probability that S_2 in the oxygen evolution center in PS II of plant photosynthesis corresponds to $Mn^{III}Mn_3^{IV}$. If so, then S_4 is $Mn_3^{IV}Mn^V$ or its equivalents: Mn_4^{IV} + oxidant or $Mn^{III}Mn_3^{IV}O_2^{2-}$. The concrete scheme of the catalytic cycle may be therefore close to the catalytic cycle in which manganese hydroxo complexes are used: the initial (reduced) state in both cases includes mainly Mn^{III}, while the oxidized state ready to evolve dioxygen (S_4 in the OEC) includes mainly Mn^{IV} with one Mn^V.

This supports the hypothesis that the photosystem producing dioxygen (later developed into PS II) when originated could use manganese hydroxo complexes, incorporating them in biological membranes. Among hydroxo complexes of various metals, those of manganese turned out to possess optimal properties and were preferred. Subsequently, the manganese hydroxo complexes could have developed in the process of evolution to the modern state of four-nuclear effective catalyst in PS II. If the hypothesis is correct, the adsorption of manganese hydroxo complexes on organic membranes not only should produce good catalysts of O_2 evolution, but these catalysts should have some advantages compared with hydroxo complexes of other metals. Comparison of the catalytic activity of different metals hydroxo complexes adsorbed on the phospholipid membrane is given in Table III.4.[156,207,209] In

TABLE III.4
Dependence of O_2 Yield[a] per Oxidant (%) on pH for Different Catalysts Adsorbed at the Lipid Membrane

	Oxidant: $Ru(bpy)_3^{3+}$					Oxidant: $(Fe(bpy)_3^{3+}$		
Catalyst (M)	pH 5.7	7.0	7.6	8.4	9.0	7.6	8.4	9.0
Cr					0			0
Mn	21.5	52.0	43.0	43.3	43.5	3.8	6.8	7.6
Fe	0	0	52.1	71.5	81.6	0	0	0
Co	0	8.1	79.5	90.5	94.5	4.4	16.6	29.8
Ni	0	5.7	33.7	36.9	32.3	3.1	17.3	31.1
Cu	0	0	4.1	22.3	27.0	0	0	0

[a] 0 means that O_2 yield does not exceed 0.1%.

$[Ru(bpy)_3^{3+}] = 5 \times 10^{-4}$ *M*; $Fe(bpy)_3^{3+} = 5 \times 10^{-4}$ *M*; $[M] = 3 \times 10^{-5}$ *M*.

the case of $Ru(bpy)_3^{3+}$ as the oxidant (E = 1.26 V) dioxygen is evolved in the presence of all the 3d elements of the periodic table, beginning with manganese.

At the same time, it can be seen in Table III.4 that at pH 7 and lower the activity of the manganese complexes is higher than those of all the other elements chosen. In the case of the weaker oxidant $(Fe(bpy)_3^{3+}$ (E = 1.06 V) only Mn, Co, and Ni complexes remain catalytically active, the activity being observed only in alkaline media. Thus manganese turns out indeed to be the most suitable in the form of a hydroxo complex supported by the organic membrane to serve as a catalyst of O_2 evolution.

If higher than Mn^{IV} oxidation state is indeed involved in the S_4 state then the supported manganese hydroxo complexes (e.g., on vesicles or cyclodextrines) are particularly close to the manganese cluster in the OEC: they both contain polynuclear manganese complexes and function at similar redox potentials of the oxidating agent at pH 7. In both cases direct formation of O_2 without free intermediate hydrogen peroxide is very likely and energy barriers in dioxygen formation from oxygen ligands at manganese atoms are low.

The activity of the hydroxo complexes is due apparently to the presence of two water ligands that produce OH and then O ligands when manganese is oxidized. There could be several kinds of active centers in freshly prepared amorphous hydroxides, including those with suitable positions of two oxygen ligands to easily form O_2. Even few very active centers in polynuclear hydroxo complexes may create an efficient catalyst with low activation energy. High activity and selectivity of manganese hydroxo complexes supported on organic material in dioxygen production from water is presumably due to the existence of some flexible bridges uniting manganese clusters containing μ-oxo structures. It is not excluded that these flexible bridges involve carboxylates produced in partial oxidation of organic material. In this case they would be particularly close to the structure that can be suggested now for the manganese cluster of the water oxidase in the oxygen evolution center of PS II.

It may be thought, therefore, that the answer to the question concerning the role of the manganese hydroxo complexes in the early evolution might be positive and the latter indeed could have been used as catalysts for O_2 evolution when photosynthetic bacteria were transforming into algae. Manganese is always present in seawater, forming iron–manganese concretions at the bottom, and it is constantly being renewed by manganese from the water of rivers. However, hydroxo complexes naturally could not be used inside living cells and they had to be transformed eventually into the smaller cofactors of the enzyme water oxidase without loss of the catalytic activity. The catalytic complex created in the process of evolution was considered in Section III.2. It is still not easy to mimic at present by a purely synthetic approach because we have yet to determine the exact structure of the OEC and the detailed mechanism of O_2 evolution.

It may be predicted that individual manganese complexes, effective catalysts of dioxygen evolution, will be prepared in the not too distant future. But even now catalytic O_2 evolution from water is not the mystery it was not very long ago: indeed, a man-made catalyst can do the job in a manner similar to that of the natural system.

III.5.5 CONCLUSION

Almost all the functions of the separate systems of the plant photosynthesis, at least its simple analogue, photocleavage of water, can be simulated now in their chemical analogues: light-induced long-distance electron transfer through a membrane, and catalytic and photocatalytic dihydrogen and dioxygen evolution. The problems that remained unanswered for many years are now not so mysterious as they appeared not very long ago. However, it is still difficult to mimic photosynthesis as a whole. Molecular engineering, although it already exists, is not yet sufficiently developed to combine all the parts of the total system, the more so because the separate parts are still far from perfection. Self-developing systems are not yet available, and these, when they are created, would hopefully provide an answer to the problem of constructing the full effective model of plant photosynthesis.

At the same time, new hybrid systems combining some principles of natural photosynthesis with the properties of solid state, e.g., semiconductors, like the Grätzel photovoltaic system described in this chapter, are being developed already and seem promising as useful devices for transforming solar energy into electricity or some other forms of energy. Thus biomimetics starts to transform itself gradually into the applied science.

REFERENCES

1. Govindjee, Ed., *Photosynthesis*, Academic Press, New York, 1982.
2. van Niel, C. B., *Arch. Microbiol.*, 3, 1, 1931.
3. Hill, R. and Scarisbrick, R., *Nature*, 146, 61, 1940.
4. Ruben, S., Randall, M., Kamen, M., and Hyde, J. I., *J. Am. Chem. Soc.*, 63, 877, 1941.
5. Bassham, J. A. and Calvin, M., *The Path of Carbon in Photosynthesis*, Prentice Hall, Englewood Cliffs, NJ, 1957.
6. Hill, R. and Bendall, F., *Nature*, 186, 136, 1960.

7. Terenin, A. N., *Photonics in Dyes and Related Organic Substances,* Leningrad, Nauka, 1967 (in Russian).
8. Krasvosky, A. A. and Voynovskaya, K. K., *Dokl. Akad. Nauk SSSR,* 81, 879, 1951.
9a. Deisenhofer, J., Epp, O., Miki, K., Huber, R., and Michel, H., *Nature,* 318, 618, 1985.
9b. Shuvalov, V. A. and Asadov, A. A., *Biochim. Biophys. Acta,* 545, 296, 1979.
10. Emerson, R., Chalmers, R., and Cederstrand, C., *Proc. Natl. Acad. Sci. USA,* 43, 133, 1957.
11. Anderson, J. M. and Boardman, N. K., *Biochim. Biophys. Acta,* 416, 191, 1975.
12. Emerson, R. and Arnold, W., *J. Gen. Phys.,* 16, 191, 1932.
13. Harriman, A., in *Energy Resources through Photochemistry and Catalysis,* Grätzel, M., Ed., Academic Press, New York, 1983, 163.
14. Govindjee, and Whitmarsh, J., in *Photosynthesis,* Govindjee, Ed., Academic Press, New York, 1982.
15. Kondratjeva, E. N., *Photosynthesizing Bacteria and Bacterial Photosynthesis,* MGU, Moscow, 1972 (in Russian).
16. Olson, J. M. and Pierson, B. K., *Int. Rev. Cytol.,* 108, 209, 1987.
17a. Prigogine, I., *From Being to Becoming. Time and Complexity in Physical Sciences,* Freeman, W. H., San Francisco, 1980.
17b. Utyanskaya, E. Z., Pavlovsky, A. G., Sosfenov, N. I., and Shilov, A. E., *Dokl. Akad. Nauk SSSR,* 301, 149, 1988.
17c. Sigel, H., *Inorg. Chim. Acta,* 198, 1, 1992.
17d. Utyanskaya, E. Z., Neihaus, M. G., Lidskii, B. V., and Shilov, A. E., *React. Kinet. Catal. Lett.,* 54, 431, 1995.
18. Wraight, C. A., in *Photosynthesis,* Govindjee, Ed., Academic Press, New York, 1982.
19. Bolton, J. R., *J. Solid State Chem.,* 22, 3, 1977; Zamaraev, K. I. and Parmon, V. N., *Usp. Khim.,* 49, 1457, 1980.
20a. Turro, N. J., *Molecular Photochemistry,* Benjamin, New York, 1965.
20b. Watanabe, T., Fujishima, A., and Honda, K., in *Energy Resources through Photochemistry and Catalysis,* Grätzel, M., Ed., Academic Press, New York, 1983, chap. 11.
21a. Shuvalov, V. A., *Primary Conversion of Light Energy in Photosynthesis,* Nauka, Moscow, 1990 (in Russian).
21b. Likhtenshtein, G. I., *Polynuclear Redox Enzymes,* Nauka, Moscow, 1979.
22. Marcus, R. A., *Annu. Rev. Phys. Chem.,* 15, 155, 1964.
23. Nadtochenko, V. A., unpublished results.
24. Gray, H. B. and Ellis, W. R., Jr., in *Bioinorganic Chemistry,* Bertini, I., Gray, H. B., Lippard, S. J., and Valantine, J. S., Eds., University Science Books, Mill Valley, CA, 1994, 315.
25. Deisenhofer, J., Epp, O., Miki, K., Huber, R., and Michel, H., *J. Mol. Biol.,* 180, 385, 1984.
26. Fendler, J. H., *J. Phys. Chem.,* 84, 1485, 1980.
27. Infelta, P. P., in *Energy Resources through Photochemistry and Catalysis,* Grätzel, M., Ed., Academic Press, New York, 1983, chap. 2.
28. Sprintschnik, G., Sprintschnik, H. W., Kirsch, P. P., and Whitten, D. G., *J. Am. Chem. Soc.,* 98, 2337, 1976.
29. Grimaldi, J. J., Boileau, S., and Lehn, J.-M., *Nature,* 265, 229, 1977.
30. Tien, H. T., *Nature,* 214, 272, 1988.
31. Tien, H. T., in *Membranes and Transport,* Martonosi, A. N., Ed., Plenum, New York, 1982, 165.
32. Fendler, J. H., *J. Photochem.,* 17, 303, 1981.
33. Matsuo, T., *Pure Appl. Chem.,* 54, 1693, 1982.
34. Matsuo, T., *J. Photochem.,* 28, 41, 1985.
35. Deanier, D. W. and Uster, P. S., in *Liposomes,* Ostro, M. J. and Dekker, M., Eds., Basel, New York, 1983, 27.
36. Huang, C. H., *Biochemistry,* 8, 344, 1962.
37. Ford, W. E., Otwos, J. W., and Calvin, M., *Proc. Natl. Acad. Sci. USA,* 76, 3590, 1979.
38. Infelta, P. P., Grätzel, M., and Fendler, J. H., *J. Am. Chem. Soc.,* 102, 1479, 1980.
39. Mangel, M., *Biochem. Biophys. Acta,* 430, 459, 1976.

40. Katagi, T., Yamamura, T., Laito, T., and Lasaki, Y., *Chem. Lett.*, 509, 1981.
41. Krasnovsky, A. A., Semenova, A. N., and Nikandrov, V. V., *Dokl. Akad, Nauk SSSR*, 262, 469, 1982.
42a. Lymar', S. V., Tsvetkov, I. M., Parmon, V. N., and Zamaraev, K. I., *Khim. Fizika*, 405, 1982.
42b. Zamaraev, K. I. and Parmon, V. N., in *Energy Resources through Photochemistry and Catalysis*, Grätzel, M., Ed., Academic Press, New York, 1983, 123.
42c. Nadtochenko, V. A., Lavrentyev, A. I., Rubtsov, I. V., Nikandrov, V. V., and Barannikova, Ya. V., *J. Photochem. Photobiol. B: Biol.*, 18, 221, 1993.
43. Yablonskaya, E. E. and Shafirovich, V. Ya., *Kinet. Katal.*, 24, 1032, 1983.
44. Parmon, V. N., Khramov, M. I., Tsvetkov, I. M., and Zamaraev, K. I., *Kinet. Katal.*, 24, 1034, 1983.
45. Yablonskaya, E. E., Klabunovskii, E. I., Shafirovich, V. Ya., and Shilov, A. E., *Dokl. Akad. Nauk SSSR*, 286, 150, 1986.
46. Knerel'man, E. I. and Shafirovich, V. Ya., *Izv. Akad. Nauk SSSR Ser. Khim.*, 2213, 1986.
47. Khannanov, N. K., Kuzmin, V. A., Levin, P. P., Shafirovich, V. Ya., and Yablonskaya, E. E., *Khim. Fiz.*, 5, 1358, 1986.
48. Yablonskaya, E. E. and Shafirovich, V. Ya., *Izv. Akad. Nauk SSSR Ser. Khim.*, 5, 1011, 1987.
49. Knerel'man, E. I., Khannanov, N. K., Shafirovich, V. Ya., and Yablonskaya, E. E., *Izv. Akad. Nauk SSSR Ser. Khim.*, 1, 185, 1989.
50. Wasielewski, M. R., *Chem. Rev.*, 92, 435, 1992.
51. Craig, D. C., Lawson, J. M., Oliver, A. M., and Paddon-Row, M. N., *J. Chem. Soc. Perkin Trans.*, 1, 3305, 1990.
52. Wasielewski, M. R., Minsek, D. W., Niemczyk, M. P., Swec, W. A., and Yang, N. C., *J. Am. Chem. Soc.*, 112, 2823, 1990.
53. Hermant, R. M., Bakker, N. A. C., Scherer, T., Krijnen, B., and Verhoeven, J. W., *J. Am. Chem. Soc.*, 112, 1214, 1990.
54. Closs, G. L., Calcaterra, L. T., Green, N. J., Penfield, K. W., and Miller, J. R., *J. Phys. Chem.*, 90, 3673, 1986.
55. Liang, N., Miller, J. R., and Closs, G. L., *J. Am. Chem. Soc.*, 112, 5353, 1990.
56. Osuka, A. and Maruyama, K., *J. Am. Chem. Soc.*, 110, 4454, 1988.
57. Nagata, T., Osuka, A., and Maruyama, K., *J. Am. Chem. Soc.*, 112, 3054, 1990.
58. Osuka, A., Nagata, T., and Maruyama, K., *Chem. Lett.*, 481, 1991.
59. Wasielewski, M. R., Johnson, D. G., Niemczyk, M. P., Gaines, G. L., O'Neil, M. P., and Svec, W. A., *J. Am. Chem. Soc.*, 112, 6482, 1990.
60. Moore, T. A., Gust, D., Mathis, P., Mialocq, J.-C., Chachaty, C., Bensasson, R. V., Land, E. J., Doizi, D., Liddell, P. A., Lehnman, W. R., Nemeth, G. A., and Moore, A. L., *Nature*, 307, 630, 1984.
61. Gust, D., Moore, T. A., Moore, A. L., Seely, G., Liddell, P., Barrett, D., Harding, L. O., Ma, X. C., Lee, S. J., and Gao, F., *Tetrahedron*, 45, 4867, 1989.
62a. Gust, D., Moore, T. A., Moore, A. L., Lee, S. J., Bittersman, E., Luttrull, D. K., Rehms, A. A., DeGraziano, J. M., Ma, X. C., Gao, F., Belford, R. E., and Trier, T. T., *Science*, 248, 199, 1990.
62b. Gust, D., Moore, T. A., and Moore, A. L., *Acc. Chem. Res.*, 26, 198, 1993.
63. Collin, J.-P., Guillerez, S., Sauvage, J.-P., Barigelletti, F., Flamigni, L., De Cola, L., and Balzani, V., *Coord. Chem. Rev.*, 111, 291, 1991.
64. Collin, J.-P., Guillerez, S., Sauvage, J.-P., Barigelletti, F., De Cola, L., Flamigni, L., and Balzani, V., *Inorg. Chem.*, 30, 4230, 1991.
65. Wasielewski, M. R., Johnson, D. G., Svec, W. A., Kersey, K. M., and Minsek, D. W., *J. Am. Chem. Soc.*, 110, 7219, 1988.
66. Gaines, G. L., O'Neil, M. P., Svec, W. A., Niemczyk, M. P., and Wasielewski, M. R., *J. Am. Chem. Soc.*, 113, 719, 1991.
67. Wasielewski, M. P., Gaines, G. L., O'Neil, M. P., Niemczyk, M. P., and Svec, W. A., *J. Am. Chem. Soc.*, 112, 4559, 1990.
68. Borovkov, V. V., Evstigneeva, R. P., Sturganova, I. A., Kamalov, V. F., and Toleytaev, B. N., *J. Phys. Chem.*, 95, 6437, 1991.
69. Lendzian, F. and von Matzan, B., *Chem. Phys. Lett.*, 180, 191, 1991.
70. Lendzian, F., Schlupmann, J., von Gersdorff, J., Mobius, K., and Kurreck, H., *Angew. Chem. Int. Ed. Engl.*, 30, 1461, 1991.

71. Aoyama, Y., Asakawa, M., Matsui, Y., and Ogoshi, H., *J. Am. Chem. Soc.*, 113, 6233, 1991.
72. Harriman, A., Kubo, Y., and Sessler, J., *J. Am. Chem. Soc.*, 114, 388, 1992.
73. Hugerat, M., Levanon, H. E., Ojadi, H., Biczok, L., and Lindschitz, H., *Chem. Phys. Lett.*, 181, 400, 1991.
74. Levin, P. P., Pluzhnikov, P. F., and Kuzmin, V. A., *Chem. Phys.*, 137, 331, 1989.
75. Miller, G. R., *New J. Chem.*, 11, 83, 1987.
76. Batova, E. E. and Shafirovich, V. Ya., *Dokl. Akad. Nauk SSSR*, 307, 1131, 1989.
77. Shafirovich, V. Ya., Amoyal, E., and Delaire, J., *Chem. Phys. Lett.*, 178, 24, 1991.
78. Batova, E. E., Levin, P. P., and Shafirovich, V. Ya., *Izv. Akad. Nauk SSSR Ser. Khim.*, 1287, 1990.
79. Levin, P. P., Batova, E. E., and Shafirovich, V. Ya., *Chem. Phys.*, 142, 279, 1990.
80. Shafirovich, V. Ya., Batova, E. E., and Levin, P. P., *Chem. Phys. Lett.*, 172, 10, 1990.
81. Shafirovich, V. Ya., Batova, E. E., and Levin, P. P., *Chem. Phys. Lett.*, 162, 155, 1992.
82. Shafirovich, V. Ya., Batova, E. E., and Levin, P. P., *J. Chem. Soc. Faraday Trans.*, 88, 935, 1992.
83. Batova, E. E., Levin, P. P., and Shafirovich, V. Ya., *New J. Chem.*, 14, 269, 1990.
84. Moore, T. A., Gust, D., Moore, A. L., Bensasson, R., Seta, P., and Bienvenue, E., in *Supramolecular Photochemistry*, Balzani, V., Ed., D. Reidel, Amsterdam, 1987, 283.
85. Momenteau, M., Loock, B., Seta, P., Bienvenue, E., and d'Epenoux, B., *Tetrahedron*, 45, 4893, 1989.
86. Shafirovich, V. Ya., Batova, E. E., and Levin, P. P., *Photochem. Photobiol.*, 55, 473, 1992.
87. Shafirovich, V. Ya., Batova, E. E., and Levin, P. P., *Chem. Phys. Lett.*, 210, 101, 1993.
88. Grätzel, M., *Platinum Met. Rev.*, 38, 151, 1994.
89a. Hagfeldt, A. and Grätzel, M., *Chem. Rev.*, 95, 49, 1995.
89b. Piccini, A. and Marino, L., *Z. Anorg. Chem.*, 32, 55, 1902.
90. Stephenson, M. and Stickland, L. H., *Biochem. J.*, 25, 204, 1931.
91. Schönheit, P., Moll, J., and Thauer, R. K., *Arch. Microbiol.*, 123, 105, 1979.
92. Graf, E. G. and Thauer, R. K., *FEBS Lett.*, 136, 156, 1981.
93. Albracht, S. P. J., Graf, E. G., and Thauer, R. K., *FEBS Lett.*, 140, 311, 1982.
94. van der Zwaan, J. W., Albracht, S. P. J., Fontijn, R. D., and Slater, E. C., *FEBS Lett.*, 179, 91, 1984.
95. Lindahl, P. A., Kojima, N., Hausinger, R. P., Fox, J. A., Teo, B. K., Walsh, C. T., and Orme-Johnson, W. H., *J. Am. Chem. Soc.*, 106, 3062, 1984.
96a. Fauque, G., Peck, H. D., Jr., Moura, J. J. G., Huinh, B. H., Berlier, Y., DerVartanian, D. V., Teixeira, M., Prrzybyla, A. E., Lespinat, P. A., Moura, I., and Legall, J., *FEMS Microbiol. Rev.*, 54, 299, 1988.
96b. Nakos, G. and Mortenson, L. E., *Biochim. Biophys. Acta*, 227, 576, 1971.
96c. Chen, J.-S. and Blanchard, D. K., *Biochim. Biophys. Res. Commun.*, 84, 1144, 1978.
97. Adams, M. W. W., *Biochim. Biophys. Acta*, 1020, 115, 1990.
98. Adams, M. W. W., Eccleston, E., and Howard, J. B., *Proc. Natl. Acad. Sci. USA*, 86, 4932, 1989.
99. Krasnovsky, A. A., Nikandrov, V. V., Brin, G. P., Gogotov, I. N., and Oschepkov, V. V., *Dokl. Akad. Nauk SSSR*, 225, 711, 1975.
100. van der Zwaan, Thesis, Laboratory of Biochemistry, B.C.P. Jansen Institute, University of Amsterdam, The Netherlands, 1987.
101. Zamaraev, K. I. and Parmon, V. N., *Usp. Khim.*, 52, 1433, 1983.
102. McLendon, G., in *Energy Resources through Photochemistry and Catalysis*, Grätzel, M., Ed., Academic Press, New York, 1983, 99.
103. Kiwi, J., in *Energy Resources through Photochemistry and Catalysis*, Grätzel, M., Ed., Academic Press, New York, 1983, 297.
104. Henglein, A., *J. Phys. Chem.*, 83, 2209, 1979.
105. Kiwi, J. and Grätzel, M., *J. Am. Chem. Soc.*, 101, 7214, 1979.
106. Amouyal, E. and Koffi, P., *J. Photochem.*, 29, 227, 1985.
107. Koryakin, B. V., Dzhabiev, T. S., and Shilov, A. E., *Dokl. Akad. Nauk SSSR*, 229, 128, 1976.
108. Krasna, A. I., *Photochem. Photobiol.*, 29, 267, 1979.
109. Lehn, J.-M. and Sauvage, J. P., *Nouv. J. Chim.*, 1, 449, 1977.
110. Moradpour, A., Amouyal, E., Keller, P., and Kagan, H., *Nouv. J. Chim.*, 2, 547, 1978.
111. Kalyanasundaram, K., Kiwi, J., and Grätzel, M., *Helv. Chim. Acta*, 61, 2720, 1978.
112. Demos, J. N. and Crosby, O. A., *J. Am. Chem. Soc.*, 93, 2841, 1971.

113. Meyer, T. J., *Acc. Chem. Res.*, 22, 163, 1989.
114. Forster, M. and Hester, R. E., *Chem. Phys. Lett.*, 81, 42, 1981.
115. Creutz, C. and Sutin, N., *Inorg. Chem.*, 15, 498, 1976.
116a. Sutin, N. and Creutz, C., *Pure Appl. Chem.*, 52, 2717, 1980.
116b. Kirsh, M., Lehn, J.-M., and Sauvage, J.-P., *Helv. Chim. Acta*, 62, 1345, 1979.
116c. Crutchley, R. J. and Lever, A. B. P., *J. Am. Chem. Soc.*, 102, 7128, 1980.
116d. Maestry, M. and Sandrini, D., *Nouv. J. Chim.*, 5, 637, 1981.
116e. Harriman, A., Porter, G., and Richoux, M.-C., *J. Chem. Soc. Faraday, Trans.*, 2, 77, 833, 1981.
116f. Kitamura, N., Kawanishi, Y., and Tazuke, S., *Chem. Lett.*, 1185, 1983.
116g. Johansen, O., Mau, A. W.-H., and Sasse, W. H. P., *Chem. Phys. Lett.*, 94, 113, 1983.
116h. Johansen, O., Mau, A. W.-H., and Sasse, W. H. P., *Chem. Phys. Lett.*, 94, 107, 1983.
116i. Nadtochenko, V. A., Rubtsov, I. V., and Dzhabiev, T. S., *Khim. Fiz.*, 1, 1515, 1983.
117. Tunuli, M. S. and Fendler, J. H., *J. Am. Chem. Soc.*, 103, 2507, 1981.
118. Kurihara, K. and Fendler, J. H., *J. Am. Chem. Soc.*, 105, 6152, 1983.
119. Tricot, Y.-M. and Fendler, J. H., *J. Am. Chem. Soc.*, 105, 7359, 1984.
120. Rafaeloff, R., Tricot, Y.-M., Nome, F., and Fendler, J. H., *J. Phys. Chem.*, 89, 533, 1985.
121. Tricot, Y.-M., Emeren, A., and Fendler, J. H., *J. Phys. Chem.*, 89, 4721, 1985.
122. Maier, V. E. and Shafirovich, V. Ya., *Dokl. Akad. Nauk SSSR*, 227, 125, 1984.
123. Maier, V. E. and Shafirovich, V. Ya., *Chem. Commun.*, 1063, 1985.
124. Maier, V. E., Levchenko, L. A., and Shafirovich, V. Ya., *Kinet. Katal.*, 27, 1378, 1986.
125. Maier, V. E. and Shafirovich, V. Ya., *Kinet. Katal.*, 28, 839, 1987.
126. Semenov, N. N., Shilov, A. E., and Likhtenshtein, G. I., *Dokl. Akad. Nauk SSSR*, 221, 1374, 1975.
127. Kok, B., Forbush, B., and McGloin, M., *Photochem. Photobiol.*, 11, 457, 1970.
128. Joliot, P. and Joliot, A., *Biochim. Biophys. Acta*, 153, 625, 1966.
129. Cheniae, G. M., *Annu. Rev. Plant. Physiol.*, 21, 467, 1970; Cheniae, G. M. and Martin, I. F., *Biochim. Biophys. Acta*, 197, 219, 1971.
130. Calvin, M., *Science*, 184, 375, 1974.
131a. Ruthherford, A. W., Zimmermann, J.-L., and Boussac, A., in *The Photosystems: Structure, Function and Molecular Biology*, Barber, J., Ed., Elsevier, 1992, 179.
131b. Wieghardt, K., *Angew. Chem. Int. Ed. Engl.*, 33, 725, 1994.
132. Wydrzynki, T. J., in *Photosynthesis*, Govindjee, Ed., Academic Press, New York, 1982, chap. 10.
133. Hansson, O. and Wydrzynki, T. J., *Photosynth. Res.*, 23, 131, 1990.
134. Andersson, B., Larsson, C., Jansson, C., Ljungberg, U., and Åkerlund, H.-E., *Biochim. Biophys. Acta*, 766, 21, 1984.
135. Åkerlund, H.-E., Hansson, C., and Andreasson, B., *Biochim. Biophys. Acta*, 681, 1, 1988.
136. Akabori, K., Imaoka, A., and Toyshima, Y., *FEBS Lett.*, 173, 36, 1984.
137. Barry, B. A. and Babcock, G. T., *Proc. Natl. Acad. Sci. USA*, 84, 7099, 1987.
138. Gerken, S., Brettel, K., Schlodder, E., and Witt, H. T., *FEBS Lett.*, 237, 69, 1988.
139a. Bouges-Bocquet, B., *Biochim. Biophys. Acta*, 594, 85, 1980.
139b. Vass, I. and Styring, S., *Biochemistry*, 30, 830, 1991.
140. Tamura, N. and Cheniae, G. M., *Biochim. Biophys. Acta*, 809, 245, 1985.
141. Nixon, P. J. and Diiner, B. A., *Biochemistry*, 31, 942, 1992.
142. Allakhverdiev, S. I., Shafiev, M. A., and Klimov, V. V., *Photobiochem. Photobiophys.*, 12, 61, 1986.
143. Pauly, S. and Witt, H. T., *Biochim. Biophys. Acta*, 1099, 211, 1992.
144. Dismukes, G. C. and Siderer, Y., *FEBS Lett.*, 121, 78, 1980.
145. Dismukes, G. C. and Siderer, Y., *Proc. Natl. Acad. Sci. USA*, 78, 274, 1981.
146. Bonvoisin, J., Blondin, G., Girerd, J.-J., and Zimmerman, J.-L., *Biophys. J.*, 61, 1076, 1992.
147. Kusunoki, M., *Chem. Phys. Lett.*, 197, 108, 1992.
148a. Proserpio, D. M., Hoffmann, R., and Dismukes, G. C., *J. Am. Chem. Soc.*, 114, 4374, 1992.
148b. Yachandra, V. K., DeRose, V. J., Latimer, M. J., Mukerji, I., Sauer, K., and Klein, M. P., *Science*, 260, 675, 1993.
149. Critchley, C., *Biochim. Biophys. Acta*, 811, 33, 1985.
150. Homann, P. H., *J. Bioenerg. Biomembr.*, 19, 105, 1987.
151. Coleman, W. J., *Photosynth. Res.*, 23, 1, 1990.
152. Brand, J. J. and Becker, D. W., *J. Bioenerg. Biomembr.*, 16, 239, 1984.

153. Homann, P. H., in *The Role of Calcium in Biological Systems,* Vol. 5, Anghileri, L. J., Ed., CRC Press, Boca Raton, 1989, 79.
154. Yocum, C. F., *Biochim. Biophys. Acta,* 1991.
155. Debus, R. C., *Biochim. Biophys. Acta,* 1991.
156a. Shilov, A. E., in *Perspectives in Catalysis,* Thomas, J. M. and Zamaraev, K. I., Eds., Blackwell, Oxford, 1992, 169.
156b. Barynin, V. V., Vagin, A. A., Melik-Adamyan, V. R., Grebenko, A. I., Khangulov, S. V., Popov, A. N., Andrianova, M. E., and Vainshtein, B. K., *Dokl. Akad. Nauk SSSR,* 288, 877, 1986.
156c. Klimov, V., Ananyev, G., Zastryzhnaya, O., Wydrzynski, T., and Renger, G., *Photosynth. Res.,* 38, 409, 1993.
157. Trasatti, S. and Buzzanca, G., *J. Electroanal. Chem.,* 29, 1, 1971.
158. Glikman, T. S. and Shwedova, I. S., *Kinet. Katal.,* 9, 461, 1968.
159. Kiwi, J. and Grätzel, M., *Chimia,* 33, 289, 1979.
160. Lehn, J.-M., Sauvage, J.-P., and Ziessel, R., *Nouv. J. Chim.,* 3, 423, 1979.
161. Mills, A., *Chem. Soc. Rev.,* 18, 285, 1989.
162. Shafirovich, V. Ya. and Strelets, V. V., *Nouv. J. Chim.,* 6, 183, 1982.
163. Nijs, H., Cruz, M. I., Fripiat, J. J., and Van Damme, H., *Nouv. J. Chim.,* 6, 551, 1982.
164. Ramaraj, R., Kira, A., and Kaneko, M., *J. Chem. Soc. D,* 227, 1987.
165. Rotzinger, F. P., Munavalli, S., Compte, P., Hurst, J. K., Grätzel, M., Pern, F.-J., and Frank, A. J., *J. Am. Chem. Soc.,* 109, 6619, 1987.
166. Gersten, S. W., Samuels, G. J., and Meyer, T. J., *J. Am. Chem. Soc.,* 104, 4029, 1982.
167. Gilbert, J. A., Gersteten, D. S., Hodgson, D. J., and Meyer, T. J., *J. Am. Chem. Soc.,* 107, 3855, 1985.
168. Ramaraj, R., Kira, A., and Kaneko, M., *Angew. Chem. Int. Ed. Engl.,* 25, 1009, 1986.
169. Ramaraj, R., Kira, A., and Kaneko, M., *J. Chem. Soc. Faraday Trans.,* 83, 1539, 1987.
170. Mills, A. and Russel, T., *J. Chem. Soc. Faraday Trans.,* 87, 313, 1991.
171. Landsberg, R. and Heckner, K. H., *Nature,* 201, 1123, 1964.
172. Veprek-Siska, J. and Ettel, V., *Chem. Ind.,* 548, 1967.
173. Veprek-Siska, J. and Ettel, V., *J. Inorg. Nucl. Chem.,* 31, 789, 1969.
174. Anbar, M. and Pecht, I., *Trans. Faraday Soc.,* 64, 744, 1968.
175. Sutin, N. and Creutz, C., *Pure Appl. Chem.,* 52, 2717, 1980.
176. Klimov, V. V., Allakhverdiev, S. I., Demeter, S., and Krasnovsky, A. A., *Dokl. Akad. Nauk SSSR,* 149, 227, 1979.
177. Blau, F., *Monatsch. Chem.,* 19, 647, 1898.
178. Brandt, W. W., Dwyer, F. P., and Gyarfas, E. C., *Chem. Rev.,* 54, 959, 1955.
179. Creutz, C. and Sutin, N., *Proc. Natl. Acad. Sci. USA,* 72, 2858, 1975.
180. Nord, G. and Wernberg, O., *J. Chem. Soc. Dalton Trans.,* 866, 1972.
181. Nord, G. and Wernberg, O., *J. Chem. Soc. Dalton Trans.,* 845, 1974.
182. Khannanov, N. K., Khramov, A. V., Moravsky, A. P., and Shafirovich, V. Ya., *Kinet. Katal.,* 24, 858, 1983.
183. Shafirovich, V. Ya., Khannanov, N. K., and Strelets, V. V., *Nouv. J. Chim.,* 4, 81, 1980.
184. Shafirovich, V. Ya. and Khannanov, N. K., *Kinet. Katal.,* 22, 248, 1981.
185. Harriman, A., Porter, G., and Walters, P., *J. Chem. Soc. Faraday Trans.,* 77, 2337, 1981.
186. Shafirovich, V. Ya. and Shilov, A. E., *Kinet. Katal.,* 19, 877, 1978.
187. Shafirovich, V. Ya., Khannanov, N. K., and Shilov, A. E., *Kinet. Katal.,* 19, 1498, 1978.
188. Shafirovich, V. Ya., Khannanov, N. K., and Shilov, A. E., *Kinet. Katal.,* 19, 1502, 1978.
189. Glikman, T. S. and Shvedova, I. S., *Kinet. Katal.,* 9, 461, 1968.
190. Shafirovich, V. Ya., Khannanov, N. K., and Shilov, A. E., *Kinet. Katal.,* 19, 1591, 1978.
191. Shafirovich, V. Ya. and Shilov, A. E., *Kinet. Katal.,* 20, 1156, 1979.
192. Luneva, N. P., Shafirovich, V. Ya., and Shilov, A. E., *J. Mol. Catal.,* 52, 49, 1989.
193. Shafirovich, V. Ya. and Shilov, A. E., *Isr. J. Chem.,* 38, 149, 1988.
194. Luneva, N. P., Knerelman, E. I., Shafirovich, V. Ya., and Shilov, A. E., *J. Chem. Soc. D,* 1504, 1987.
195. Luneva, N. P., Knerelman, E. I., Shafirovich, V. Ya., and Shilov, A. E., *Dokl. Akad. Nauk SSSR,* 299, 388, 1988.
196. Dekker, J. P., Gorkom, H. J., Wensik, J., and Ouwehand, Z., *Biochim. Biophys. Acta,* 776, 1, 1984.

197a. Kritschmann, H., Dekker, J. P., Saygin, O., and Witt, H. T., *Biochim. Biophys. Acta*, 932, 358, 1988.
197b. Luneva, N. P., Shafirovich, V. Ya., and Shilov, A. E., *Kinet. Katal.*, 30, 250, 1989.
198a. Armstrong, W. H., in *Manganese Redox Enzymes*, Pecoraro, V. L., Ed., VCH, New York, 1992, chap. 12.
198b. Brudvig, G. V. and Crabtree, R. H., *Prog. Inorg. Chem.*, 37, 99, 1989.
198c. Christou, G., *Acc. Chem. Res.*, 22, 328, 1989.
199. Chan, M. K. and Armstrong, W. H., *J. Am. Chem. Soc.*, 113, 5055, 1991.
200. Cooper, S. R. and Calvin, M., *Science*, 185, 376, 1974.
201. Ramaraj, R., Kira, A., and Kaneko, M., *Angew. Chem. Int. Ed. Engl.*, 25, 825, 1986.
202. Ramaraj, R., Kira, A., and Kaneko, M., *Chem. Lett.*, 261, 1987.
203a. Luneva, N. P., Shilov, A. E., Crabtree, R., and Sarneski, J. E., *Kinet. Katal.*, 34, 870, 1993.
203b. Mandal, S. K. and Armstrong, W. H., *Inorg. Chem. Acta*, 229, 361, 1995.
204. Harriman, A., Porter, G., and Walters, P., *J. Chem. Soc. Faraday Trans.*, 79, 1335, 1983.
205. Lehn, J.-M., Sauvage, J.-P., and Ziessel, R., *Nouv. J. Chim.*, 4, 355, 1980.
206. Harriman, A., *Platinum Met. Res.*, 27, 102, 1984.
207. Luneva, N. P., Shafirovich, V. Ya., and Shilov, A. E., *J. Mol. Catal.*, 52, 49, 1989.
208. Shafirovich, V. Ya., Khannanov, N. K., and Shilov, A. E., *J. Inorg. Biochem.*, 15, 113, 1981.
209. Luneva, N. P., Knerelman, E. I., Shafirovich, V. Ya., and Shilov, A. E., *New J. Chem.*, 13, 107, 1989.
210. Navon, G. and Sutin, N., *Inorg. Chem.*, 13, 2159, 1974.
211. Laurence, G. S. and Balzani, V., *Inorg. Chem.*, 13, 2976, 1974.
212. Bolletta, F., Juris, A., Maestri, M., and Sandrini, D., *Inorg. Chem. Acta Lett.*, 44, L175, 1980.

INDEX